handbook

Key Notes
Terms
Definitions
Formulae

Physics

Highly Important for Class XI & XII Students, Engineering & Medical Entrances and Other Competitions Aspirants

A Multi-Purpose Quick *Revision* Resource

Keshav M Agarwal

ARIHANT PRAKASHAN, MEERUT

arihant
ARIHANT PRAKASHAN, MEERUT

Administrative & Production Offices

Corporate Office: 'Ramchhaya' 4577/15, Agarwal Road, Darya Ganj, New Delhi -110002
Tele: 011- 47630600, 43518550; Fax: 011- 23280316

Head Office: Kalindi, TP Nagar, Meerut (UP) - 250002
Tele: 0121-2401479, 2512970, 4004199; Fax: 0121-2401648
All disputes subject to Meerut (UP) jurisdiction only.

Sales & Support Offices
Agra, Ahmedabad, Bengaluru, Bhubaneswar, Bareilly, Chennai, Delhi, Guwahati, Haldwani Hyderabad, Jaipur, Jalandhar, Jhansi, Kolkata, Kota, Lucknow, Meerut, Nagpur & Pune

ISBN 978-93-5203-603-5

Typeset by Arihant DTP Unit at Meerut

For further information about the products from Arihant
log on to **www.arihantbooks.com** or email to **info@arihantbooks.com**

PREFACE

Handbook means reference book listing brief facts on a subject. So, to facilitate the students in this we have released this **Handbook of Physics** this book has been prepared to serve the special purpose of the students, to rectify any query or any concern point of a particular subject.

This book will be of highly use whether students are looking for a quick revision before the board exams or just before other examinations like Engineering Entrances, Medical Entrances, NDA Entrance etc or Competitive Exam like CDS exam or any similar examination, they will find that this handbook will answer their needs admirably.

This handbook can even be used for revision of a subject in the time between two shift of the exams, even this handbook can be used while travelling to Examination Centre or whenever you have time, less sufficient or more.

The format of this handbook has been developed particularly so that it can be carried around by the students conveniently.

The objectives of publishing these handbooks are :

- To support students in their revision of a subject just before an examination.
- To provide a focus to students to clear up their doubts about particular concepts which were not clear to them earlier.
- To give confidence to the students just before they attempt important examinations.

However, we have put our best efforts in preparing this book, but if any error or what so ever has been skipped out, we will by heart welcome your suggestions. A part from all those who helped in the compilation of this book a special note of thanks to Priyanshi Garg.

Author

CONTENTS

1

Measurement

The comparison of any physical quantity with its standard unit is called **measurement.**

Physical Quantities

All the quantities in terms of which laws of physics are described, and whose measurement is necessary are called physical quantities.

Units

A definite amount of a physical quantity is taken as its standard unit. The standard unit should be easily reproducible, internationally accepted.

Fundamental Units

Those physical quantities which are independent to each other are called fundamental quantities and their units are called fundamental units.

S.No.	Fundamental Quantities	Fundamental Units	Symbol
1.	Length	metre	m
2.	Mass	kilogram	kg
3.	Time	second	s
4.	Temperature	kelvin	K
5.	Electric current	ampere	A
6.	Luminous intensity	candela	cd
7.	Amount of substance	mole	mol

Supplementary Fundamental Units

Radian and steradian are two supplementary fundamental units. It measures plane angle and solid angle respectively.

S.No.	Supplementary Fundamental Quantities	Supplementary Unit	Symbol
1.	Plane angle	radian	rad
2.	Solid angle	steradian	sr

Derived Units

Those physical quantities which are derived from fundamental quantities are called **derived quantities** and their units are called **derived units.**
e.g. velocity, acceleration, force, work etc.

Definitions of Fundamental Units

The seven fundamental units of SI have been defined as under :

(i) **1 kilogram** A cylindrical prototype mass made of platinum and iridium alloys of height 39 mm and diameter 39 mm. It is mass of 5.0188×10^{25} atoms of carbon-12.

(ii) **1 metre** 1 metre is the length of the path travelled by light in vacuum during a time interval of 1/ 299792, 458 of a second 1983.

(iii) **1 second** 1 second is the time in which cesium atom vibrates 9192631770 times in an atomic clock.

(iv) **1 kelvin** 1 kelvin is the $\frac{1}{273.16}$ part of the thermodynamics temperature of the triple point of water.

(v) **1 candela** 1 candela is the luminous intensity in a given direction of a source which emits monochromatic radiation of frequency 540×10^{12} Hz and whose radiant intensity is 1/683 watt per steradian.

(vi) **1 ampere** 1 ampere is the electric current which it maintained in two straight parallel conductor of infinite length and of negligible circular cross-section area placed one metre apart in vacuum would produce between them a force equal to 2×10^{-7} N per metre of length.

(vii) **1 mole** 1 mole is the amount of substance of a system which contains a many elementry entities (may be atoms, molecules, ions, electrons or group of particles, as this and atoms in 0.012 kg of carbon isotope ${}_6C^{12}$.

Systems of Units

A system of units is the complete set of units, both fundamental and derived, for all kinds of physical quantities. The common system of units which is used in mechanics are given below:

(i) **CGS System** In this system, the unit of length is centimetre, the unit of mass is gram and the unit of time is second.

(ii) **FPS System** In this system, the unit of length is foot, the unit of mass is pound and the unit of time is second.

(iii) **MKS System** In this system, the unit of length is metre, the unit of mass is kilogram and the unit of time is second.

(iv) **SI System** This system contain seven fundamental units and two supplementary fundamental units.

Relationship between Some Mechanical SI Unit and Commonly Used Units

S.No.	Physical Quantity		Unit
1.	Length	(a)	1 micrometre $= 10^{-6}$ m,
		(b)	1 nanometre $= 10^{-9}$ m
		(c)	1 angstrom $= 10^{-10}$ m
2.	Mass	(a)	1 metric ton $= 10^3$ kg
		(b)	1 pound = 0.4537 kg
		(c)	1 amu $= 1.66 \times 10^{-23}$ kg
3.	Volume		1 litre $= 10^{-3}$ m^3
4.	Force	(a)	1 dyne $= 10^{-5}$ N
		(b)	1 kgf = 9.81 N
5.	Pressure	(a)	1 kgf-m^2 = 9.81 Nm^{-2}
		(b)	1 mm of Hg = 133 Nm^{-2}
		(c)	1 pascal = 1 Nm^{-2}
		(d)	1 atmosphere pressure = 76 cm of Hg $= 1.01 \times 10^5$ pascal
6.	Work and energy	(a)	1 erg $= 10^{-7}$ J
		(b)	1 kgf-m = 9.81 J
		(c)	1 kWh $= 3.6 \times 10^6$ J
		(d)	1 eV $= 1.6 \times 10^{-19}$ J
7.	Power	(a)	1 kgf- ms^{-1} = 9.81 W
		(b)	1 horse power = 746 W

Some Practical Units

(i) 1 fermi $= 10^{-15}$ m

(ii) 1 X-ray unit $= 10^{-13}$ m

(iii) 1 astronomical unit $= 1.49 \times 10^{11}$ m (average distance between sun and earth)

(iv) 1 light year $= 9.46 \times 10^{15}$ m

(v) 1 parsec = 3.08×10^{16} m = 3.26 light year

Some Approximate Masses

Object	Kilogram
Our galaxy	2×10^{41}
Sun	2×10^{30}
Moon	7×10^{22}
Asteroid Eros	5×10^{15}

Dimensions

Dimensions of any physical quantity are those powers which are raised on fundamental units to express its unit. The expression which shows how and which of the base quantites represent the dimensions of a physical quantity, is called the **dimensional formula**.

Dimensional Formula of Some Physical Quantities

S.No.	Physical Quantity	Dimensional Formula	MKS Unit
1.	Area	$[L^2]$	metre^2
2.	Volume	$[L^3]$	metre^3
3.	Velocity	$[LT^{-1}]$	ms^{-1}
4.	Acceleration	$[LT^{-2}]$	ms^{-2}
5.	Force	$[MLT^{-2}]$	newton (N)
6.	Work or energy	$[ML^2T^{-2}]$	joule (J)
7.	Power	$[ML^2T^{-3}]$	$J\ s^{-1}$ or watt
8.	Pressure or stress	$[ML^{-1}T^{-2}]$	Nm^{-2}
9.	Linear momentum or Impulse	$[MLT^{-1}]$	$kg\ ms^{-1}$
10.	Density	$[ML^{-3}]$	$kg\ m^{-3}$
11.	Strain	Dimensionless	unitless
12.	Modulus of elasticity	$[ML^{-1}T^{-2}]$	Nm^{-2}
13.	Surface tension	$[MT^{-2}]$	Nm^{-1}
14.	Velocity gradient	$[T^{-1}]$	second^{-1}
15.	Coefficient of velocity	$[ML^{-1}T^{-1}]$	$kg\ m^{-1}s^{-1}$
16.	Gravitational constant	$[M^{-1}L^3T^{-2}]$	Nm^2/kg^2
17.	Moment of inertia	$[ML^2]$	$kg\ m^2$
18.	Angular velocity	$[T^{-1}]$	rad/s
19.	Angular acceleration	$[T^{-2}]$	rad/s^2

S.No.	Physical Quantity	Dimensional Formula	MKS Unit
20.	Angular momentum	$[ML^2T^{-1}]$	kg m^2s^{-1}
21.	Specific heat	$[L^2T^{-2}\theta^{-1}]$	kcal $kg^{-1}K^{-1}$
22.	Latent heat	$[L^2T^{-2}]$	kcal/kg
23.	Planck's constant	$[ML^2T^{-1}]$	J-s
24.	Universal gas constant	$[ML^2T^{-2}\theta^{-1}]$	J/mol-K

Homogeneity Principle

If the dimensions of left hand side of an equation are equal to the dimensions of right hand side of the equation, then the equation is dimensionally correct. This is known as **homogeneity principle**.

Mathematically [LHS] = [RHS]

Applications of Dimensions

(i) To check the accuracy of physical equations.

(ii) To change a physical quantity from one system of units to another system of units.

(iii) To obtain a relation between different physical quantities.

Significant Figures

In the measured value of a physical quantity, the number of digits about the correctness of which we are sure plus the next doubtful digit, are called the significant figures.

Rules for Finding Significant Figures

(i) All non-zeros digits are significant figures, e.g. 4362 m has 4 significant figures.

(ii) All zeros occuring between non-zero digits are significant figures, e.g. 1005 has 4 significant figures.

(iii) All zeros to the right of the last non-zero digit are not significant, e.g. 6250 has only 3 significant figures.

(iv) In a digit less than one, all zeros to the right of the decimal point and to the left of a non-zero digit are not significant, e.g. 0.00325 has only 3 significant figures.

(v) All zeros to the right of a non-zero digit in the decimal part are significant, e.g. 1.4750 has 5 significant figures.

Significant Figures in Algebraic Operations

(i) **In Addition or Subtraction** In addition or subtraction of the numerical values the final result should retain the least decimal place as in the various numerical values. e.g.

If $l_1 = 4.326$ m and $l_2 = 1.50$ m

Then, $$l_1 + l_2 = (4.326 + 1.50)\text{ m} = 5.826\text{ m}$$

As l_2 has measured upto two decimal places, therefore

$$l_1 + l_2 = 5.83\text{ m}$$

(ii) **In Multiplication or Division** In multiplication or division of the numerical values, the final result should retain the least significant figures as the various numerical values. e.g. If length $l = 12.5$ m and breadth $b = 4.125$ m.

Then, area $$A = l \times b = 12.5 \times 4.125 = 51.5625\text{ m}^2$$

As l has only 3 significant figures, therefore, $A = 51.6\text{ m}^2$

Rules of Rounding Off Significant Figures

(i) If the digit to be dropped is less than 5, then the preceding digit is left unchanged. e.g. 1.54 is rounded off to 1.5.

(ii) If the digit to be dropped is greater than 5, then the preceding digit is raised by one. e.g. 2.49 is rounded off to 2.5.

(iii) If the digit to be dropped is 5 followed by digit other than zero, then the preceding digit is raised by one. e.g. 3.55 is rounded off to 3.6.

(iv) If the digit to be dropped is 5 or 5 followed by zeros, then the preceding digit is raised by one, if it is odd and left unchanged if it is even. e.g. 3.750 is rounded off to 3.8 and 4.650 is rounded off to 4.6.

Error

The lack in accuracy in the measurement due to the limit of accuracy of the instrument or due to any other cause is called an error.

1. Absolute Error

The difference between the true value and the measured value of a quantity is called absolute error.

If $a_1, a_2, a_3, \dots, a_n$ are the measured values of any quantity a in an experiment performed n times, then the arithmetic mean of these values is called the true value (a_m) of the quantity.

$$a_m = \frac{a_1 + a_2 + a_3 + \dots + a_n}{n}$$

The absolute error in measured values is given by

$$\Delta a_1 = a_m - a_1$$
$$\Delta a_2 = a_m - a_2$$
$$\dots\dots\dots\dots\dots\dots$$
$$\Delta a_n = a_m - a_n$$

2. Mean Absolute Error

The arithmetic mean of the magnitude of absolute errors in all the measurement is called mean absolute error.

$$\overline{\Delta a} = \frac{|\Delta a_1| + |\Delta a_2| + \dots + |\Delta a_n|}{n}$$

3. Relative Error

The ratio of mean absolute error to the true value is called relative error. Relative error $= \dfrac{\text{Mean absolute error}}{\text{True value}} = \dfrac{\overline{\Delta a}}{a_m}$

4. Percentage Error

The relative error expressed in percentage is called percentage error.

$$\text{Percentage error} = \frac{\overline{\Delta a}}{a_m} \times 100\%$$

Propagation of Error

(i) **Error in Addition or Subtraction** Let $x = a + b$ or $x = a - b$

If the measured values of two quantities a and b are $(a \pm \Delta a)$ and $(b \pm \Delta b)$, then maximum absolute error in their addition or subtraction. $\Delta x = \pm (\Delta a + \Delta b)$

(ii) **Error in Multiplication or Division** Let $x = a \times b$ or $x = \dfrac{a}{b}$

If the measured values of a and b are $(a \pm \Delta a)$ and $(b \pm \Delta b)$, then maximum relative error $\dfrac{\Delta x}{x} = \pm \left(\dfrac{\Delta a}{a} + \dfrac{\Delta b}{b}\right)$

(iii) **Error in Raised to a Power** Let $Z = A^p B^q / C^r$

If the measured values of A, B and C are $(A \pm \Delta A)$ and $(B \pm \Delta B)$, then maximum error $\dfrac{\Delta Z}{Z} = p\left(\dfrac{\Delta A}{A}\right) + q\left(\dfrac{\Delta B}{B}\right) + r\left(\dfrac{\Delta C}{C}\right)$

2

Vectors

Those physical quantities which require magnitude as well as direction for their complete representation and follows vector laws are called **vectors.**

Vector can be divided into two types :

1. **Polar Vectors**

These are those vectors which have a starting point or a point of application as a displacement, force etc.

2. **Axial Vectors**

These are those vectors which represent rotational effect and act along the axis of rotation in accordance with right hand screw rule as angular velocity, torque, angular momentum etc.

Scalars

Those physical quantities which require only magnitude but no direction for their complete representation are called scalars.

Distance, speed, work, mass, density, etc are the examples of scalars. Scalars can be added, subtracted, multiplied or divided by simple algebraic laws.

Tensors

Tensors are those physical quantities which have different values in different directions at the same point.

Moment of inertia, radius of gyration, modulus of elasticity, pressure, stress, conductivity, resistivity, refractive index, wave velocity and density, etc are the examples of tensors. Magnitude of tensor is not unique.

Different Types of Vectors

(i) **Equal Vectors** Two vectors of equal magnitude, in same direction are called equal vectors.

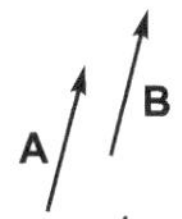

(ii) **Negative Vectors** Two vectors of equal magnitude but in opposite directions are called negative vectors.

The negative vector of $\mathbf{A}$ is represented as $-\mathbf{A}$.

(iii) **Zero Vector or Null Vector** A vector whose magnitude is zero, known as a zero or null vector. Its direction is not defined. It is denoted by **0**.

Velocity of a stationary object, acceleration of an object moving with uniform velocity and resultant of two equal and opposite vectors are the examples of null vector.

(iv) **Unit Vector** A vector having unit magnitude is called a unit vector.

A unit vector in the direction of vector $\mathbf{A}$ is given by

$$\hat{\mathbf{A}} = \frac{\mathbf{A}}{A}$$

A unit vector is unitless and dimensionless vector and represents direction only.

(v) **Orthogonal Unit Vectors** The unit vectors along the direction of orthogonal axis, i.e. X-axis, Y-axis and Z-axis are called orthogonal unit vectors. They are represented by $\hat{\mathbf{i}}$, $\hat{\mathbf{j}}$ and $\hat{\mathbf{k}}$.

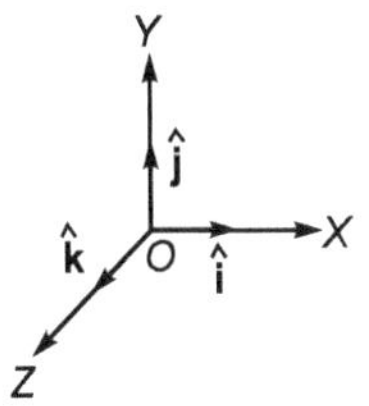

(vi) **Co-initial Vectors** Vectors having a common initial point, are called co-initial vectors.

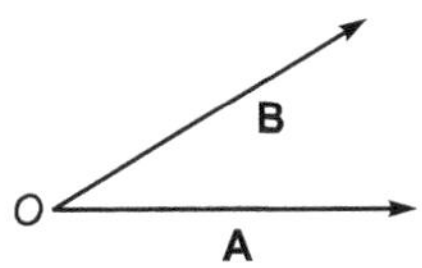

(vii) **Collinear Vectors** Vectors having equal or unequal magnitudes but acting along the same or parallel lines are called collinear vectors.

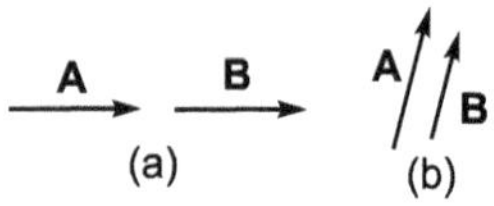

(viii) **Coplanar Vectors** Vectors acting in the same plane are called coplanar vectors.

(ix) **Localised Vector** A vector whose initial point is fixed, is called a localised vector.

(x) **Non-localised or Free Vector** A vector whose initial point is not fixed is called a non-localised or a free vector.

(xi) **Position Vector** A vector which gives position of an object with reference to the origin of a coordinate system is called position vector. It is represented by a symbol **r**.

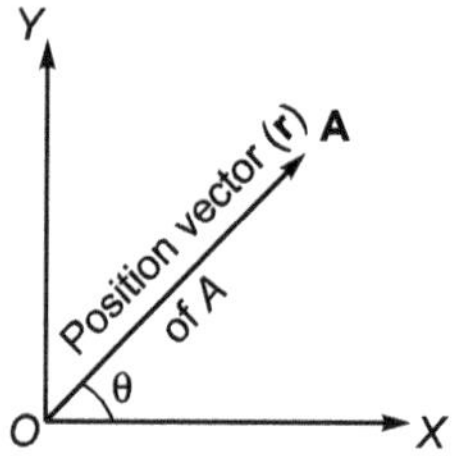

(xii) **Displacement Vector** The vector which tells how much and in which direction an object has changed its position in a given interval of time is called **displacement vector.**

Displacement vector is the straight line joining the initial and final positions and does not depend on the actual path undertaken by the object between the two positions.

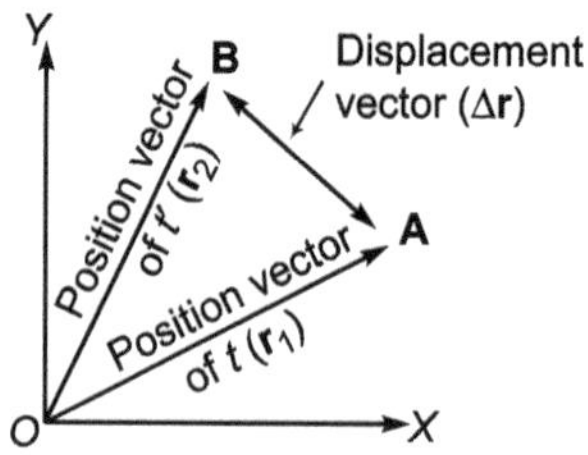

The displacement vector for **AB** is

$$\Delta \mathbf{r} = \mathbf{r}_2 - \mathbf{r}_1$$

Addition of Vectors

1. Triangle Law of Vectors Addition

If two vectors acting at a point are represented in magnitude and direction by the two sides of a triangle taken in one order, then their resultant is represented by the third side of the triangle taken in the opposite order.

If two vectors **A** and **B** acting at a point are inclined at an angle θ, then their resultant

$$R = \sqrt{A^2 + B^2 + 2AB\cos\theta}$$

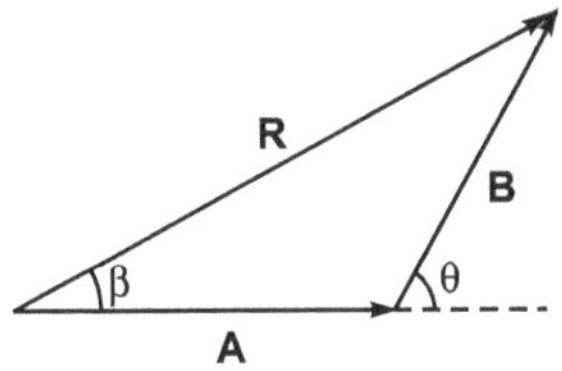

If the resultant vector **R** subtends an angle β with vector **A**, then

$$\tan\beta = \frac{B\sin\theta}{A + B\cos\theta}$$

2. Parallelogram Law of Vectors Addition

If two vectors acting at a point are represented in magnitude and direction by the two adjacent sides of a parallelogram draw from a point, then their resultant is represented in magnitude and direction by the diagonal of the parallelogram draw from the same point.

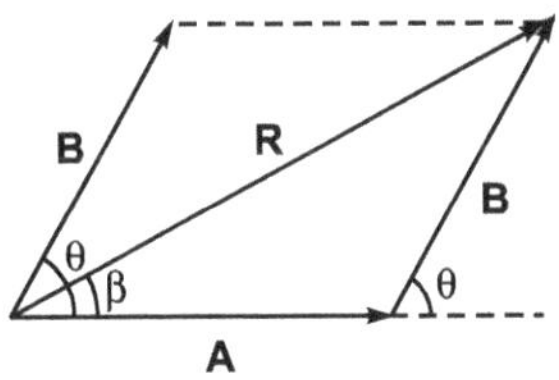

Resultant of vectors **A** and **B** is given by

$$R = \sqrt{A^2 + B^2 + 2AB\cos\theta}$$

If the resultant vector **R** subtends an angle β with vector **A**, then

$$\tan\beta = \frac{B\sin\theta}{A + B\cos\theta}$$

3. Polygon Law of Vectors Addition

It states that if number of vectors acting on a particle at a time are represented in magnitude and direction by the various sides of an open polygon taken in same order, their resultant vector is represented in magnitude and direction by the closing side of polygon taken in opposite order. In fact, polygon law of vectors is the outcome of triangle law of vectors.

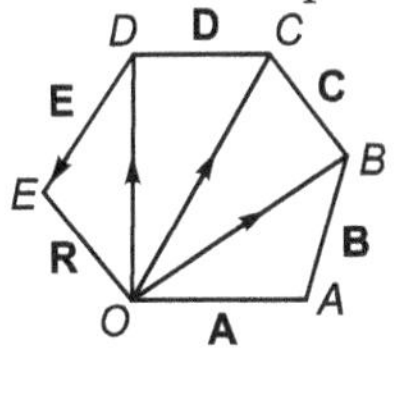

$$\mathbf{R} = \mathbf{A} + \mathbf{B} + \mathbf{C} + \mathbf{D} + \mathbf{E}$$

$$\mathbf{OE} = \mathbf{OA} + \mathbf{AB} + \mathbf{BC} + \mathbf{CD} + \mathbf{DE}$$

Properties of Vector Addition

(i) Vector addition is commutative, i.e. $\mathbf{A} + \mathbf{B} = \mathbf{B} + \mathbf{A}$

(ii) Vector addition is associative, i.e.

$$\mathbf{A} + (\mathbf{B} + \mathbf{C}) = \mathbf{B} + (\mathbf{C} + \mathbf{A}) = \mathbf{C} + (\mathbf{A} + \mathbf{B})$$

(iii) Vector addition is distributive, i.e. $m(\mathbf{A} + \mathbf{B}) = m\mathbf{A} + m\mathbf{B}$

(iv) $\mathbf{A} + \mathbf{0} = \mathbf{A}$

Rotation of a Vector

(i) If a vector is rotated through an angle θ, which is not an integral multiple of 2π, the vector changes.

(ii) If the frame of reference is rotated or translated, the given vector does not change. The components of the vector may, however, change.

Resolution of Vectors into Two Components

If two component vectors of **R** are **OP** and **PQ** in the direction of A and B, respectively.

Suppose $\mathbf{OP} = \lambda\mathbf{A}$ and $\mathbf{PQ} = \mu\mathbf{B}$, where λ and μ are two real numbers.

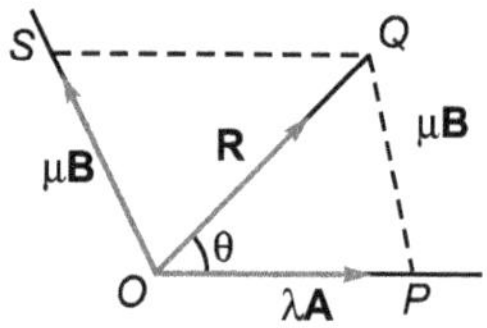

The resultant vector, $\mathbf{R} = \lambda\mathbf{A} + \mu\mathbf{B}$

Resolution of a Vector into Rectangular Components

If any vector **A** subtends an angle θ with X-axis, then its

Horizontal component $A_x = A\cos\theta$

Vertical component $A_y = A\sin\theta$

Magnitude of vector $A = \sqrt{A_x^2 + A_y^2}$

$$\tan\theta = \frac{A_y}{A_x}$$

Angle, $$\theta = \tan^{-1}\left(\frac{A_y}{A_x}\right)$$

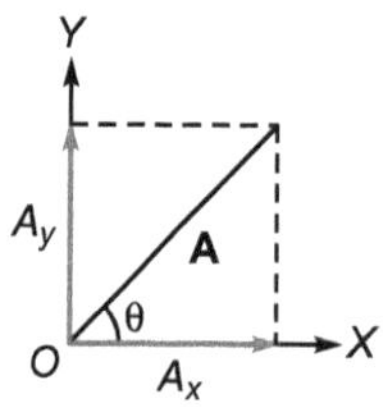

Direction Cosines of a Vector

If any vector **A** subtend angles α, β and γ with X-axis, Y-axis and Z-axis respectively and its components along these axes are A_x, A_y and A_z, then

$$\cos\alpha = \frac{A_x}{A}, \quad \cos\beta = \frac{A_y}{A},$$

$$\cos\gamma = \frac{A_z}{A}$$

and $$\cos^2\alpha + \cos^2\beta + \cos^2\gamma = 1$$

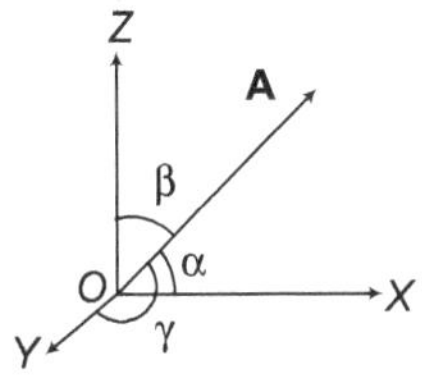

Subtraction of Vectors

Subtraction of a vector **B** from a vector **A** is defined as the addition of vector $-\mathbf{B}$ (negative of vector **B**) to vector **A**.

Thus, $$\mathbf{A} - \mathbf{B} = \mathbf{A} + (-\mathbf{B})$$

Multiplication of a Vector

1. By a Real Number

When a vector **A** is multiplied by a real number n, then its magnitude becomes n times but direction and unit remains unchanged.

2. By a Scalar

When a vector **A** is multiplied by a scalar S, then its magnitude becomes S times, and unit is the product of units of **A** and S but direction remains same as that of vector **A**.

Scalar or Dot Product of Two Vectors

The scalar product of two vectors is equal to the product of their magnitudes and the cosine of the smaller angle between them. It is denoted by · (dot).

$$\mathbf{A} \cdot \mathbf{B} = AB\cos\theta$$

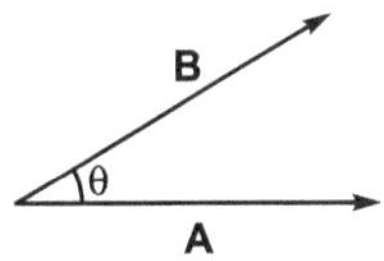

The scalar or dot product of two vectors is a scalar.

Properties of Scalar Product

(i) Scalar product is commutative,

i.e., $$\mathbf{A} \cdot \mathbf{B} = \mathbf{B} \cdot \mathbf{A}$$

(ii) Scalar product is distributive,

i.e., $$\mathbf{A} \cdot (\mathbf{B} + \mathbf{C}) = \mathbf{A} \cdot \mathbf{B} + \mathbf{A} \cdot \mathbf{C}$$

(iii) Scalar product of two perpendicular vectors is zero.

$$\mathbf{A} \cdot \mathbf{B} = AB\cos 90^\circ = 0$$

(iv) Scalar product of two parallel vectors or anti-parallel vectors is equal to the product of their magnitudes, i.e.

$\mathbf{A} \cdot \mathbf{B} = AB \cos 0° = AB$ (for parallel)

$\mathbf{A} \cdot \mathbf{B} = AB \cos 180° = -AB$ (for anti-parallel)

(v) Scalar product of a vector with itself is equal to the square of its magnitude, i.e. $\mathbf{A} \cdot \mathbf{A} = AA \cos 0° = A^2$

(vi) Scalar product of orthogonal unit vectors

$$\hat{\mathbf{i}} \cdot \hat{\mathbf{i}} = \hat{\mathbf{j}} \cdot \hat{\mathbf{j}} = \hat{\mathbf{k}} \cdot \hat{\mathbf{k}} = 1$$

and $$\hat{\mathbf{i}} \cdot \hat{\mathbf{j}} = \hat{\mathbf{j}} \cdot \hat{\mathbf{k}} = \hat{\mathbf{k}} \cdot \hat{\mathbf{i}} = 0$$

(vii) Scalar product in cartesian coordinates

$$\mathbf{A} \cdot \mathbf{B} = (A_x \hat{\mathbf{i}} + A_y \hat{\mathbf{j}} + A_z \hat{\mathbf{k}}) \cdot (B_x \hat{\mathbf{i}} + B_y \hat{\mathbf{j}} + B_z \hat{\mathbf{k}})$$
$$= A_x B_x + A_y B_y + A_z B_z$$

Vector or Cross Product of Two Vectors

The vector product of two vectors is equal to the product of their magnitudes and the sine of the smaller angle between them. It is denoted by × (cross).

$$\mathbf{A} \times \mathbf{B} = AB \sin \theta \, \hat{\mathbf{n}}$$

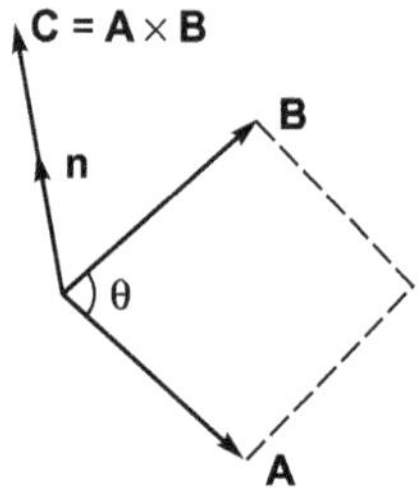

The direction of unit vector $\hat{\mathbf{n}}$ can be obtained from right hand thumb rule.

If fingers of right hand are curled from **A** to **B** through smaller angle between them, then thumb will represent the direction of vector $(\mathbf{A} \times \mathbf{B})$.

The vector or cross product of two vectors is also a vector.

Properties of Vector Product

(i) Vector product is not commutative, i.e.

$$\mathbf{A} \times \mathbf{B} \neq \mathbf{B} \times \mathbf{A} \qquad [\because (\mathbf{A} \times \mathbf{B}) = -(\mathbf{B} \times \mathbf{A})]$$

(ii) Vector product is distributive, i.e.

$$\mathbf{A} \times (\mathbf{B} + \mathbf{C}) = \mathbf{A} \times \mathbf{B} + \mathbf{A} \times \mathbf{C}$$

(iii) Vector product of two parallel vectors is zero, *i.e.*,

$$\mathbf{A} \times \mathbf{B} = AB \sin 0° = 0$$

(iv) Vector product of any vector with itself is zero.

$$\mathbf{A} \times \mathbf{A} = AA \sin 0° = 0$$

(v) Vector product of orthogonal unit vectors

$$\hat{\mathbf{i}} \times \hat{\mathbf{i}} = \hat{\mathbf{j}} \times \hat{\mathbf{j}} = \hat{\mathbf{k}} \times \hat{\mathbf{k}} = 0$$

and $$\hat{\mathbf{i}} \times \hat{\mathbf{j}} = \hat{\mathbf{k}}, \quad \hat{\mathbf{j}} \times \hat{\mathbf{k}} = \hat{\mathbf{i}}, \quad \hat{\mathbf{k}} \times \hat{\mathbf{i}} = \hat{\mathbf{j}}$$

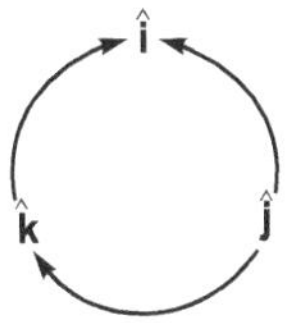

(vi) Vector product in cartesian coordinates

$$\mathbf{A} \times \mathbf{B} = (A_x\hat{\mathbf{i}} + A_y\hat{\mathbf{j}} + A_z\hat{\mathbf{k}}) \times (B_x\hat{\mathbf{i}} + B_y\hat{\mathbf{j}} + B_z\hat{\mathbf{k}})$$

$$= \begin{vmatrix} \hat{\mathbf{i}} & \hat{\mathbf{j}} & \hat{\mathbf{k}} \\ A_x & A_y & A_z \\ B_x & B_y & B_z \end{vmatrix}$$

$$= (A_yB_z - A_zB_y)\,\hat{\mathbf{i}} - (A_xB_z - B_xA_z)\,\hat{\mathbf{j}} + (A_xB_y - A_yB_x)\,\hat{\mathbf{k}}$$

Direction of Vector Cross Product

When $\mathbf{C} = \mathbf{A} \times \mathbf{B}$, the direction of **C** is at right angles to the plane containing the vectors **A** and **B**. The direction is determined by the right hand screw rule and right hand thumb rule.

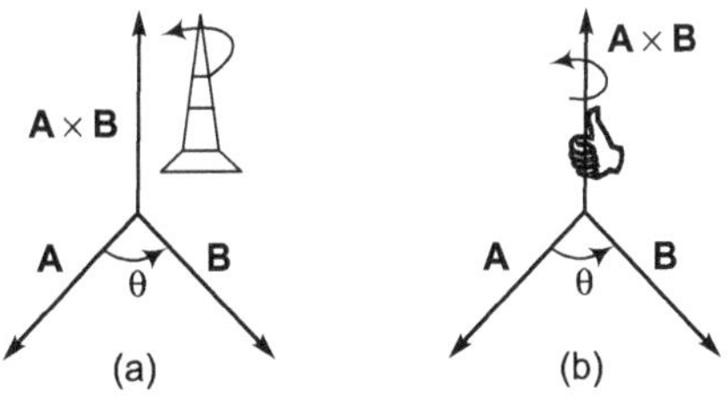

(a) (b)

(i) **Right Hand Screw Rule** Rotate a right handed screw from first vector (**A**) towards second vector (**B**). The direction in which the right handed screw moves gives the direction of vector (**C**).

(ii) **Right Hand Thumb Rule** Curl the fingers of your right hand from **A** to **B**. Then, the direction of the erect thumb will point in the direction of $\mathbf{A} \times \mathbf{B}$.

3

Motion in a Straight Line

Motion

If an object changes its position with respect to its surroundings with time, then it is called in motion.

Rest

If an object does not change its position with respect to its surroundings with time, then it is called at rest.

Rest and motion are relative states. It means an object which is at rest in one frame of reference can be in motion in another frame of reference at the same time.

Point Object An object can be considered as a point object, if the distance travelled by it in motion is very large in comparison to its dimensions.

Types of Motion

1. One Dimensional Motion

If only one out of three coordinates specifying the position of the object changes with respect to time, then the motion is called one dimensional motion. or rectilinear motion.

For instance, motion of a block in a straight line, motion of a train along a straight track, a man walking on a level and narrow road and object falling under gravity etc.

2. Two Dimensional Motion

If only two out of three coordinates specifying the position of the object changes with respect to time, then the motion is called two dimensional motion.

A circular motion is an instance of two dimensional motion.

3. Three Dimensional Motion

If all the three coordinates specifying the position of the object changes with respect to time, then the motion is called three dimensional motion. A few instances of three dimension are flying bird, a flying kite, a flying aeroplane, the random motion of gas molecule etc.

Frame of Reference

The most convenient system is a rectangular coordinate system of three mutually perpendicular axes as x, y, and zi. This point of intersection of these three axes is called origin (o) and considered as the reference point. The x, y, z-coordinates describe the position of the object w.r.t the coordinate system. This coordinate system along with a clock constitutes a frame of reference.

Distance or Path Length Covered

The length of the actual path by an object is called the distance.

It is a scalar quantity and it can never be zero or negative during the motion of an object.

Its unit is metre.

Displacement

The shortest distance between the initial and final positions of any object during motion is called displacement. The displacement of an object in a given time can be positive, zero or negative.

It is a vector quantity.

Its unit is metre.

Uniform motion

If an object moving along the straight line covers equal distances in equal intervals of time, it is said to be in uniform motion along a straight line.

Speed

The time rate of change of position of the object in any direction is called speed of the object.

$$\text{Speed } (v) = \frac{\text{Distance travelled } (s)}{\text{Time taken } (t)}$$

Its unit is m/s.
It is a scalar quantity.
Its dimensional formula is $[M^0LT^{-1}]$.

Uniform Speed

If an object covers equal distances in equal intervals of time, then its speed is called uniform speed.

Non-uniform or Variable Speed

If an object covers unequal distances in equal intervals of time, then its speed is called non-uniform or variable speed.

Average Speed

The ratio of the total distance travelled by the object to the total time taken is called average speed of the object.

$$\text{Average speed} = \frac{\text{Total distance travelled}}{\text{Total time taken}}$$

If a particle travels distances $s_1, s_2, s_3, \ldots$ with speeds $v_1, v_2, v_3, \ldots$, then

$$\text{Average speed} = \frac{s_1 + s_2 + s_3 + \ldots}{\left(\frac{s_1}{v_1} + \frac{s_2}{v_2} + \frac{s_3}{v_3} + \ldots\right)}$$

If particle travels equal distances ($s_1 = s_2 = s$) with velocities v_1 and v_2, then,

$$\text{Average speed} = \frac{2v_1v_2}{(v_1 + v_2)}$$

If a particle travels with speeds $v_1, v_2, v_3, \ldots$ during time intervals $t_1, t_2, t_3, \ldots$, then

$$\text{Average speed} = \frac{v_1t_1 + v_2t_2 + v_3t_3 + \ldots}{t_1 + t_2 + t_3 + \ldots}$$

If particle travels with speeds v_1 and v_2 for equal time intervals, i.e. $t_1 = t_2 = t$, then

$$\text{Average speed} = \frac{v_1 + v_2}{2}$$

When a body travels equal distance with speeds V_1 and V_2, the average speed (v) is the harmonic mean of two speeds.

$$\frac{2}{v} = \frac{1}{v_1} + \frac{1}{v_2}$$

Instantaneous Speed

When an object is travelling with variable speed, then its speed at a given instant of time is called its instantaneous speed.

$$\text{Instantaneous speed} = \lim_{\Delta t \to 0} \frac{\Delta s}{\Delta t} = \frac{ds}{dt}$$

Velocity

The rate of change of displacement of an object in a particular direction is called its velocity.

$$\text{Velocity} = \frac{\text{Displacement}}{\text{Time taken}}$$

Its unit is m/s.

Its dimensional formula is $[M^0LT^{-1}]$.

It is a vector quantity, as it has both, the magnitude and direction.

The velocity of an object can be positive, zero and negative.

Uniform Velocity

If an object undergoes equal displacements in equal intervals of time, then it is said to be moving with a uniform velocity.

Non-uniform or Variable Velocity

If an object undergoes unequal displacements in equal intervals of time, then it is said to be moving with a non-uniform or variable velocity.

Relative Velocity

Relative velocity of one object with respect to another object is the time rate of change of relative position of one object with respect to another object.

Relative velocity of object A with respect to object B

$$v_{AB} = \mathbf{v}_A - \mathbf{v}_B$$

When two objects are moving in the same direction, then

$$\mathbf{v}_{AB} = \mathbf{v}_A - \mathbf{v}_B$$

or $$v_{AB} = v_A - v_B$$

When two objects are moving in opposite direction, then

$$\mathbf{v}_{AB} = \mathbf{v}_A + \mathbf{v}_B$$

or $$v_{AB} = v_A + v_B$$

When two objects are moving at an angle, then

$$v_{AB} = \sqrt{v_A^2 + v_B^2 - 2v_A v_B \cos\theta}$$

and $$\tan\beta = \frac{v_B \sin\theta}{v_A - v_B \cos\theta}$$

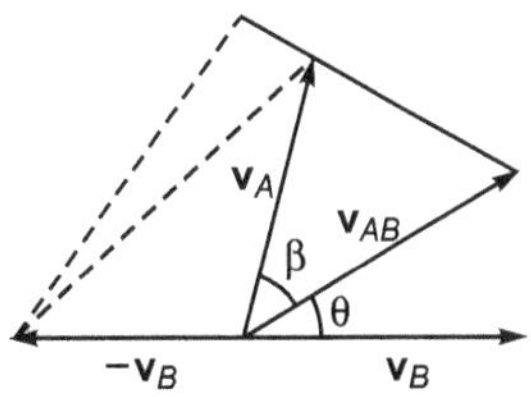

Average Velocity

The ratio of the total displacement to the total time taken is called average velocity.

$$\text{Average velocity} = \frac{\text{Total displacement}}{\text{Total time taken}}$$

Acceleration

The rate of change of velocity with time is called acceleration.

$$\text{Acceleration } (a) = \frac{\text{Change in velocity } (\Delta v)}{\text{Time interval } (\Delta t)}$$

Its unit is m/s^2.

Its dimensional formula is $[M^0 LT^{-2}]$.

It is a vector quantity.

Acceleration can be positive, zero or negative. Positive acceleration means velocity increasing with time, zero acceleration means velocity is uniform while negative acceleration (retardation) means velocity is decreasing with time.

If a particle is accelerated for a time t_1 with acceleration a_1 and for a time t_2 with acceleration a_2, then average acceleration

$$a_{av} = \frac{a_1 t_1 + a_2 t_2}{t_1 + t_2}$$

Different Graphs of Motion

Displacement-Time Graph

S. No.	Condition	Graph
(a)	For a stationary body	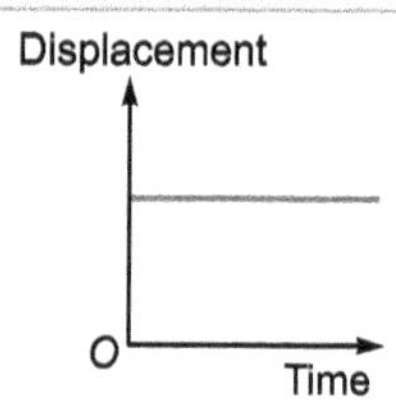
(b)	Body moving with a constant velocity	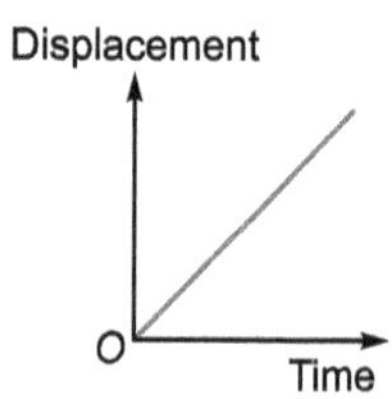

S. No.	Condition	Graph
(c)	Body moving with a constant acceleration	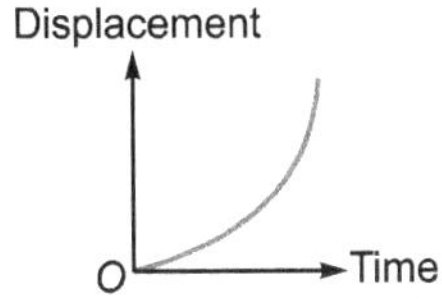
(d)	Body moving with a constant retardation	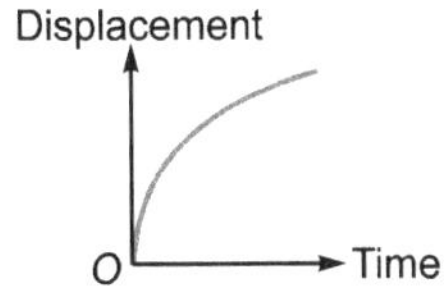
(e)	Body moving with infinite velocity. But such motion of a body is never possible.	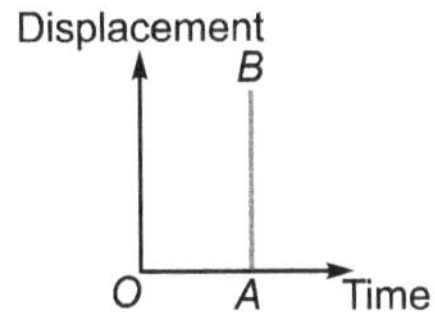

Note *Slope of displacement-time graph gives average velocity.*

Velocity-Time Graph

S. No.	Condition	Graph
(a)	Moving with a constant velocity	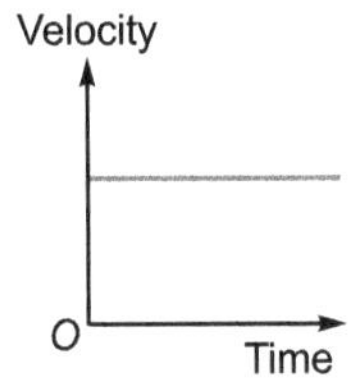
(b)	Moving with a constant acceleration having zero initial velocity	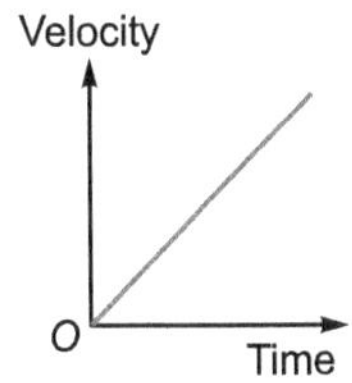
(c)	Body moving with a constant retardation and its initial velocity is not zero.	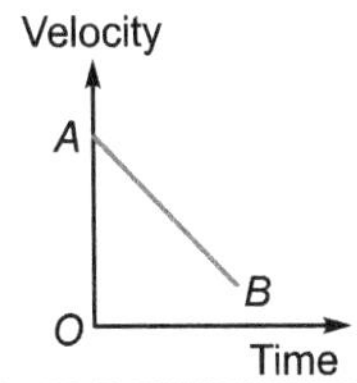

S. No.	Condition	Graph
(d)	Moving with a constant retardation	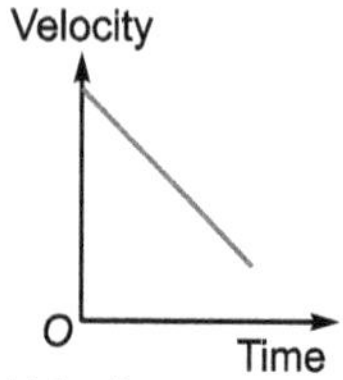
(e)	Moving with increasing acceleration	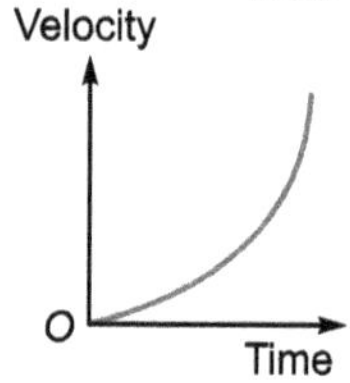
(f)	Moving with decreasing acceleration	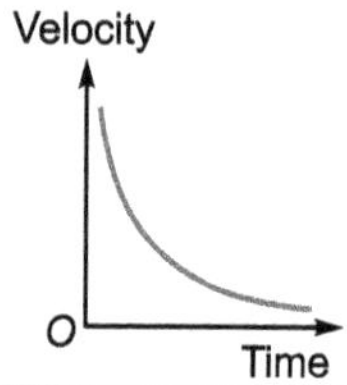

Note *Slope of velocity-time graph gives average acceleration.*

Acceleration-Time Graph

S. No.	Condition	Graph
(a)	When object is moving with constant acceleration	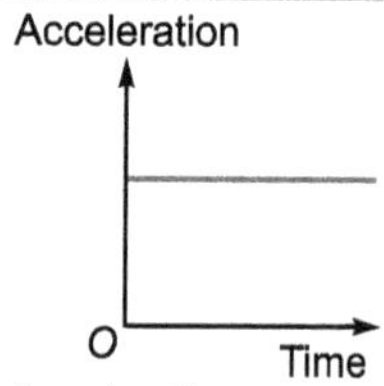
(b)	When object is moving with constant increasing acceleration	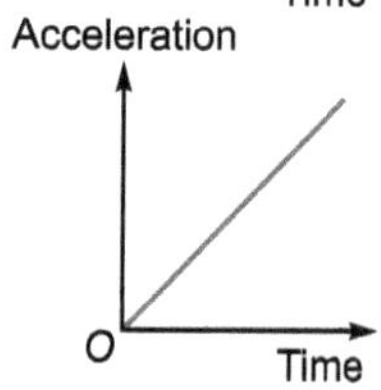
(c)	When object is moving with constant decreasing acceleration	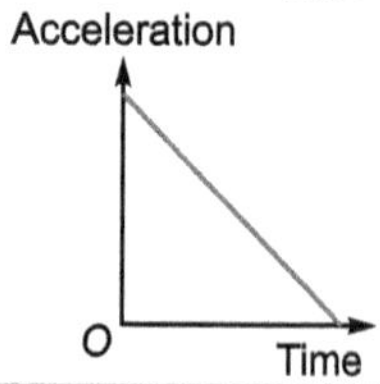

Equations of Uniformly Accelerated Motion

If a body starts with velocity (u) and after time t its velocity changes to v, if the uniform acceleration is a and the distance travelled in time t in s, then the following relations are obtained, which are called equations of uniformly accelerated motion.

(i) $v = u + at$ (ii) $s = ut + \frac{1}{2}at^2$ (iii) $v^2 = u^2 + 2as$

(iv) Distance travelled in nth second.

$$S_n = u + \frac{a}{2}(2n - 1)$$

If a body moves with uniform acceleration and velocity changes from u to v in a time interval, then the velocity at the mid point of its path

$$= \frac{\sqrt{u^2 + v^2}}{2}$$

Motion Under Gravity

If an object is falling freely ($u = 0$) under gravity, then equations of motion

(i) $v = u + gt$ (ii) $h = ut + \frac{1}{2}gt^2$ (iii) $v^2 = u^2 + 2gh$

Note *If an object is thrown upward then g is replaced by – g in above three equations.*

It thus follows that:

(i) Time taken to reach maximum height, $t_A = \frac{u}{g} = \sqrt{\frac{2h}{g}}$

(ii) Maximum height reached by the body, $h_{max} = \frac{u^2}{2g}$

(iii) A ball is dropped from a building of height h and it reaches after t seconds on earth. From the same building if two ball are thrown (one upwards and other downwards) with the same velocity u and they reach the earth surface after t_1 and t_2 seconds respectively, then

$$t = \sqrt{t_1 t_2}$$

(iv) When a body is dropped freely from the top of the tower and another body is projected horizontally from the same point, both will reach the ground at the same time.

4

Projectile Motion and Circular Motion

Projectile Motion

When any object is thrown from horizontal at an angle θ except 90°, then the path followed by it is called **trajectory,** the object is called **projectile** and its motion is called projectile motion.

If any object is thrown with velocity u, making an angle θ, from horizontal, then

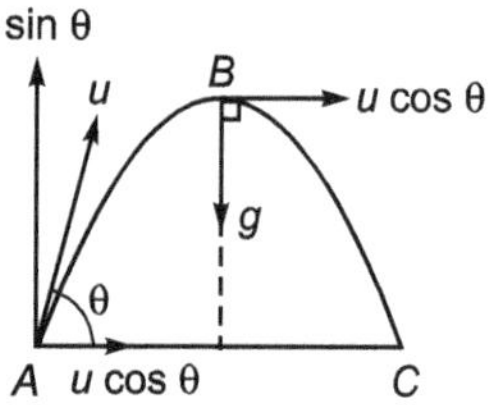

Horizontal component of initial velocity = $u \cos\theta$.

Vertical component of initial velocity = $u \sin\theta$.

Horizontal component of velocity ($u \cos\theta$) remains same during the whole journey as no acceleration is acting horizontally.

Vertical component of velocity ($u \sin\theta$) decreases gradually and becomes zero at highest point of the path.

At highest point, the velocity of the body is $u \cos\theta$ in horizontal direction and the angle between the velocity and acceleration is 90°.

Important Points and Formulae of Projectile Motion

(i) At highest point, the linear momentum is $mu\cos\theta$ and the kinetic energy is $\frac{1}{2}m(u\cos\theta)^2$.

(ii) The horizontal displacement of the projectile after t seconds

$$x = (u\cos\theta)\,t$$

(iii) The vertical displacement of the projectile after t seconds

$$y = (u\sin\theta)\,t - \frac{1}{2}gt^2$$

(iv) Equation of the path of projectile

$$y = x\tan\theta - \frac{g}{2u^2\cos^2\theta}x^2$$

(v) The path of a projectile is parabolic.

(vi) Kinetic energy at lowest point $= \frac{1}{2}mu^2$

(vii) Linear momentum at lowest point $= mu$

(viii) Acceleration of projectile is constant throughout the motion and it acts vertically downwards being equal to g.

(ix) Angular momentum of projectile $= mu\cos\theta \times h$, where h denotes the height.

(x) In case of angular projection, the angle between velocity and acceleration varies from $0° < \theta < 180°$.

(xi) The maximum height occurs when the projectile covers a horizontal distance equal to half of the horizontal range, *i.e.*, $R/2$.

(xii) When the maximum range of projectile is R, then its maximum height is $R/4$.

Time of flight It is defined as the total time for which the projectile remains in air.

$$T = \frac{2u\sin\theta}{g}$$

Maximum height It is defined as the maximum vertical distance covered by projectile.

$$H = \frac{u^2\sin^2\theta}{2g}$$

Horizontal range It is defined as the maximum distance covered in horizontal distance.

$$R = \frac{u^2 \sin 2\theta}{g}$$

(i) Horizontal range is maximum when it is thrown at an angle of 45° from the horizontal

$$R_{max} = \frac{u^2}{g}$$

(ii) For angle of projection θ and (90° – θ) the horizontal range is same.

Projectile Projected from Some Heights

1. When Projectile is Projected Horizontally

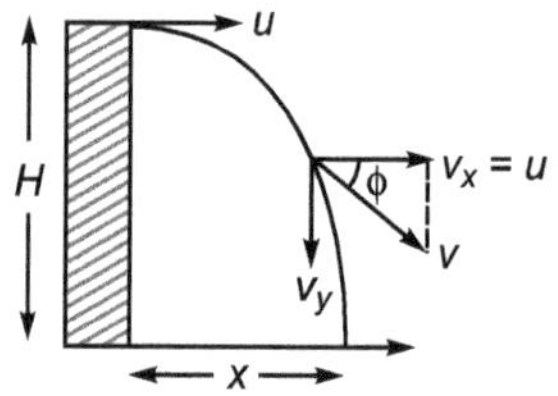

Initial velocity in vertical direction = 0

Time of flight $T = \sqrt{\frac{2H}{g}}$

Horizontal range $x = uT = u\sqrt{\frac{2H}{g}}$

Vertical velocity after t seconds

$$v_y = gt \qquad (\because u_y = 0)$$

Velocity of projectile after t seconds

$$v = \sqrt{v_x^2 + v_y^2} = \sqrt{u^2 + (gt)^2}$$

If velocity makes an angle φ from horizontal, then

$$\tan \phi = \frac{v_y}{v_x} = \frac{gt}{u}$$

Equation of the path of the projectile

$$y = -\frac{g}{2u^2}x^2$$

2. When Projectile Projected Downward at an Angle with Horizontal

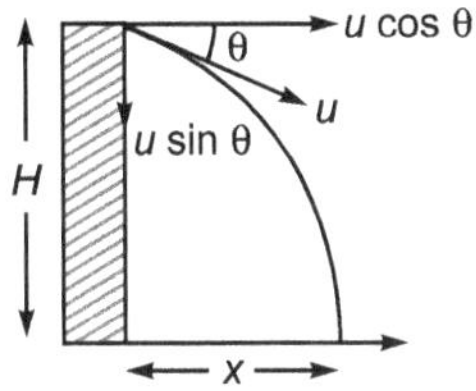

Initial velocity in horizontal direction = $u \cos\theta$

Initial velocity in vertical direction = $u \sin\theta$

Time of flight can be obtained from the equation,

$$H = u \sin\theta t + \frac{1}{2} gt^2$$

Horizontal range $x = (u \cos\theta)\, t$

Vertical velocity after t seconds

$$v_y = u \sin\theta + gt$$

Velocity of projectile after t seconds

$$v = \sqrt{v_x^2 + v_y^2} = \sqrt{(u \cos\theta)^2 + (u \sin\theta + gt)^2}$$

$$= \sqrt{u^2 + (gt)^2 + 2ugt \sin\theta}$$

3. When Projectile Projected Upward at an Angle with Horizontal

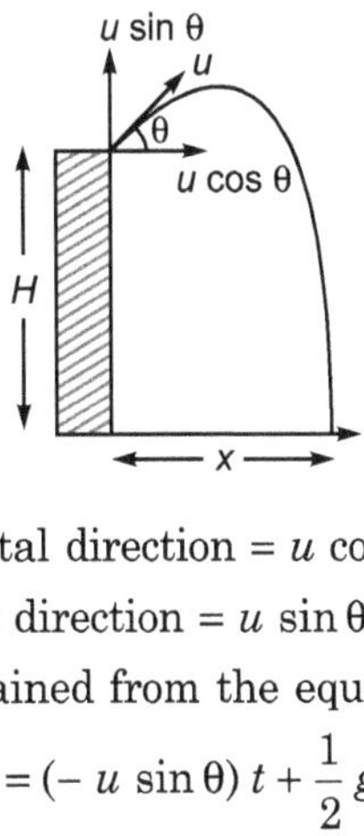

Initial velocity in horizontal direction = $u \cos\theta$

Initial velocity in vertical direction = $u \sin\theta$

Time of flight can be obtained from the equation

$$H = (- u \sin\theta)\, t + \frac{1}{2} gt^2$$

Horizontal range $x = (u \cos \theta) t$

Vertical velocity after t seconds, $v_y = (-u \sin \theta) + gt$

Velocity of projectile after t seconds

$$v = \sqrt{v_x^2 + v_y^2} = \sqrt{u^2 + (gt - u \sin \theta)^2}$$

$$= \sqrt{u^2 + (gt)^2 - 2ugt \sin \theta}$$

4. Projectile Motion on an Inclined Plane

When any object is thrown with velocity u making an angle α from horizontal, at a plane inclined at an angle β from horizontal, then

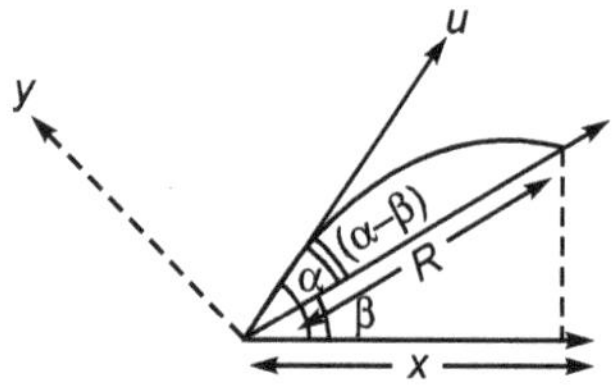

Initial velocity along the inclined plane $= u \cos(\alpha - \beta)$

Initial velocity perpendicular to the inclined plane $= u \sin(\alpha - \beta)$

Acceleration along the inclined plane $= g \sin \beta$

Acceleration perpendicular to the inclined plane $= g \cos \beta$

Time of flight $T = \dfrac{2u \sin(\alpha - \beta)}{g \cos \beta}$

Horizontal range $x = \dfrac{2u^2 \sin(\alpha - \beta) \cos \alpha}{g \cos \beta}$

Range on inclined plane

$$R = \frac{x}{\cos \beta} = \frac{2u^2 \sin(\alpha - \beta) \cos \alpha}{g \cos^2 \beta}$$

Range on inclined plane will be maximum, when

$$\alpha = 45° + \frac{\beta}{2}$$

$$R_{\max} = \frac{u^2}{g(1 + \sin \beta)}$$

For angle of projections α and $(90° - \alpha + \beta)$, the range on inclined plane are same.

Circular Motion

Circular motion is the movement of an object in a circular path.

1. Uniform Circular Motion

If the magnitude of the velocity of the particle in circular motion remains constant, then it is called uniform circular motion.

2. Non-uniform Circular Motion

If the magnitude of the velocity of the body in circular motion is not constant, then it is called non-uniform circular motion.

Note *A special kind of circular motion is when an object rotates around itself. This can be called* ***spinning motion.***

Variables in Circular Motion

(i) **Angular Displacement** Angular displacement is the angle subtended by the position vector at the centre of the circular path.

$$\text{Angular displacement } (\Delta\theta) = \frac{\Delta s}{r}$$

where Δs is the linear displacement and r is the radius.

Its unit is radian.

(ii) **Angular Velocity** The time rate of change of angular displacement $(\Delta\theta)$ is called angular velocity.

$$\text{Angular velocity } (\omega) = \frac{\Delta\theta}{\Delta t}$$

Angular velocity is a vector quantity and its unit is rad/s.

Relation between linear velocity (v) and angular velocity (ω) is given by

$$v = r\omega$$

(iii) **Angular Acceleration** The rate of change of angular velocity $(d\omega)$ is called angular acceleration.

$$\text{Angular acceleration } (\alpha) = \frac{d\omega}{dt} = \frac{d^2\theta}{dt^2}$$

Its unit is rad/s^2 and dimensional formula is $[T^{-2}]$.

Relation between linear acceleration (a) and angular acceleration (α).

$$a = r\alpha$$

where, r = radius

Centripetal Acceleration

In circular motion, an acceleration acts on the body, whose direction is always towards the centre of the path. This acceleration is called centripetal acceleration.

$$\text{Centripetal acceleration } a = \frac{v^2}{r} = r\omega^2$$

Centripetal acceleration is also called radial acceleration as it acts along radius of circle.

Its unit is m/s^2 and it is a vector quantity.

Centripetal Force

It is that force which complex a body to move in a circular path.

It is directed along radius of the circle towards its centre.

For circular motion a centripetal force is required, which is not a new force but any force present there can act as centripetal force.

$$\text{Centripetal force } F = \frac{mv^2}{r} = mr\omega^2$$

where, m = mass of the body, v = linear velocity,
ω = angular velocity and r = radius.

Work done by the centripetal force is zero because the centripetal force and displacement are at right angles to each other.

Examples of some incidents and the cause of centripetal force involved.

S. No.	Incidents	Force Providing Centripetal Force
1.	Orbital motion of planets.	Gravitational force between planet and sun.
2.	Orbital motion of electron.	Electrostatic force between electron and nucleus.
3.	Turning of vehicles at turn.	Frictional force acting between tyres of vehicle and road.
4.	Motion of a stone in circular path, tied with a string.	Tension in the string.

Kinematical Equations in Circular Motion

Relations between different variables for an object executing circular motion are called kinematical equations in circular motion.

(i) $\omega = \omega_0 + \alpha t$ (ii) $\theta = \omega_0 t + \frac{1}{2}\alpha t^2$ (iii) $\omega^2 = \omega_0^2 + 2\alpha\theta$

where, ω_0 = initial angular velocity, ω = final angular velocity,
α = angular acceleration, θ = angular displacement and
t = time.

Centrifugal Force

Centrifugal force is equal and opposite to centripetal force.

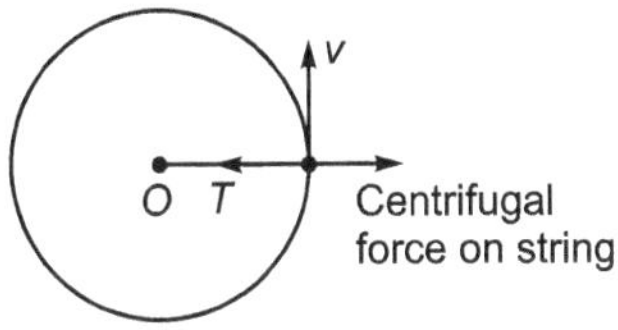

Under centrifugal force, body moves only along a straight line.

It appears when centripetal force ceases to exist.

Centrifugal force does not act on the body in an inertial frame but arises as pseudo forces in non-inertial frames and need to be considered.

Turning at Roads

If centripetal force is obtained only by the force of friction between the tyres of the vehicle and road, then for a safe turn, the coefficient of friction (μ_s) between the road and tyres should be,

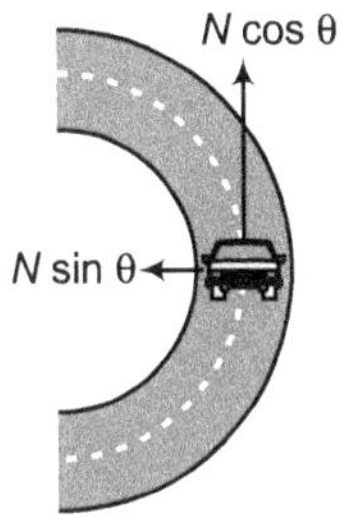

$$\mu_s \geq \frac{v^2}{rg} \quad \text{or} \quad v \leq \sqrt{\mu_s rg}$$

where, v = the velocity of the vehicle and r = radius of the circular path.

If centripetal force is obtained only by the banking of roads, then the speed (v) of the vehicle for a safe turn

$$v = \sqrt{rg \tan \theta}$$

If speed of the vehicle is less than $\sqrt{rg \tan \theta}$, than it will move inward (down) and r will decrease and if speed is more than $\sqrt{rg \tan \theta}$, then it will move outward (up) and r will increase.

In normal life, the centripetal force is obtained by the friction force between the road and tyres as well as by the banking of the roads.

Therefore, the maximum permissible speed for the vehicle is much greater than the optimum value of the speed on a banked road.

When centripetal force is obtained from friction force as well as banking of roads, then maximum safe value of speed of vehicle

$$v_{\text{max}} = \sqrt{\frac{rg(\tan \theta + \mu)}{(1 - \mu \tan \theta)}}$$

When a cyclist takes turn at road, he inclined himself from the vertical, slower down his speed and move on a circular path of larger radius.

If a cyclist inclined at an angle θ, then $\tan\theta = \frac{v^2}{rg}$

where, v = speed of the cyclist, r = radius of path and
g = acceleration due to gravity.

Motion in a Vertical Circle

(i) Minimum value of velocity at the highest point is $\sqrt{gr}$

(ii) The minimum velocity at the bottom required to complete the circle

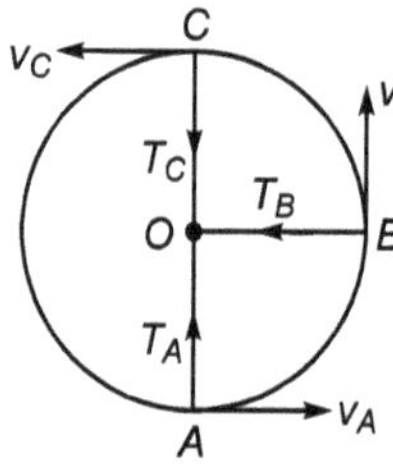

$$v_A = \sqrt{5gr}$$

(iii) Velocity of the body when string is in horizontal position

$$v_B = \sqrt{3gr}$$

(iv) **Tension in the string**

At the top $T_C = 0$,

At the bottom $T_A = 6\ mg$

When string is horizontal $T_B = 3\ mg$

(v) When a vehicle is moving over a convex bridge, then at the maximum height, reaction (N_1) is

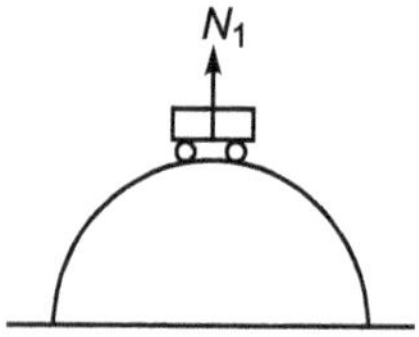

$$N_1 = mg - \frac{mv^2}{r}$$

(vi) When a vehicle is moving over a concave bridge, then at the lowest point, reaction (N_2) is

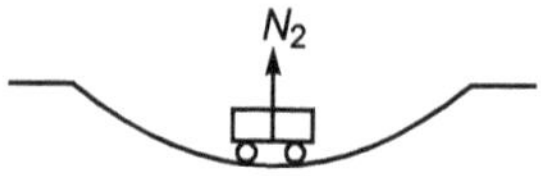

$$N_2 = mg + \frac{mv^2}{r}$$

(vii) When a car takes a turn, sometimes it overturns. During the overturning, it is the inner wheel which leaves the ground first.

(viii) A driver sees a child in front of him during driving a car, then it better to apply brake suddenly rather than taking a sharp turn to avoid an accident.

Non-uniform Horizontal Circular Motion

In non-uniform horizontal circular motion, the magnitude of the velocity of the body changes with time.

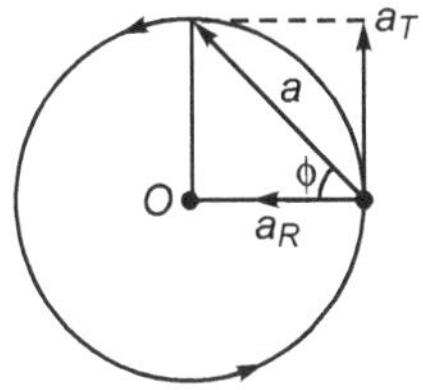

In this condition, centripetal (radial) acceleration (a_R) acts towards centre and a tangential acceleration (a_T) acts towards tangent. Both acceleration acts perpendicular to each other.

Resultant acceleration

$$a = \sqrt{a_R^2 + a_T^2} = \sqrt{\left(\frac{v^2}{r}\right)^2 + (r\alpha)^2}$$

and
$$\tan \phi = \frac{a_T}{a_R} = \frac{r^2\alpha}{v^2}$$

where, α is angular acceleration, r = radius and v = velocity.

Conical Pendulum

It consists of a string OA whose upper end O is fixed and bob is tied at the other free end. The string traces the surface of the cone, the arrangement is called a conical pendulum.

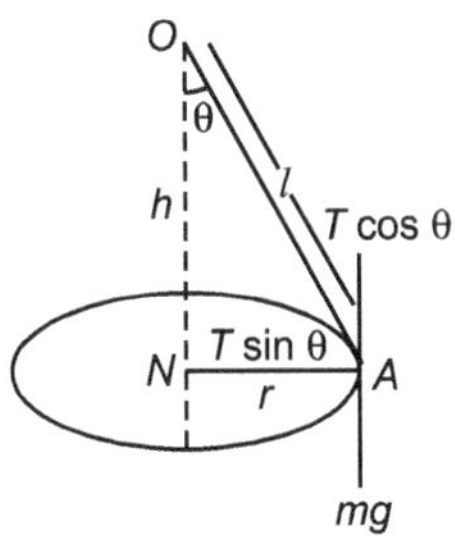

Time period of conical pendulum, $T = 2\pi \sqrt{\dfrac{l \cos \theta}{g}}$

5
Laws of Motion

Inertia

The property of an object by virtue of which it cannot change its state of rest or of uniform motion along a straight line its own, is called **inertia**.

Inertia is a measure of mass of a body. Greater the mass of a body greater will be its inertia or *vice-versa.*

Inertia is of three types:

(i) **Inertia of Rest** When a bus or train starts to move suddenly, the passangers sitting in it falls backward due to inertia of rest.

(ii) **Inertia of Motion** When a moving bus or train stops suddenly, the passangers sitting in it jerks in forward direction due to inertia of motion.

(iii) **Inertia of Direction** We can protect ourself from rain by an umbrella because rain drops can not change its direction its own due to inertia of direction.

Force

Force is a push or pull which changes or tries to change the state of rest, the state of uniform motion, size or shape of a body.

Its SI unit is newton (N) and its dimensional formula is $[MLT^{-2}]$.

Forces can be categorized into two types:

(i) **Contact Forces** Frictional force, tensional force, spring force, normal force, etc are the contact forces.

(ii) **Action at a Distance Forces** Electrostatic force, gravitational force, magnetic force, etc are action at a distance forces.

Impulsive Force

A force which acts on a body for a short interval of time, and produces a large change in momentum is called an impulsive force.

Linear Momentum

The total amount of motion present in a body is called its momentum. Linear momentum of a body is equal to the product of its mass and velocity. It is denoted by p.

Linear momentum $p = mv$.

Its SI unit is kg-m/s and dimensional formula is $[MLT^{-1}]$.

It is a vector quantity and its direction is in the direction of velocity of the body.

Impulse

The product of impulsive force and time for which it acts is called impulse.

Impulse = Force × Time = Change in momentum

Its SI unit is newton-second or kg-m/s and its dimension is $[MLT^{-1}]$.

It is a vector quantity and its direction is in the direction of force.

Newton's Laws of Motion

1. Newton's First Law of Motion

A body continues to be in its state of rest or in uniform motion along a straight line unless an external force is applied on it.
This law is also called **law of inertia**.

Examples

(i) When a carpet or a blanket is beaten with a stick then the dust particles separate out from it.

(ii) If a moving vehicle suddenly stops then the passengers inside the vehicle bend outward.

2. Newton's Second Law of Motion

The rate of change of linear momentum is proportional to the applied force and change in momentum takes place in the direction of applied force.

Mathematically, $$\mathbf{F} \propto \frac{d\mathbf{p}}{dt}$$

$$\mathbf{F} = k\frac{d}{dt}(m\mathbf{v})$$

where, k is a constant of proportionality and its value is one in SI and CGS system.

$$\mathbf{F} = \frac{md\mathbf{v}}{dt} = m\mathbf{a}$$

Examples

(i) It is easier for a strong adult to push a full shopping cart than it is for a baby to push the same cart. (This is depending on the net force acting on the object).

(ii) It is easier for a person to push an empty shopping cart than a full one (This is depending on the mass of the object).

3. Newton's Third Law of Motion

For every action there is an equal and opposite reaction and both acts on two different bodies.

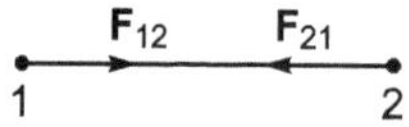

Mathematically $\mathbf{F}_{12} = -\mathbf{F}_{21}$

Examples

(i) Swimming becomes possible because of third law of motion.

(ii) Jumping of a man from a boat onto the bank of a river.

(iii) Jerk is produced in a gun when bullet is fired from it.

(iv) Pulling of cart by a horse.

Note *Newton's second law of motion is called real law of motion because first and third laws of motion can be obtained by it.*

The modern version of these laws is

(i) A body continues in its initial state of rest or motion with uniform velocity unless acted on by an unbalanced external force.

(ii) Forces always occur in pairs. If body A exerts a force on body B, an equal but opposite force is exerted by body B on body A.

Law of Conservation of Linear Momentum

If no external force acts on a system, then its total linear momentum remains conserved.

Linear momentum depends on the frame of reference but law of conservation of linear momentum is independent of frame of reference.

Newton's laws of motion are valid only in inertial frame of reference.

Equilibrium of a Particle

When the vector sum of the forces acting on a body is zero, the body is said to be in equilibrium.

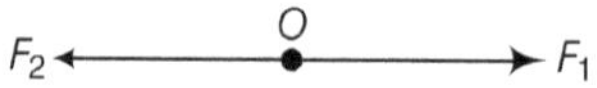

If two forces F_1 and F_2 act on a particle, then they will be in equilibrium if $F_1 = -F_2$.

Weight (w)

It is a field force. The force with which a body is pulled towards the centre of the earth due to gravity. It has the magnitude mg, where m is the mass of the body and g is the acceleration due to gravity.

$$w = mg$$

Apparent Weight in a Lift

(i) When a lift is at rest or moving with a constant speed, then

$$R = mg$$

The weighing machine will read the actual weight.

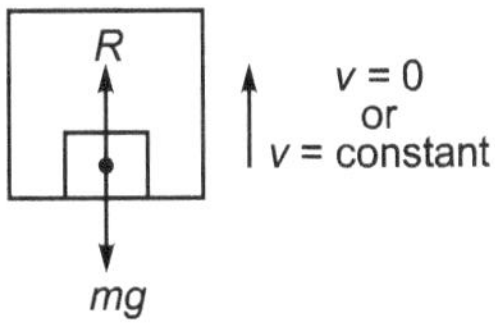

(ii) When a lift is accelerating upward, then apparent weight

$$R_1 = m(g + a)$$

The weighing machine will read the apparent weight, which is more than the actual weight.

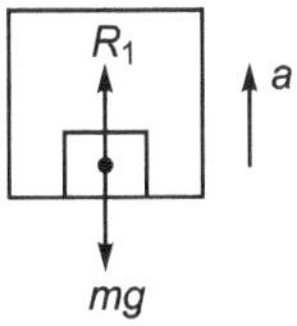

(iii) When a lift is accelerating downward, then apparent weight

$$R_2 = m\,(g - a)$$

The weighing machine will read the apparent weight, which is less than the actual weight.

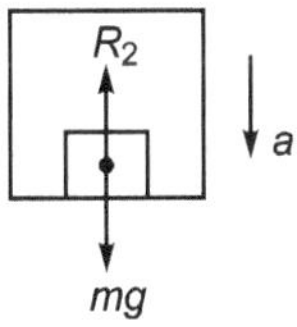

(iv) When lift is falling freely under gravity, then

$$R_2 = m\,(g - g) = 0$$

The apparent weight of the body becomes zero.

(v) If lift is accelerating downward with an acceleration greater than g, then body will lift from floor to the ceiling of the lift.

Rocket

Rocket is an example of variable mass following law of conservation of momentum.

Thrust on the rocket at any instant

$$F = -\,u\,\frac{dM}{dt}$$

where u = exhaust speed of the burnt gases and $\frac{dM}{dt}$ = rate of combustion of fuel.

Velocity of rocket at any instant is given by $v = v_0 + u \log_e \left(\frac{M_0}{M}\right)$

where, v_0 = initial velocity of the rocket,
M_0 = initial mass of the rocket and
M = present mass of the rocket.

If effect of gravity is taken into account then speed of rocket

$$v = v_0 + u \log_e \left(\frac{M_0}{M}\right) - gt$$

Friction

A force acting on the point of contact of the objects, which opposes the relative motion is called **friction.**

It acts parallel to the contact surfaces.

Frictional forces are produced due to intermolecular interactions acting between the molecules of the bodies in contact.

Friction is of three types:

1. Static Friction

It is an opposing force which comes into play when one body tends to move over the surface of the other body but actual motion is not taking place.

Static friction is a self adjusting force which increases as the applied force is increased. Static friction opposes impending motion.

Impending Motion

The motion that would take place under the applied force, if friction were absent.

2. Limiting Friction

It is the maximum value of static friction when body is at the verge of starting motion.

$$\text{Limiting friction } (f_s) = \mu_s R$$

where μ_s = coefficient of limiting friction and R = normal reaction.

Limiting friction do not depend on area of contact surfaces but depends on their nature, *i.e.*, smoothness or roughness.

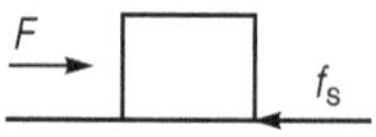

If angle of friction is θ, then coefficient of limiting friction

$$\mu_s = \tan\theta$$

3. **Kinetic Friction**

If the body begins to slide on the surface, the magnitude of the frictional force rapidly decreases to a constant value f_k kinetic friction.

$$\text{Kinetic friction, } f_k = \mu_k N$$

where $\mu_k =$ coefficient of kinetic friction and N = normal force.

Kinetic friction is of two types:

(a) Sliding friction

(b) Rolling friction

As, rolling friction < sliding friction, therefore it is easier to roll a body than to slide.

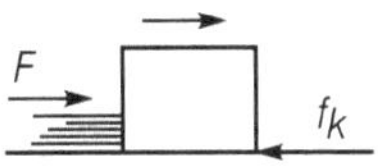

$$\text{Kinetic friction } (f_k) = \mu_k R$$

where μ_k = coefficient of kinetic friction and R = normal reaction.

Angle of repose or angle of sliding It is the minimum angle of inclination of a plane with the horizontal, such that a body placed on it, just begins to slide down.

If angle of repose is α and coefficient of limiting friction is μ, then

$$\mu_s = \tan \alpha$$

Motion on an Inclined Plane

When an object moves along an inclined plane then different forces act on it like normal reaction of plane, friction force acting in opposite direction of motion etc. Different relations for the motion are given below.

Normal reaction of plane

$$R = mg \cos \theta$$

and net force acting downward on the block.

$$F = mg \sin \theta - f$$

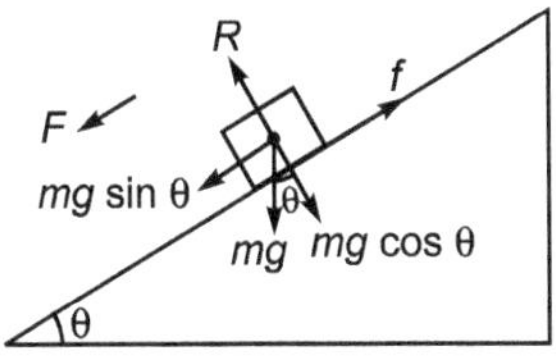

Acceleration on inclined plane $a = g(\sin \theta - \mu \cos \theta)$

When angle of inclination of the plane from horizontal is less than the angle of repose (α), then

(i) minimum force required to move the body up the inclined plane

$$f_1 = mg\,(\sin\theta + \mu\cos\theta)$$

(ii) minimum force required to push the body down the inclined plane

$$f_2 = mg\,(\mu\cos\theta - \sin\theta)$$

Tension

Tension force always pulls a body.

Tension is a reactive force. It is not an active force.

Tension across a massless pulley or frictionless pulley remain constant.

Rope becomes slack when tension force becomes zero.

Motion of Bodies in Contact

(i) **Two Bodies in Contact** If F force is applied on object of mass m_1, then acceleration of the bodies

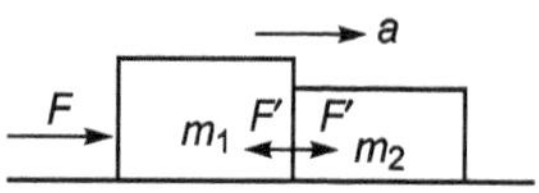

$$a = \frac{F}{(m_1 + m_2)}$$

Contact force on $m_1 = m_1 a = \dfrac{m_1 F}{(m_1 + m_2)}$

Contact force on $m_2 = m_2 a = \dfrac{m_2 F}{(m_1 + m_2)}$

(ii) **Three Bodies in Contact** If F force is applied an object of mass m_1, then acceleration of the bodies $= \dfrac{F}{(m_1 + m_2 + m_3)}$

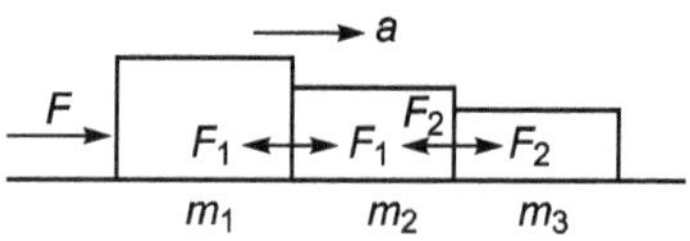

Contact force between m_1 and m_2

$$F_1 = \frac{(m_2 + m_3)\,F}{(m_1 + m_2 + m_3)}$$

Contact force between m_2 and m_3

$$F_2 = \frac{m_3 F}{(m_1 + m_2 + m_3)}$$

(iii) **Motion of Two Bodies, One Resting on the Other**

(a) The coefficient of friction between surface of A and B be μ. If a force F is applied on the lower body A, then common acceleration of two bodies

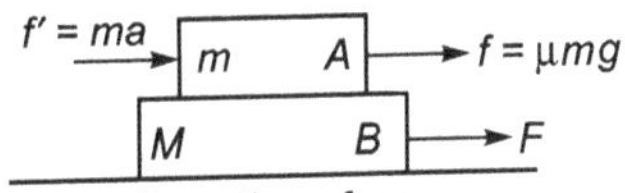

$$a = \frac{F}{(M + m)}$$

Pseudo force acting on block B due to the accelerated motion

$$f' = ma$$

The pseudo force tends to produce a relative motion between bodies A and B and consequently a frictional force

$f = \mu N = \mu mg$ is developed. For equilibrium

$$ma \le \mu mg \quad \text{or} \quad a \le \mu g$$

(b) Let friction is also present between the ground surface and body A. Let the coefficient of friction between the given surface and body A is μ_1 and the coefficient of friction between the surfaces of bodies A and B is μ_2. If a force F is applied on the lower body A.

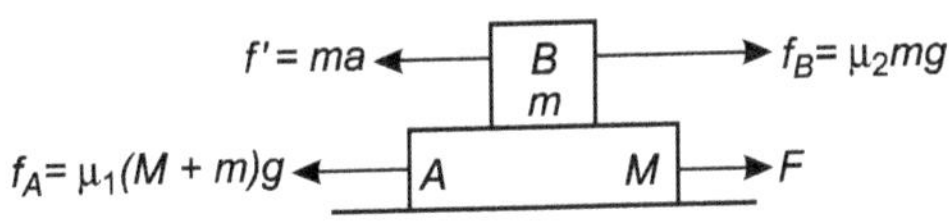

Net accelerating force $= F - f_A = F - \mu_1(M + m)g$

$\therefore$ Net acceleration

$$a = \frac{F - \mu_1(M + m)g}{(M + m)}$$

$$= \frac{F}{(M + m)} - \mu_1 g$$

Pseudo force acting on the block B

$$f' = ma$$

The pseudo force tends to produce a relative motion between the bodies A and B are consequently a frictional force $f_B = \mu_2\, mg$ is developed. For equilibrium

$$ma \le \mu_2\, mg$$

or

$$a \le \mu_2\, g$$

If acceleration produced under the the effect of force F is more than $\mu_2 g$, then two bodies will not move together.

(iv) **Motion of Bodies Connected by Strings**

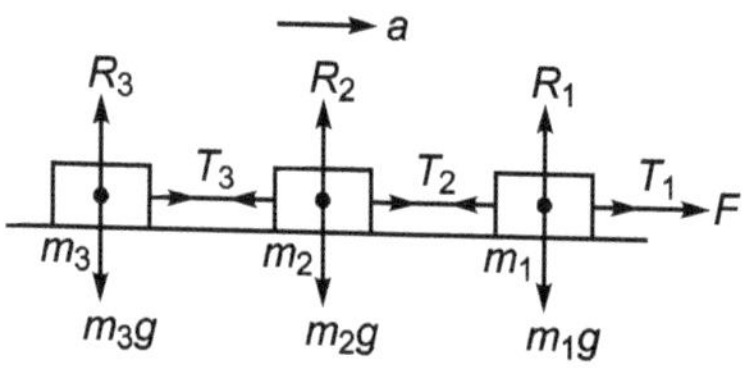

Acceleration of the system $a = \dfrac{F}{(m_1 + m_2 + m_3)}$

Tension in string $T_1 = F$

$$T_2 = (m_2 + m_3)\, a = \frac{(m_2 + m_3)\, F}{(m_1 + m_2 + m_3)}$$

$$T_3 = m_3 a = \frac{m_3 F}{(m_1 + m_2 + m_3)}$$

Pulley Mass System

(i) When unequal masses m_1 and m_2 are suspended from a pulley $(m_1 > m_2)$

$$m_1 g - T = m_1 a,$$

and

$$T - m_2 g = m_2 a$$

On solving equations, we get

$$a = \frac{(m_1 - m_2)}{(m_1 + m_2)}\, g,$$

$$T = \frac{2m_1 m_2}{(m_1 + m_2)}\, g$$

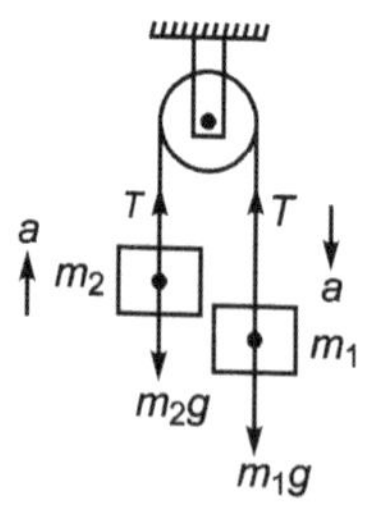

(ii) When a body of mass m_2 is placed on a frictionless horizontal surface, then

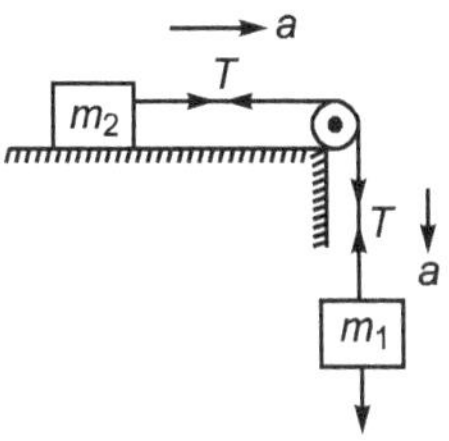

Acceleration $$a = \frac{m_1 g}{(m_1 + m_2)}$$

Tension in string $$T = \frac{m_1 m_2 g}{(m_1 + m_2)}$$

(iii) When a body of mass m_2 is placed on a rough horizontal surface, then

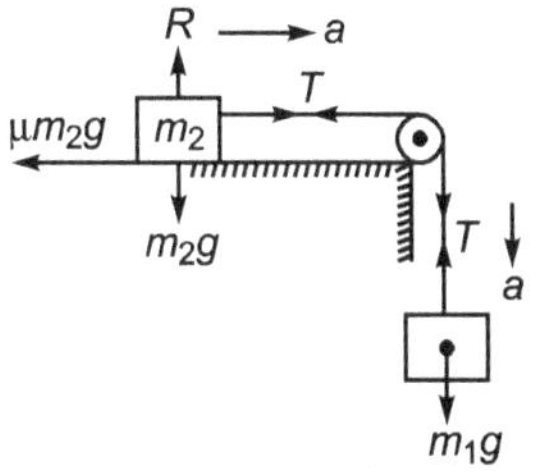

Acceleration $$a = \frac{(m_1 - \mu m_2) g}{(m_1 + m_2)}$$

Tension $$T = \frac{m_1 m_2 (1 + \mu) g}{(m_1 + m_2)}$$

(iv) When two masses m_1 and m_2 are connected to a single mass M as shown in figure, then

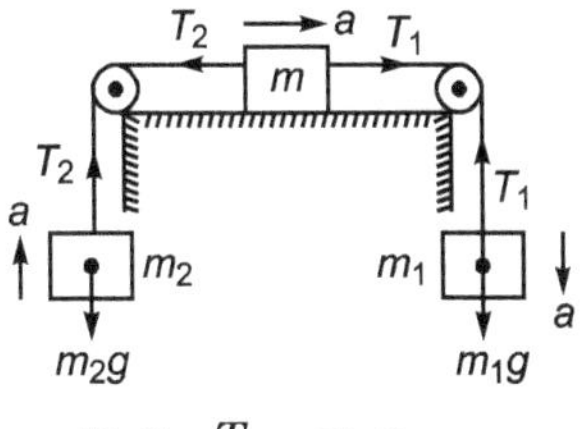

$$m_1 g - T_1 = m_1 a \quad \text{...(i)}$$

$$T_2 - m_2 g = m_2 a \quad \text{...(ii)}$$

$$T_1 - T_2 = Ma \quad \text{...(iii)}$$

Acceleration $$a = \frac{(m_1 - m_2)\,g}{(m_1 + m_2 + M)}$$

Tension $$T_1 = \left(\frac{2m_2 + M}{m_1 + m_2 + M}\right) m_1 g,$$

$$T_2 = \left(\frac{2m_1 + M}{m_1 + m_2 + M}\right) m_2 g$$

(v) Motion on a smooth inclined plane, then

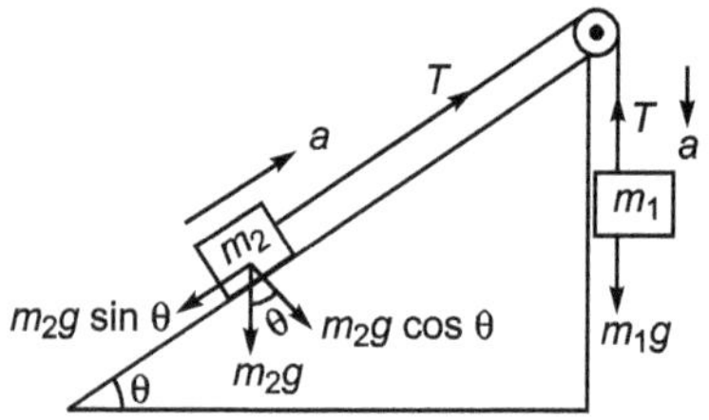

$$m_1 g - T = m_1 a \quad \text{...(i)}$$

$$T - m_2 g \sin\theta = m_2 a \quad \text{...(ii)}$$

Acceleration $$a = \left(\frac{m_1 - m_2 \sin\theta}{m_1 + m_2}\right) g$$

Tension $$T = \frac{m_1 m_2 (1 + \sin\theta)\,g}{(m_1 + m_2)}$$

(vi) Motion of two bodies placed on two inclined planes having different angle of inclination, then

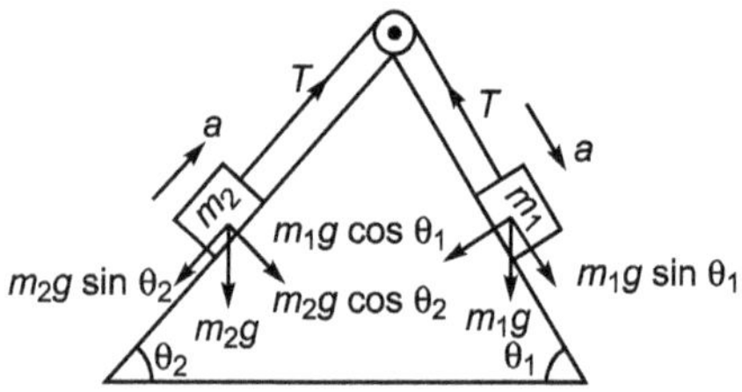

Acceleration $$a = \frac{(m_1 \sin\theta_1 - m_2 \sin\theta_2)\,g}{m_1 + m_2}$$

Tension $$T = \frac{m_1 m_2}{m_1 + m_2} (\sin\theta_1 + \sin\theta_2)\,g$$

6

Work, Power and Energy

Work

When a force acts on an object and the object actually moves in the direction of force, then the work is said to be done by the force.

Work done by the force is equal to the product of the force and the displacement of the object in the direction of force.

If under a constant force **F** the object displaced through a distance **s**, then work done by the force

$$W = \mathbf{F} \cdot \mathbf{s} = Fs \cos \theta$$

where θ is the smaller angle between **F** and **s**.

Work is a scalar quantity. Its SI unit is joule and CGS unit is erg.

$$\therefore \quad 1 \text{ joule} = 10^7 \text{ erg}$$

Its dimensional formula is $[ML^2T^{-2}]$.

Work done by a force is zero, if

(a) body is not displaced actually, *i.e.*, $\mathbf{s} = 0$

(b) body is displaced perpendicular to the direction of force, *i.e.*, $\theta = 90^\circ$

Work done by a force is **positive** if angle between **F** and **s** is acute angle.

Work done by a force is **negative** if angle between **F** and **s** is obtuse angle.

Work done by a constant force depends only on the initial and final positions of the object and not on the actual path followed between initial and final positions.

Work done in different conditions

(i) Work done by a variable force is given by

$$W = \int \mathbf{F} \cdot \mathbf{ds}$$

It is equal to the area under the force-displacement graph, along with proper sign.

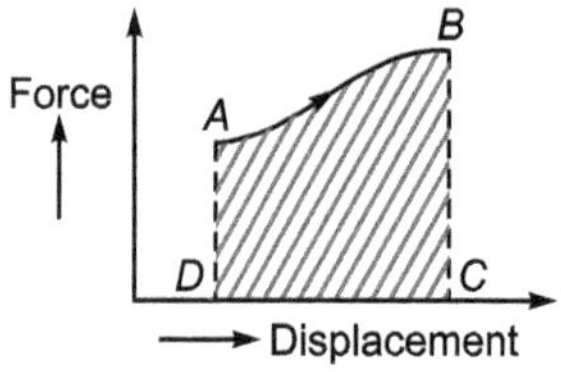

Work done = Area $ABCDA$

(ii) Work done in displacing any body under the action of a number of forces is equal to the work done by the resultant force.

(iii) In equilibrium (static or dynamic), the resultant force is zero, therefore resultant work done is zero.

(iv) If work done by a force during a rough trip of a system is zero, then the force is **conservative,** otherwise it is called **non-conservative** force.

- Gravitational force, electrostatic force, magnetic force, etc are conservative forces. All the central forces are conservative forces.
- Frictional force, viscous force, etc are non-conservative forces.

(v) Work done by the force of gravity on a particle of mass m is given by $W = mgh$

where g is acceleration due to gravity and h is height through particle one displaced.

(vi) Work done in compressing or stretching a spring is given by

$$W = -\frac{1}{2} kx^2$$

where k is spring constant and x is displacement from mean position.

(vii) When on end of a spring is attached to a fixed vertical support and a block attached to the free end moves on a horizontal table from $x = x_1$ to $x = x_2$ then $W = \frac{1}{2} k(x_2^2 - x_1^2)$

(viii) Work done by the couple for an angular displacement θ is given by $W = \tau \cdot \theta$

where τ is the torque of the couple.

Power

The rate of work done by a body is called its power.

$$\text{Power} = \text{Rate of doing work} = \frac{\text{Work done}}{\text{Time taken}}$$

If under a constant force **F** a body is displaced through a distance **s** in time t, the power, $P = \frac{W}{t} = \frac{\mathbf{F} \cdot \mathbf{s}}{t}$

But $\frac{\mathbf{s}}{t} = \mathbf{v}$, uniform velocity with which body is displaced.

$$\therefore \qquad P = \mathbf{F} \cdot \mathbf{v} = F\, v \cos\theta$$

where θ is the smaller angle between **F** and **v**.

Power is a scalar quantity. Its SI unit is watt and its dimensional formula is $[ML^2T^{-3}]$.

Its other units are kilowatt and horse power,

$$1 \text{ kilowatt} = 1000 \text{ watt}$$
$$1 \text{ horse power} = 746 \text{ watt}$$
$$1 \text{ kWh} = 3.6 \times 10^6 \text{J}$$

Energy

Energy of a body is its capacity of doing work.

It is a scalar quantity.

Its SI unit is joule and CGS unit is erg. Its dimensional formula is $[ML^2T^{-2}]$.

There are several types of energies, such as mechanical energy (kinetic energy and potential energy), chemical energy, light energy, heat energy, sound energy, nuclear energy, electric energy etc.

Mechanical Energy

The sum of kinetic and potential energies at any point remains constant throughout the motion. It does not depend upon time. This is known as law of conservation of mechanical energy.

Mechanical energy is of two types:

1. Kinetic Energy

The energy possessed by any object by virtue of its motion is called its kinetic energy.

Kinetic energy of an object is given by, $K = \frac{1}{2} mv^2 = \frac{p^2}{2m}$

where m = mass of the object, v = velocity of the object and $p = mv$ = momentum of the object.

2. Potential Energy

The energy possessed by any object by virtue of its position or configuration is called its potential energy.

There are three important types of potential energies:

(i) **Gravitational Potential Energy** If a body of mass m is raised through a height h against gravity, then its gravitational potential energy = mgh.

(ii) **Elastic Potential Energy** If a spring of spring constant k is stretched through a distance x, then elastic potential energy of the spring $= \frac{1}{2} kx^2$.

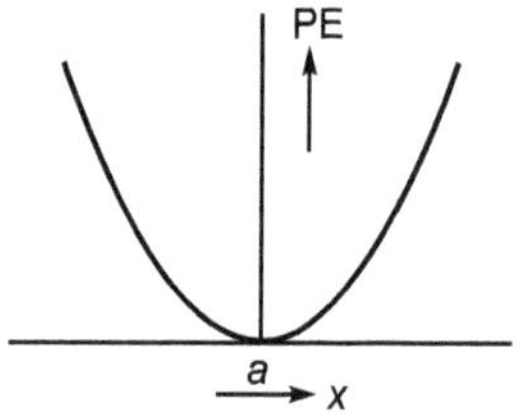

The variation of potential energy with distance is shown in figure.

Potential energy is defined only for conservative forces. It does not exist for non-conservative forces.

Potential energy depends upon frame of reference.

(iii) **Electric Potential Energy** The electric potential energy of two point charges q_1 and q_2 separated by a distance r in vacuum is given by

$$U = \frac{1}{4\pi\varepsilon_0} \cdot \frac{q_1 q_2}{r}$$

Here $$\frac{1}{4\pi\varepsilon_0} = 9.0 \times 10^9 \frac{\text{N-m}^2}{\text{C}^2} = \text{constant}$$

Various Forms of Energy

Heat Energy

A body possess heat energy due to the disorderly motion of its molecules. Heat energy is also related to the internal energy of the body.

Chemical Energy

A stable chemical compound has lesser energy than its constituent atoms, the difference being in arrangement and motion of the electron in the compound. This difference is called chemical energy.

If the total energy of the reactant is more than the product of the reaction, then heat is released and the reaction is said to be an **exothermic reaction.** If the reverse is true, then heat is absorbed and the reaction is **endothermic.**

Electrical Energy

It is the energy which is associated with the flow of electric current or with charging or discharging of a body.

Nuclear Energy

It is the binding energy of the nucleus of an atom. It is used in nuclear reactors, nuclear fission etc.

Work-Energy Theorem

Work done by a force in displacing a body is equal to change in its kinetic energy.

$$W = \int_{v_1}^{v_2} F \cdot ds = \frac{1}{2} mv_2^2 - \frac{1}{2} mv_1^2 = K_f - K_i = \Delta \text{KE}$$

where, K_i = initial kinetic energy

and K_f = final kinetic energy.

Regarding the work-energy theorem it is worth noting that

(*i*) If W_{net} is positive, then $K_f - K_i$ = positive, *i.e.*, $K_f > K_i$ or kinetic energy will increase and *vice-versa*.

(*ii*) This theorem can be applied to non-inertial frames also. In a non-inertial frame it can be written as:

Work done by all the forces (including the Pseudo force) = change in kinetic energy in non-inertial frame.

Mass-Energy Equivalence

According to Einstein, the mass can be transformed into energy and *vice-versa*.

When Δm mass disappears, then produced energy, $E = \Delta mc^2$

where c is the speed of light in vacuum.

Principle of Conservation of Energy

The sum of all kinds of energies in an isolated system remains constant at all times.

Principle of Conservation of Mechanical Energy

For conservative forces the sum of kinetic and potential energies of any object remains constant throughout the motion.

According to the quantum physics, mass and energy are not conserved separately but are conserved as a single entity called 'mass-energy'.

Collisions

Collision between two or more particles is the interaction for a short interval of time in which they apply relatively strong forces on each other.

In a collision physical contact of two bodies is not necessary.

There are two types of collisions:

1. Elastic Collision

The collision in which both the momentum and the kinetic energy of the system remains conserved are called elastic collisions.

In an elastic collision all the involved forces are conservative forces.

Total energy remains conserved.

2. Inelastic Collision

The collision in which only the momentum remains conserved but kinetic energy does not remain conserved are called inelastic collisions.

The collision in which two particles move together after the collision, is called a completely inelastic collision.

In an inelastic collision some or all the involved forces are non-conservative forces.

Total energy of the system remains conserved.

If after the collision two bodies stick to each other, then the collision is said to be perfectly inelastic.

Coefficient of Restitution or Resilience

The ratio of relative velocity of separation after collision to the relative velocity of approach before collision is called coefficient of restitution or resilience.

It is represented by e and it depends upon the material of the colliding bodies.

For a perfectly elastic collision, $e = 1$

For a perfectly inelastic collision, $e = 0$

For all other collisions, $0 < e < 1$

One Dimensional or Head-on Collision

If the initial and final velocities of colliding bodies lie along the same line, then the collision is called one dimensional or head-on collision.

Inelastic One Dimensional Collision

Applying Newton's experimental law, we have

$$v_2 - v_1 = u_2 - u_1$$

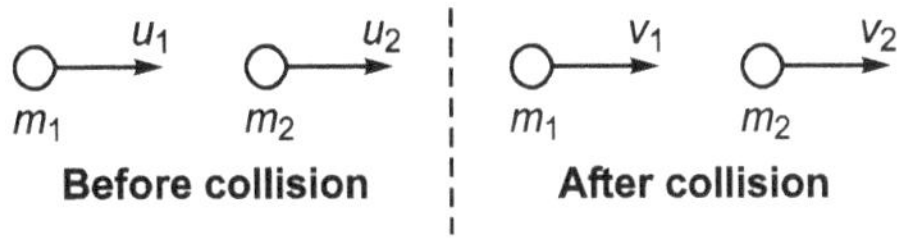

Velocities after collision

$$v_1 = \frac{(m_1 - m_2)\,u_1 + 2m_2u_2}{(m_1 + m_2)}$$

and $$v_2 = \frac{(m_2 - m_1)\,u_2 + 2m_1u_1}{(m_1 + m_2)}$$

When masses of two colliding bodies are equal, then after the collision, the bodies exchange their velocities.

$$v_1 = u_2 \quad \text{and} \quad v_2 = u_1$$

If second body of same mass ($m_1 = m_2$) is at rest, then after collision first body comes to rest and second body starts moving with the initial velocity of first body.

$$v_1 = 0 \quad \text{and} \quad v_2 = u_1$$

If a light body of mass m_1 collides with a very heavy body of mass m_2 at rest, then after collision.

$$v_1 = -u_1 \quad \text{and} \quad v_2 = 0$$

It means light body will rebound with its own velocity and heavy body will continue to be at rest.

If a very heavy body of mass m_1 collides with a light body of mass $m_2 (m_1 >> m_2)$ at rest, then after collision

$$v_1 = u_1 \quad \text{and} \quad v_2 = 2u_1$$

In Inelastic One Dimensional Collision

Loss of kinetic energy

$$\Delta E = \frac{m_1 m_2}{2\,(m_1 + m_2)}(u_1 - u_2)^2\,(1 - e^2)$$

In Perfectly Inelastic One Dimensional Collision

Velocity of separation after collision = 0.

$$\text{Loss of kinetic energy} = \frac{m_1 m_2\,(u_1 - u_2)^2}{2\,(m_1 + m_2)}$$

If a body is dropped from a height h_0 and it strikes the ground with velocity v_0 and after inelastic collision it rebounds with velocity v_1 and rises to a height h_1, then

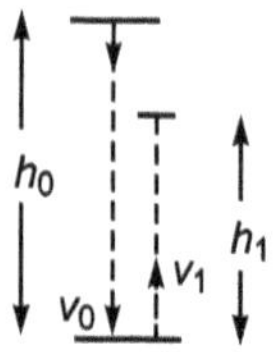

$$e = \frac{v_1}{v_0} = \sqrt{\frac{2gh_1}{2gh_0}} = \sqrt{\frac{h_1}{h_0}}$$

If after n collisions with the ground, the body rebounds with a velocity v_n and rises to a height h_n, then

$$e^n = \frac{v_n}{v_0} = \sqrt{\frac{h_n}{h_0}}$$

Two Dimensional or Oblique Collision

If the initial and final velocities of colliding bodies do not lie along the same line, then the collision is called two dimensional or oblique collision.

In horizontal direction,

$$m_1u_1 \cos \alpha_1 + m_2u_2 \cos \alpha_2 = m_1v_1 \cos \beta_1 + m_2v_2 \cos \beta_2$$

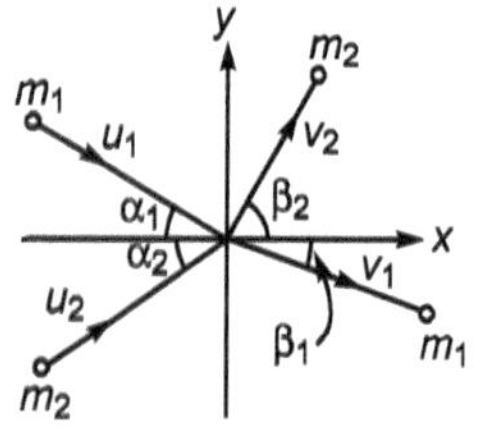

In vertical direction,

$$m_1u_1 \sin \alpha_1 - m_2u_2 \sin \alpha_2 = m_1v_1 \sin \beta_1 - m_2v_2 \sin \beta_2$$

If $m_1 = m_2$ and $\alpha_1 + \alpha_2 = 90°$

then $\beta_1 + \beta_2 = 90°$

If a particle A of mass m_1 moving along x-axis with a speed u and makes an elastic collision with another stationary body B of mass m_2.

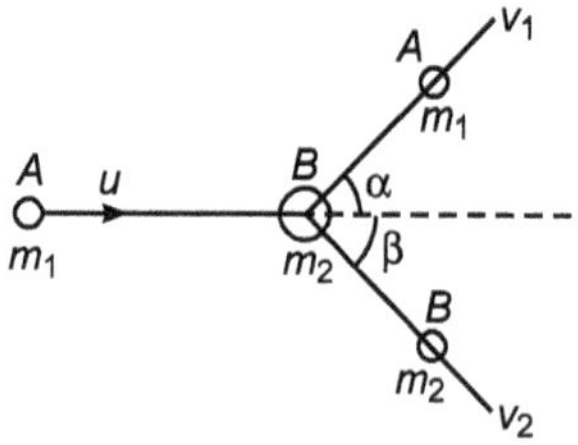

From conservation law of momentum

$$m_1u = m_1v_1 \cos \alpha + m_2v_2 \cos \beta$$

$$0 = m_1v_1 \sin \alpha - m_2v_2 \sin \beta$$

7

Rotational Motion

Centre of Mass

Centre of mass of a system is the point that behaves as whole mass of the system is concentrated at it and all external forces are acting on it.

For rigid bodies, centre of mass is independent of the state of the body, i.e. whether it is in rest or in accelerated motion centre of mass will remain same.

Centre of Mass of System of *n* Particles

If a system consists of n particles of masses $m_1, m_2, m_3, \ldots, m_n$ having position vectors $\mathbf{r}_1, \mathbf{r}_2, \mathbf{r}_3, \ldots, \mathbf{r}_n$, then position vector of centre of mass of the system,

$$\mathbf{r}_{CM} = \frac{m_1\mathbf{r}_1 + m_2\mathbf{r}_2 + m_3\mathbf{r}_3 + \ldots + m_n\mathbf{r}_n}{m_1 + m_2 + m_3 + \ldots + m_n} = \frac{\sum_{i=1}^{n} m_i\mathbf{r}_i}{\sum m_i}$$

In terms of coordinates,

$$x_{CM} = \frac{m_1x_1 + m_2x_2 + \ldots + m_nx_n}{m_1 + m_2 + \ldots + m_n} = \frac{\sum_{i=1}^{n} m_i\mathbf{x}_i}{\sum m_i}$$

$$y_{CM} = \frac{m_1y_1 + m_2y_2 + \ldots + m_ny_n}{m_1 + m_2 + \ldots + m_n} = \frac{\sum_{i=1}^{n} m_i\mathbf{y}_i}{\sum m_i}$$

$$z_{CM} = \frac{m_1z_1 + m_2z_2 + \ldots + m_nz_n}{m_1 + m_2 + \ldots + m_n} = \frac{\sum_{i=1}^{n} m_i\mathbf{z}_i}{\sum m_i}$$

Centre of Mass of Two Particle System

Choosing O as origin of the coordinate axis.

(i) Then, position of centre of mass from $m_1 = \dfrac{m_2d}{m_1 + m_2}$

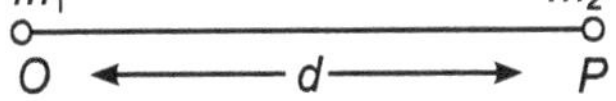

(ii) Position of centre of mass from $m_2 = \dfrac{m_1 d}{m_1 + m_2}$

(iii) If position vectors of particles of masses m_1 and m_2 are $\mathbf{r}_1$ and $\mathbf{r}_2$, respectively, then

$$\mathbf{r}_{CM} = \frac{m_1\mathbf{r}_1 + m_2\mathbf{r}_2}{m_1 + m_2}$$

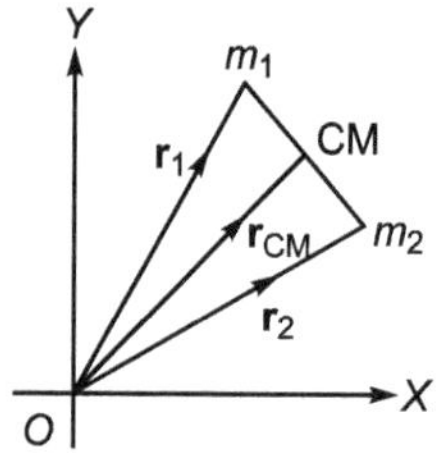

(iv) If in a two particle system, particles of masses m_1 and m_2 are moving with velocities $\mathbf{v}_1$ and $\mathbf{v}_2$ respectively, then velocity of the centre of mass

$$\mathbf{v}_{CM} = \frac{m_1\mathbf{v}_1 + m_2\mathbf{v}_2}{m_1 + m_2}$$

(v) If accelerations of the particles are $\mathbf{a}_1$ and $\mathbf{a}_2$ respectively, then acceleration of the centre of mass

$$\mathbf{a}_{CM} = \frac{m_1\mathbf{a}_1 + m_2\mathbf{a}_2}{m_1 + m_2}$$

(vi) Centre of mass of an isolated system has a constant velocity. It means isolated system will remain at rest if it is initially at rest or will move with a same velocity if it is in motion initially.

(vii) The position of centre of mass depends upon the shape, size and distribution of the mass of the body.

(viii) The centre of mass of an object need not to lie with in the object.

(ix) In symmetrical bodies having homogeneous distribution of mass the centre of mass coincides with the geometrical centre of the body.

(x) The position of centre of mass of an object changes in translatory motion but remains unchanged in rotatory motion.

Linear Momentum of a System of Particles

For a system of n particles, the total momentum of a system of particles is equal to the product of the total mass and velocity of its centre of mass.

$$p = Mv_{CM}$$

Newton's Second Law for System of Particles

Net external force, $F_{ext} = \frac{dp}{dt}$.

Rigid Body A body is said to be a rigid body when it has perfectly definite shape and size. The distance between all points of particles of such a body do not change, while applying any force on it.

General motion of a rigid body consists of both the translational motion and the rotational motion.

Translational Motion

A rigid body performs a pure translational motion, if each particle of the body undergoes the same displacement in the same direction in a given interval of time.

Rotational Motion

A rigid body performs a pure rotational motion, if each particle of the body moves in a circle, and the centre of all the circles lie on a straight line called the axes of rotation.

Equations of Rotational Motion

(i) $\omega = \omega_0 + \alpha t$ (ii) $\theta = \omega_0 t + \frac{1}{2}\alpha t^2$ (iii) $\omega^2 = \omega_0^2 + 2\alpha\theta$

where θ is displacement in rotational motion, ω_0 is initial velocity, ω is final velocity and α is acceleration.

Moment of Inertia

The inertia of rotational motion is called **moment of inertia**. It is denoted by I.

Moment of inertia is the property of an object by virtue of which it opposes any change in its state of rotation about an axis.

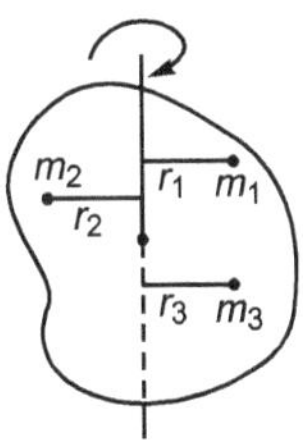

The moment of inertia of a body about a given axis is equal to the sum of the products of the masses of its constituent particles and the square of their respective distances from the axis of rotation.

Moment of inertia of a body

$$I = m_1 r_1^2 + m_2 r_2^2 + m_3 r_3^2 + \ldots = \sum_{i=1}^{n} m_i r_i^2$$

Its unit is kg-m^2 and its dimensional formula is [ML^2].

The moment of inertia of a body depends upon

(a) position of the axis of rotation
(b) orientation of the axis of rotation
(c) shape and size of the body
(d) distribution of mass of the body about the axis of rotation.

The physical significance of the moment of inertia is same in rotational motion as the mass in linear motion.

The Radius of Gyration

The root mean square distance of its constituent particles from the axis of rotation is called the radius of gyration of a body.

It is denoted by K.

Radius of gyration $K = \sqrt{\dfrac{r_1^2 + r_2^2 + \ldots + r_n^2}{n}}$

The product of the mass of the body (M) and square of its radius of gyration (K) gives the same moment of inertia of the body about the rotational axis.

Therefore, moment of inertia $I = MK^2 \Rightarrow K = \sqrt{\dfrac{I}{M}}$

Parallel Axes Theorem

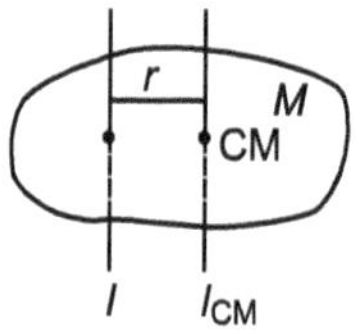

The moment of inertia of any object about any arbitrary axes is equal to the sum of moment of inertia about a parallel axis passing through the centre of mass and the product of mass of the body and the square of the perpendicular distance between the two axes.

Mathematically $I = I_{CM} + Mr^2$

where I is the moment of inertia about the arbitrary axis, I_{CM} is the moment of inertia about the parallel axis through the centre of mass, M is the total mass of the object and r is the perpendicular distance between the axis.

Perpendicular Axes Theorem

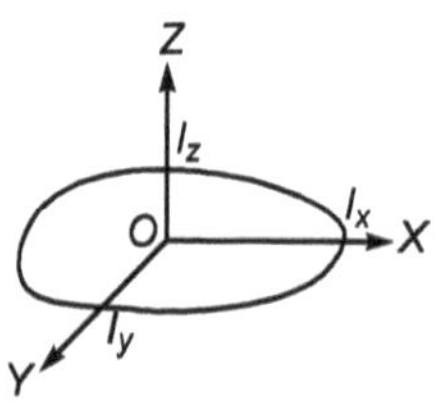

The moment of inertia of any two dimensional body about an axis perpendicular to its plane (I_z) is equal to the sum of moments of inertia of the body about two mutually perpendicular axes lying in its own plane and intersecting each other at a point, where the perpendicular axis passes through it.

Mathematically $I_z = I_x + I_y$

where I_x and I_y are the moments of inertia of plane lamina about the perpendicular axes X and Y respectively which lie in the plane of lamina an intersect each other.

> Theorem of parallel axes is applicable for any type of rigid body whether it is a two dimensional or three dimensional, while the theorem of perpendicular is applicable for laminar type or two dimensional bodies only.

Moment of Inertia of Homogeneous Rigid Bodies

For a Thin Circular Ring

S. No.	Axis of Rotation	Moment of Inertia
(a)	About an axis passing through its centre and perpendicular to its plane	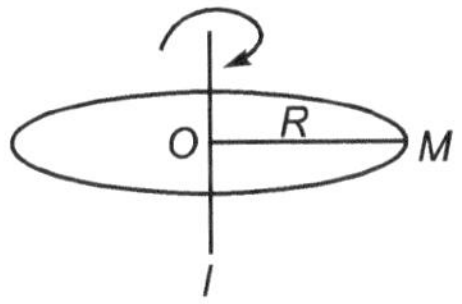 $I = MR^2$
(b)	About a tangent perpendicular to its plane	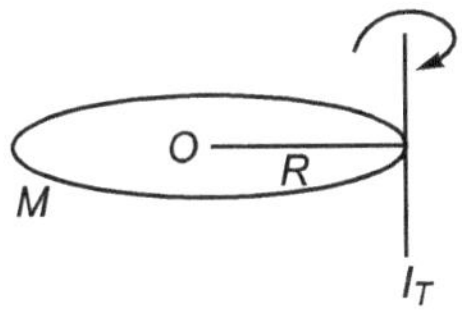 $I_T = 2MR^2$
(c)	About a tangent in the plane of ring	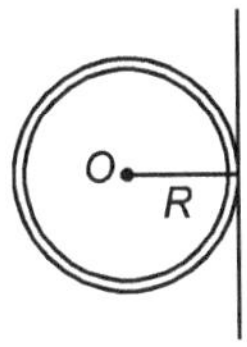 $I_{T'} = \frac{3}{2} MR^2$
(d)	About a diameter	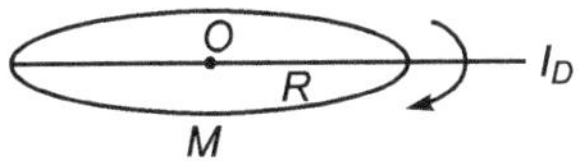 $I_D = \frac{1}{2} MR^2$

For a Circular Disc

S. No.	Axis of Rotation	Moment of Inertia
(a)	About an axis passing through its centre and perpendicular to its plane	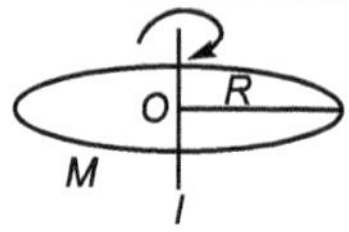 $I = \frac{1}{2}MR^2$
(b)	About a tangent perpendicular to its plane	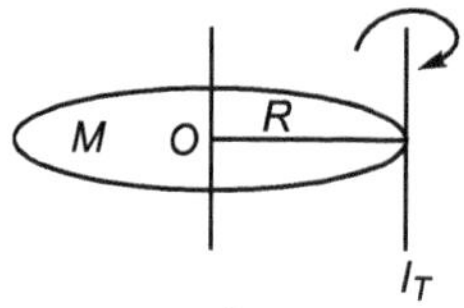 $I_T = \frac{3}{2}MR^2$
(c)	About a tangent in its plane	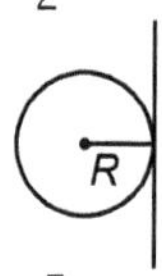 $I_{T'} = \frac{5}{4}MR^2$
(d)	About a diameter	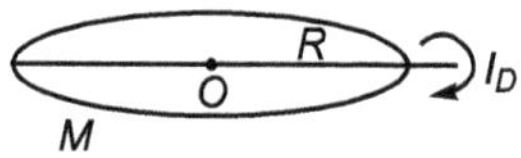 $I_D = \frac{1}{4}MR^2$

For a Thin Rod

S. No.	Axis of Rotation	Moment of Inertia
(a)	About an axis passing through its centre and perpendicular to its length	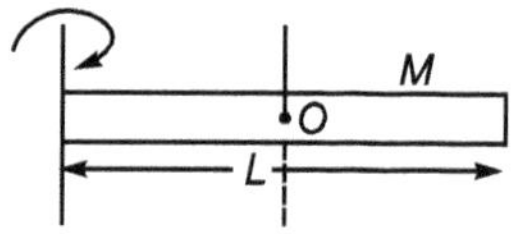 $I = \frac{ML^2}{12}$
(b)	About an axis passing through its one end and perpendicular to its length	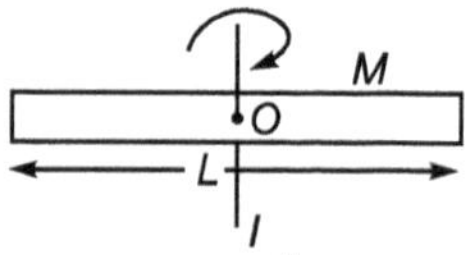 $I = \frac{ML^2}{3}$

For a Solid Cylinder

S. No.	Axis of Rotation	Moment of Inertia
(a)	About its geometrical axis	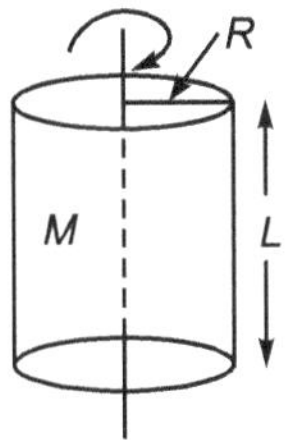 $I = \frac{MR^2}{2}$
(b)	About an axis passing through its outer face along its length	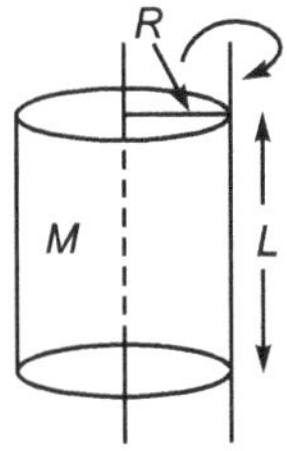 $I = \frac{3}{2} MR^2$
(c)	About an axis passing through its centre and perpendicular to its length	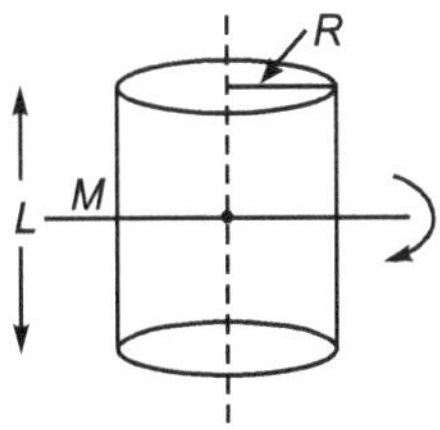 $I = \left(\frac{ML^2}{12} + \frac{MR^2}{4}\right)$
(d)	About an axis passing through its diameter of circular surface.	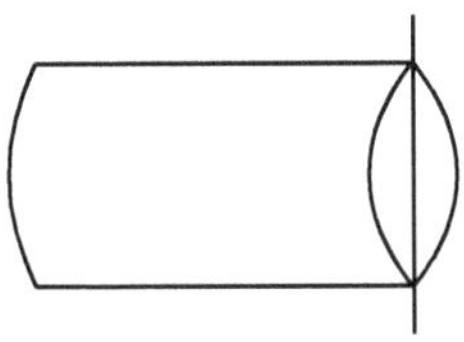 $I = \left(\frac{ML^2}{3} + \frac{MR^2}{4}\right)$

For a Rectangular Plate

Axis of Rotation	Moment of Inertia
About an axis passing through its centre and perpendicular to its plane $I = \frac{M(l^2 + b^2)}{12}$	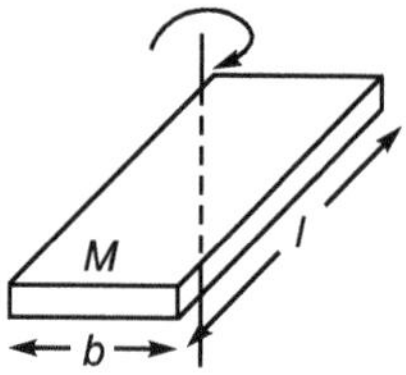

For a Thin Spherical Shell

S. No.	Axis of Rotation	Moment of Inertia
(a)	About its any diameter $I_D = \frac{2}{3}MR^2$	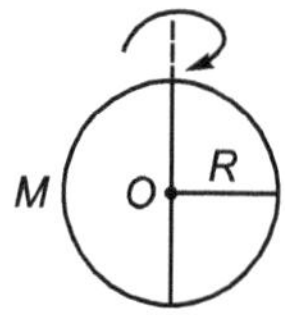
(b)	About its any tangent $I_T = \frac{5}{3}MR^2$	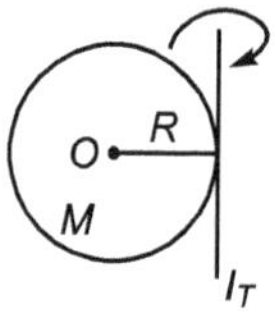

For a Solid Sphere

S. No.	Axis of Rotation	Moment of Inertia
(a)	About its any diameter $I_D = \frac{2}{5}MR^2$	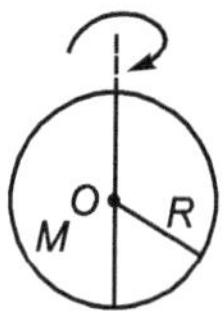
(b)	About its any tangent $I_T = \frac{7}{5}MR^2$	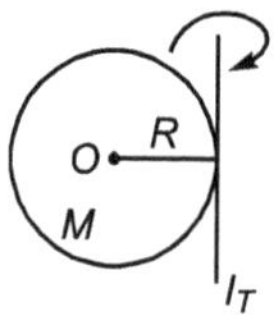

Centre of Gravity

If a body is supported on a point such that total gravitational torque about this point is zero, then this point is called centre of gravity of the body.

Torque

Torque or moment of a force about the axis of rotation

$$\tau = \mathbf{r} \times \mathbf{F} = rF \sin\theta \, \hat{\mathbf{n}}$$

It is a vector quantity. It is also known as moment of force or couple.

If the nature of the force is to rotate the object clockwise, then torque is called **negative** and if rotate the object anticlockwise, then it is called **positive.**

Its SI unit is 'newton-metre' and its dimension is $[ML^2T^{-2}]$.

In rotational motion, torque, $\tau = I\alpha$

where α is angular acceleration and I is moment of inertia.

Angular Momentum

The moment of linear momentum is called angular momentum.

It is denoted by L.

Angular momentum, $L = I\omega = mvr$

In vector form, $L = I\omega = \mathbf{r} \times m\mathbf{v}$

Its unit is 'joule-second' and its dimensional formula is $[ML^2T^{-1}]$.

Torque, $$\tau = \frac{d\mathbf{L}}{dt}$$

Torque and Angular Momentum for a System of Particles

The rate of the total angular momentum of a system of particles about a point is equal to the sum of the external torques acting on the system taken about the same point.

$$\frac{d\mathbf{L}}{dt} = \tau_{ext}$$

Equilibrium of Rigid Body

A rigid body is said to be in equilibrium, if both of its linear momentum and angular momentum are not changing with time. Thus, for equilibrium body does not possess linear acceleration or angular acceleration.

Principle of Moment

When an object is in rotational equilibrium, then algebraic sum of all torques acting on it is zero. Clockwise torques are taken negative and anti-clockwise torques are taken positive.

Conservation of Angular Momentum

If the external torque acting on a system is zero, then its angular momentum remains conserved.

If $\tau_{ext} = 0$, then $L = I\omega = \text{constant} \Rightarrow I_1\omega_1 = I_2\omega_2$

Angular Impulse

Total effect of a torque applied on a rotating body in a given time is called angular impulse. Angular impulse is equal to total change in angular momentum of the system in given time. Thus, angular impulse

$$J = \int_0^{\Delta L} \tau \, dt = L_f - L_i$$

Rotational Kinetic Energy

Rotational kinetic energy of a body is equal to the sum of kinetic energies of its constituent particles.

Rotational kinetic energy $K = \frac{1}{2} I\omega^2$

Motion of a Body Rolling Down Without Slipping on an Inclined Plane

Acceleration of the body

$$a = \frac{mg \sin\theta}{m + \frac{I}{r^2}} = \frac{g \sin\theta}{1 + \frac{K^2}{r^2}}$$

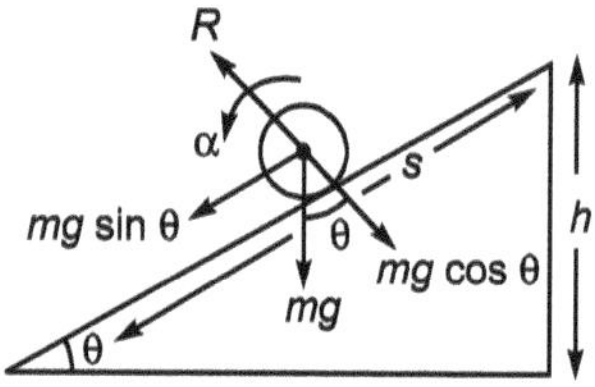

where, K = radius of gyration, m = mass of the body, r = radius of the body and θ = inclination of the plane.

Velocity attained at the bottom

$$v = \sqrt{\frac{2gh}{1 + \frac{K^2}{r^2}}} = \sqrt{\frac{2gs \sin\theta}{1 + \frac{K^2}{r^2}}} \qquad \left(\because s = \frac{h}{\sin\theta}\right)$$

where, h = height of slope and s = length of slope

$$\text{Time} = \sqrt{\frac{2s\left(1 + \frac{k^2}{r^2}\right)}{g \sin\theta}}$$

where, s = Length of slope, r = Radius of rolling body

If a cylinder, ring, disc and sphere roll on inclined plane, then the sphere will reach the bottom first with greater velocity while ring will reach the bottom with least velocity.

If a solid and hollow body of same shape are allowed to roll down an inclined plane, then solid body will reach the bottom first with greater velocity.

For rolling without slipping the minimum value of coefficient of friction

$$\mu = \frac{F}{R} = \frac{(Ia / r^2)}{mg\cos\theta}$$

Total kinetic energy of a rolling object

$$= \text{Kinetic energy of translation} + \text{Kinetic energy of rotation.}$$

$$= \frac{1}{2}mv^2 + \frac{1}{2}I\omega^2$$

Power delivered by torque $P = \tau \cdot \omega$

Work done by torque $W = \int_{\theta_1}^{\theta_2} \rho \, d\theta$

If ρ is constant, then $W = \rho(\theta_2 - \theta_1)$

or $W = \rho \times$ angle moved by the particle.

In rolling motion, all points of the body have same angular speed but different linear speeds.

In pure translational motion

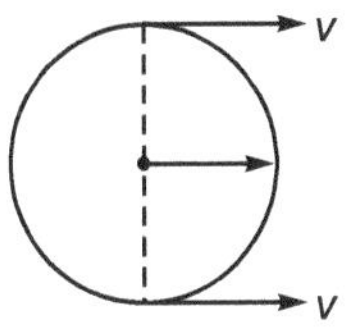

In pure rotational motion

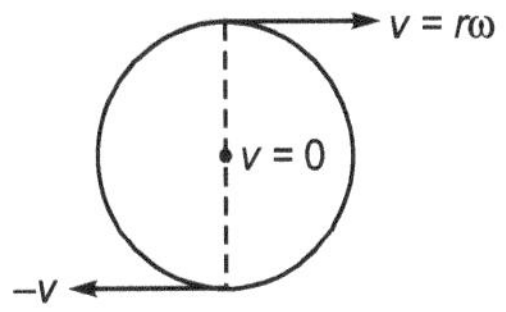

In combined motion, i.e. translational as well as rotational motion.

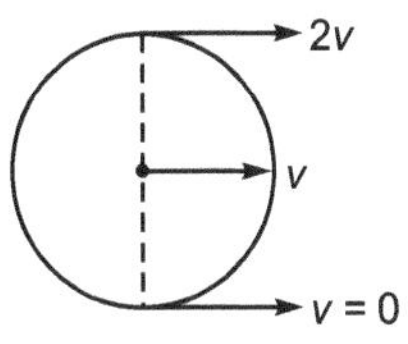

8 Gravitation

Every object in the universe attracts every other object with a force, which is called the force of **gravitation.**

Gravitation is one of the four classes of interactions found in nature. These are

(i) the gravitational force

(ii) the electromagnetic force

(iii) the strong nuclear force (also called the hadronic force).

(iv) the weak nuclear forces.

Although, of negligible importance in the interactions of elementary particles, gravity is of primary importance in the interactions of large objects. It is gravity that holds the universe together.

Newton's Law of Gravitation

Gravitational force is a attractive force between two masses m_1 and m_2 separated by a distance r.

The gravitational force acting between two point objects is directly proportional to the product of their masses and inversely proportional to the square of the distance between them.

Gravitational force, $F = \dfrac{Gm_1 m_2}{r^2}$

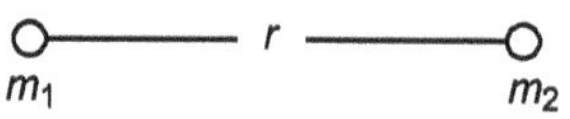

where G is universal gravitational constant.

The value of G is 6.67×10^{-11} Nm^2 kg^{-2} and is same throughout the universe.

The value of G is independent of the nature and size of the bodies as well as the nature of the medium between them.

Dimensional formula of G is $[M^{-1}L^3T^{-2}]$.

Important Points about Gravitational Force

(i) Gravitational force is a central as well as conservative force.

(ii) It is the weakest force in nature.

(iii) It is 10^{36} times smaller than electrostatic force and 10^{38} times smaller than nuclear force.

(iv) The law of gravitational is applicable for all bodies, irrespective of their size, shape and position.

(v) Gravitational force acting between sun and planet provide it centripetal force for orbital motion.

(vi) Gravitational pull of the earth is called **gravity**.

(vii) Newton's third law of motion holds good for the force of gravitation. It means the gravitational forces between two bodies are action-reaction pairs.

Following three points are important regarding the gravitational force

(i) Unlike the electrostatic force, it is independent of the medium between the particles.

(ii) It is conservative in nature.

(iii) It expresses the force between two point masses (of negligible volume). However, for external points of spherical bodies the whole mass can be assumed to be concentrated at its centre of mass.

Note *Newton's law of gravitation holds goods for object lying at very large distances and also at very short distances. It fails when the distance between the objects is less than* 10^{-9} *m i.e. of the order of intermolecular distances.*

Central Forces

Central force is that force which acts along the line joining towards the centres of two interacting bodies. A central force is always directed towards the centre as away from a fixed point.

Acceleration Due to Gravity

The uniform acceleration produced in a freely falling object due to the gravitational pull of the earth is known as **acceleration due to gravity**.

It is denoted by g and its unit is m/s^2. It is a vector quantity and its direction is towards the centre of the earth.

The value of g is independent of the mass of the object which is falling freely under gravity.

The value of g changes slightly from place to place. The value of g is taken to be 9.8 m/s^2 for all practical purposes. The value of acceleration due to gravity on the moon is about one sixth of that on the earth and on the sun is about 27 times of that on the earth.

Among the planets, the acceleration due to gravity is minimum on the mercury.

Relation between g and G is given by

$$g = \frac{GM}{R^2}$$

where, M = mass of the earth $= 6.4 \times 10^{24}$ kg and R = radius of the earth $= 6.38 \times 10^6$ m.

Factors Affecting Acceleration Due to Gravity

(i) **Shape of Earth** Acceleration due to gravity $g \propto \frac{1}{R^2}$.

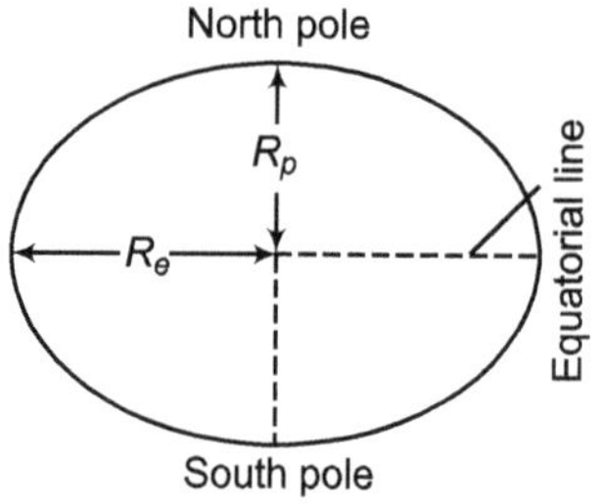

Earth is elliptical in shape. Its diameter at poles is approximately 42 km less than its diameter at equator.

Therefore, g is minimum at equator and maximum at poles.

(ii) **Rotation of Earth about Its Own Axis** If ω is the angular velocity of rotation of earth about its own axis, then acceleration due to gravity at a place having latitude λ is given by

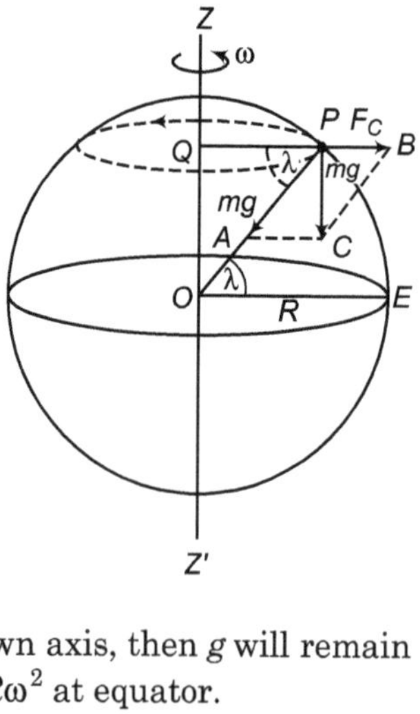

$$g' = g - R\omega^2 \cos^2 \lambda$$

At poles $\lambda = 90°$ and $g' = g$

Therefore, there is no effect of rotation of earth about its own axis at poles.

At equator $\lambda = 0°$ and $g' = g - R\omega^2$

The value of g is minimum at equator.

If earth stopes its rotation about its own axis, then g will remain unchanged at poles but increases by $R\omega^2$ at equator.

(iii) **Effect of Altitude** The value of g at height h from earth's surface

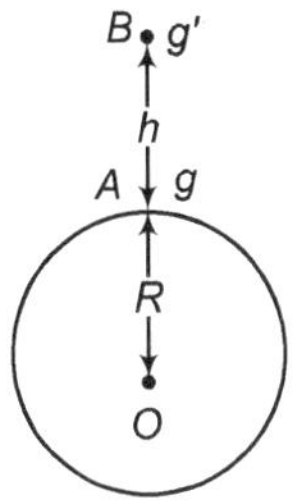

$$g' = \frac{g}{\left(1 + \frac{h}{R}\right)^2}$$

Therefore g decreases with altitude.

(iv) **Effect of Depth** The value of g at depth h from earth's surface

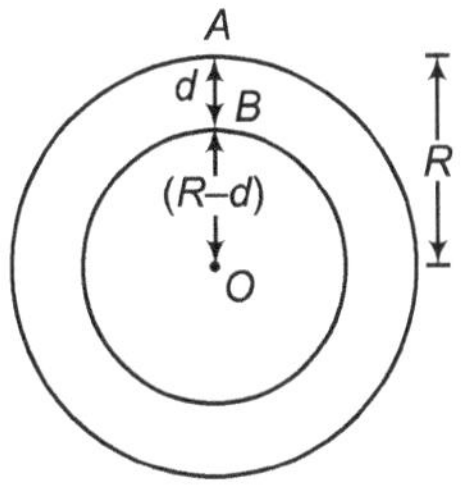

$$g' = g\left(1 - \frac{h}{R}\right)$$

Therefore g decreases with depth from earth's surface.

The value of g becomes zero at earth's centre.

Gravitational Field

The space in the surrounding of any body in which its gravitational pull can be experienced by other bodies is called **gravitational field**.

Intensity of Gravitational Field

The gravitational force acting per unit mass at any point in gravitational field is called intensity of gravitational field at that point.

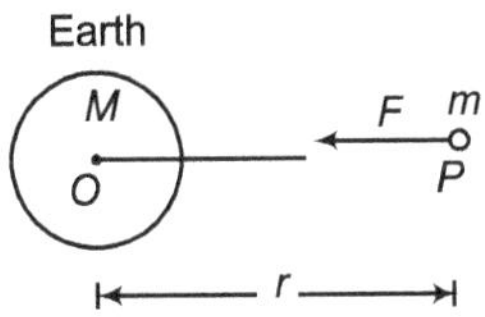

It is denoted by E_g or I.

$$E_g \text{ or } I = \frac{F}{m}$$

Intensity of gravitational field at a distance r from a body of mass M is given by

$$E_g \quad \text{or} \quad I = \frac{GM}{r^2}$$

It is a vector quantity and its direction is towards the centre of gravity of the body.

Its SI unit is N/m and its dimensional formula is $[LT^{-2}]$.

Gravitational mass M_g is defined by Newton's law of gravitation.

$$M_g = \frac{F_g}{g} = \frac{W}{g} = \frac{\text{Weight of body}}{\text{Acceleration due to gravity}}$$

$$\therefore \qquad \frac{(M_1)_g}{(M_2)_g} = \frac{Fg_1g_2}{Fg_2g_1}$$

Gravitational Potential

Gravitational potential at any point in gravitational field is equal to the work done per unit mass in bringing a very light body from infinity to that point.

It is denoted by V_g.

Gravitational potential, $V_g = \frac{W}{m} = -\frac{GM}{r}$

Its SI unit is J/kg and it is a scalar quantity.

Its dimensional formula is $[L^2T^{-2}]$.

Since work W is obtained, that is, it is negative, the gravitational potential is always negative.

Gravitational Potential Energy

Gravitational potential energy of any object at any point in gravitational field is equal to the work done in bringing it from infinity to that point. It is denoted by U.

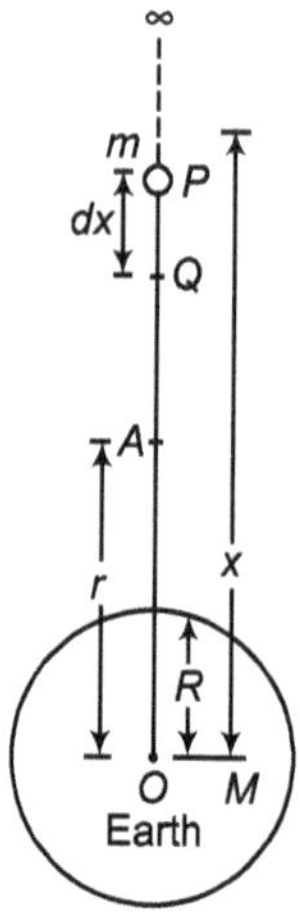

Gravitational potential energy $U = -\frac{GMm}{r}$

The negative sign shows that the gravitational potential energy decreases with increase in distance.

Gravitational potential energy at height h from surface of earth

$$U_h = -\frac{GMm}{R+h} = \frac{mgR}{1+\frac{h}{R}}$$

Satellite

A heavenly object which revolves around a planet is called a satellite.

Natural satellites are those heavenly objects which are not man made and revolve around the earth. Artificial satellites are those heavenly objects which are man made and launched for some purposes and revolve around the earth.

Time period of satellite

$$T = 2\pi\sqrt{\frac{r^3}{GM}} = \frac{2\pi}{R}\sqrt{\frac{(R+h)^3}{g}} \qquad \left[\because g = \frac{GM}{R^2}\right]$$

Near the earth surface, time period of the satellite

$$T = 2\pi\sqrt{\frac{R^3}{GM}} = \sqrt{\frac{3\pi}{G\rho}}$$

$$T = 2\pi\sqrt{\frac{R}{g}} = 5.08 \times 10^3 \text{ s} = 84 \text{ min}$$

where, ρ is the average density of earth.
Artificial satellites are of two types :

1. Geostationary or Parking Satellites

A satellite which appears to be at a fixed position at a definite height to an observer on earth is called geostationary or parking satellite.

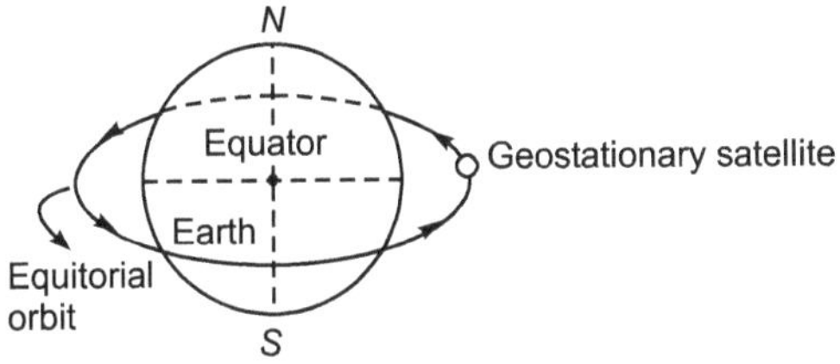

$$\text{Height from earth's surface} = 36000 \text{ km}$$

$$\text{Radius of orbit} = 42400 \text{ km}$$

$$\text{Time period} = 24 \text{ h}$$

$$\text{Orbital velocity} = 3.1 \text{ km/s}$$

$$\text{Angular velocity} = \frac{2\pi}{24} = \frac{\pi}{12} \text{ rad/h}$$

These satellites revolve around the earth in equitorial orbits.

The angular velocity of the satellite is same in magnitude and direction as that of angular velocity of the earth about its own axis.

These satellites are used in communication purpose.

INSAT 2B and INSAT 2C are geostationary satellites of India.

2. Polar Satellites

These are those satellites which revolve in polar orbits around earth. A polar orbit is that orbit whose angle of inclination with equatorial plane of earth is 90°.

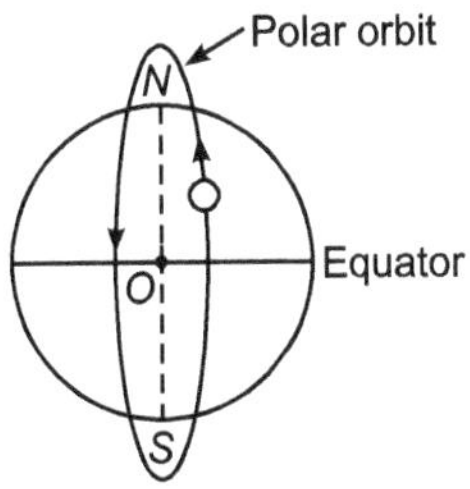

Height from earth's surface ≈ 880 km

Time period ≈ 84 min

Orbital velocity = 8 km/s

$$\text{Angular velocity} = \frac{2\pi}{84} = \frac{\pi}{42} \text{ rad/min}$$

These satellites revolve around the earth in polar orbits.

These satellites are used in forecasting weather, studying the upper region of the atmosphere, in mapping, etc.

PSLV series satellites are polar satellites of India.

Orbital Velocity

Orbital velocity of a satellite is the minimum velocity required to put the satellite into a given orbit around earth.

Orbital velocity of a satellite is given by

$$v_o = \sqrt{\frac{GM}{r}} = R\sqrt{\frac{g}{R+h}}$$

where, M = mass of the planet, R = radius of the planet and h = height of the satellite from planet's surface.

If satellite is revolving near the earth's surface, then $r = (R + h) \approx R$

Now orbital velocity,

$$v_o = \sqrt{gR} \approx 7.92 \text{ km/h}$$

If v is the speed of a satellite in its orbit and v_o is the required orbital velocity to move in the orbit, then

(i) If $v < v_o$, then satellite will move on a parabolic path and satellite falls back to earth.

(ii) If $v = v_o$, then satellite revolves in circular path/orbit around earth.

(iii) If $v_o < v < v_e$, then satellite shall revolve around earth in elliptical orbit.

Energy of a Satellite in Orbit

Total energy of a satellite

$$E = \text{KE} + \text{PE}$$

$$= \frac{GMm}{2r} + \left(-\frac{GMm}{r}\right)$$

$$= -\frac{GMm}{2r}$$

Binding Energy

The energy required to remove a satellite from its orbit around the earth (planet) to infinity is called **binding energy** of the satellite.

Binding energy of the satellite of mass m is given by

$$BE = + \frac{GMm}{2r}$$

Escape Velocity

Escape velocity on earth is the minimum velocity with which a body has to be projected vertically upwards from the earth's surface so that it just crosses the earth's gravitational field and never returns.

Escape velocity of any object

$$v_e = \sqrt{\frac{2GM}{R}}$$

$$= \sqrt{2gR} = \sqrt{\frac{8\pi\rho GR^2}{3}}$$

Escape velocity does not depend upon the mass or shape or size of the body as well as the direction of projection of the body.

Escape velocity at earth is 11.2 km/s.

Some Important Escape Velocities

Heavenly body	Escape velocity
Moon	2.3 km/s
Mercury	4.28 km/s
Earth	11.2 km/s
Jupiter	60 km/s
Sun	618 km/s
Neutron star	2×10^5 km/s

Relation between escape velocity and orbital velocity of the satellite

$$v_e = \sqrt{2}\, v_o$$

If velocity of projection v is equal the escape velocity ($v = v_e$), then the satellite will escape away following a parabolic path.

If velocity of projection v of satellite is greater than the escape velocity ($v > v_e$), then the satellite will escape away following a hyperbolic path.

Weightlessness

It is a situation in which the effective weight of the body becomes zero. Weightlessness is achieved

(i) during freely falling under gravity

(ii) inside a space craft or satellite

(iii) at the centre of the earth

(iv) when a body is lying in a freely falling lift.

Kepler's Laws of Planetary Motion

(i) **Law of orbit** Every planet revolves around the sun in an elliptical orbit and sun is at its one focus.

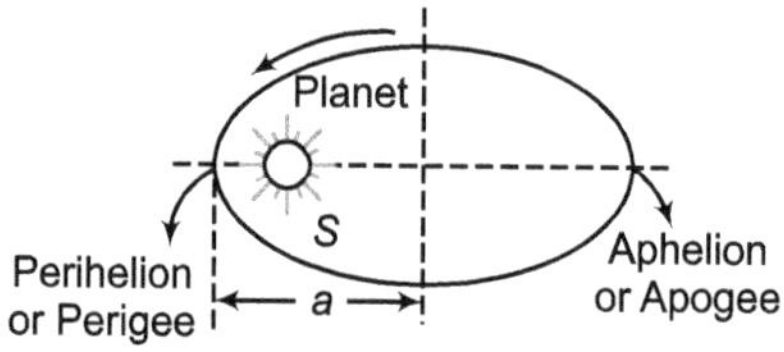

(ii) **Law of area** The radius vector drawn from the sun to a planet sweeps out equal areas in equal intervals of time, i.e. the areal velocity of the planet around the sun is constant.

Areal velocity of a planet

$$\frac{d\mathbf{A}}{dt} = \frac{\mathbf{L}}{2m} = \text{constant}$$

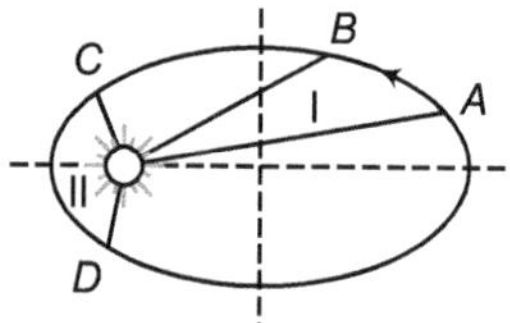

where L = angular momentum and m = mass of the planet.

(iii) **Law of period** The square of the time period of revolution of a planet around the sun is directly proportional to the cube of semi-major axis of its elliptical orbit.

$$T^2 \propto a^3 \quad \text{or} \quad \left(\frac{T_1}{T_2}\right)^2 = \left(\frac{a_1}{a_2}\right)^3$$

where, a = semi-major axis of the elliptical orbit.

Important Points

(i) A missile is launched with a velocity less than the escape velocity. The sum of its kinetic energy and potential energy is negative.

(ii) The orbital speed of jupiter is less than the orbital speed of earth.

(iii) A bomb explodes on the moon. You cannot hear the sound of the explosion on earth.

(iv) A bottle filled with water at 30°C and fitted with a cork is taken to the moon. If the cork is opened at the surface of the moon then water will boil.

(v) For a satellite orbiting near earth's surface

(a) Orbital velocity = 8 km/s

(b) Time period = 84 min approximately

(c) Angular speed $\omega = \frac{2\pi}{84}$ rad/min

$= 0.00125$ rad/s

(vi) Inertial mass and gravitational mass

(a) Inertial mass $= \frac{\text{force}}{\text{acceleration}}$

(b) Gravitational mass $= \frac{\text{weight of body}}{\text{acceleration due to gravity}}$

(c) They are equal to each other in magnitude.

(d) Gravitational mass of a body is affected by the presence of other bodies near it. Inertial mass of a body remains unaffected by the presence of other bodies near it.

9

Elasticity

Deforming Force

A force which produces a change in configuration of the object on applying it, is called a deforming force.

Elasticity

Elasticity is that property of the object by virtue of which it regain its original configuration after the removal of the deforming force.

Elastic Limit

Elastic limit is the upper limit of deforming force upto which, if deforming force is removed, the body regains its original form completely and beyond which if deforming force is increased the body loses its property of elasticity and get permanently deformed.

Perfectly Elastic Bodies

Those bodies which regain its original configuration immediately and completely after the removal of deforming force are called perfectly elastic bodies. e.g. quartz, phospher bronze etc.

Perfectly Plastic Bodies

Those bodies which does not regain its original configuration at all on the removal of deforming force are called perfectly plastic bodies, *e.g.*, putty, paraffin, wax etc.

Stress

The internal restoring force acting per unit area of a deformed body is called stress.

$$\text{Stress} = \frac{\text{Restoring force}}{\text{Area}}$$

Its unit is N/m^2 or pascal and dimensional formula is $[ML^{-1}T^{-2}]$.

Stress is a tensor quantity.

Types of Stress

Stress is of two types :

(i) **Normal Stress** If deforming force is applied normal to the area, then the stress is called normal stress.

If there is an increase in length, then stress is called **tensile stress.**

If there is a decrease in length, then stress is called **compression stress.**

(ii) **Tangential Stress** If deforming force is applied tangentially, then the stress is called tangential stress.

Strain

The fractional change in configuration is called strain.

$$\text{Strain} = \frac{\text{Change in configuration}}{\text{Original configuration}}$$

It has no unit and it is a dimensionless quantity.

According to the change in configuration, the strain is of three types

(i) $\text{Longitudinal strain} = \frac{\text{Change in length}}{\text{Original length}}$

(ii) $\text{Volumetric strain} = \frac{\text{Change in volume}}{\text{Original volume}}$

(iii) Shearing strain = Angular displacement of the plane perpendicular to the fixed surface.

Hooke's Law

Within the limit of elasticity, the stress is proportional to the strain.

$$\text{Stress} \propto \text{Strain}$$

or

$$\text{Stress} = E \times \text{Strain}$$

where, E is the **modulus of elasticity** of the material of the body.

Elastic Modulii

The ratio of stress and strain, called modulus of elasticity.

Stress and Strain Curve

When a wire is stretched by a load as in Fig. (a) it is seen that for small value of load, the extension produced in the wire is proportional to the load.

$$\text{Stress} \propto \text{Strain}$$

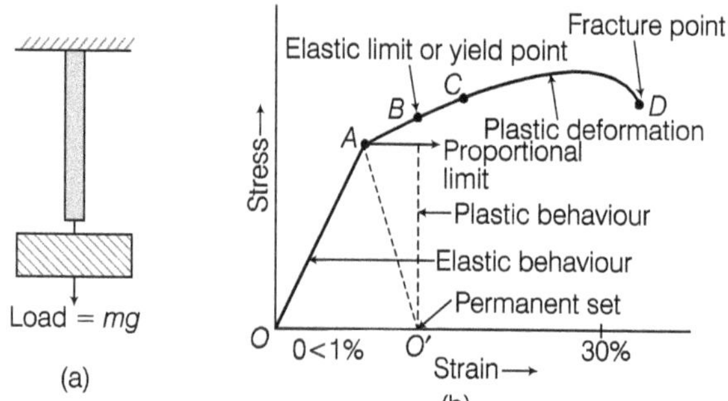

(a) (b)

Types of Modulus of Elasticity

1. Young's Modulus of Elasticity

It is defined as the ratio of normal stress to the longitudinal strain within the elastic limit.

$$Y = \frac{\text{Normal stress}}{\text{Longitudinal strain}}$$

$$Y = \frac{F\Delta l}{Al} = \frac{Mg\,\Delta l}{\pi r^2 l}$$

Its unit is N/m^2 or Pascal and its dimensional formula is $[ML^{-1}T^{-2}]$.

2. Bulk Modulus of Elasticity

It is defined as the ratio of normal stress to the volumetric strain within the elastic limit.

$$K = \frac{\text{Normal stress}}{\text{Volumetric strain}}$$

$$K = \frac{FV}{A\Delta V} = \frac{\Delta p V}{\Delta V}$$

where, $\Delta p = \frac{F}{A}$ = Change in pressure.

Its unit is N/m^2 or Pascal and its dimensional formula is $[ML^{-1}T^{-2}]$.

3. **Modulus of Rigidity (η)** (Shear Modulus)

It is defined as the ratio of tangential stress to the shearing strain, within the elastic limit.

$$\eta = \frac{\text{Tangential stress}}{\text{Shearing strain}}$$

$$\eta = \frac{F}{A\theta}$$

Its unit is N/m^2 or Pascal and its dimensional formula is $[ML^{-1}T^{-2}]$.

Compressibility

Compressibility of a material is the reciprocal of its bulk modulus of elasticity.

$$\text{Compressibility } (C) = \frac{1}{K}$$

Its SI unit is $N^{-1}m^2$ and CGS unit is dyne^{-1} cm^2.

Steel is more elastic than rubber. Solids are more elastic and gases are least elastic.

For liquids, modulus of rigidity is zero.

Young's modulus (Y) and modulus of rigidity (η) are possessed by solid materials only.

Limit of Elasticity

The maximum value of deforming force for which elasticity is present in the body is called its limit of elasticity.

Breaking Stress

The minimum value of stress required to break a wire, is called breaking stress.

Breaking stress is fixed for a material but breaking force varies with area of cross-section of the wire.

$$\text{Safety factor} = \frac{\text{Breaking stress}}{\text{Working stress}}$$

Elastic Relaxation Time

The time delay in restoring the original configuration after removal of deforming force is called elastic relaxation time.

For quartz and phospher bronze this time is negligible.

Elastic After Effect

The temporary delay in regaining the original configuration by the elastic body after the removal of deforming force, is called elastic after effect.

Elastic Fatigue

The property of an elastic body by virtue of which its behaviour becomes less elastic under the action of repeated alternating deforming force is called elastic fatigue.

Ductile Materials

The materials which show large plastic range beyond elastic limit are called ductile materials, e.g. copper, silver, iron, aluminum, etc.

Ductile materials are used for making springs and sheets.

Brittle Materials

The materials which show very small plastic range beyond elastic limit are called brittle materials, e.g. glass, cast iron, etc.

Elastomers

The materials for which strain produced is much larger than the stress applied, with in the limit of elasticity are called elastomers, e.g. rubber, the elastic tissue of arota, the large vessel carrying blood from heart, etc.

Elastomers have no plastic range.

Elastic Potential Energy in a Stretched Wire

The work done in stretching a wire is stored in form of potential energy of the wire.

Potential energy

$$U = \text{Average force} \times \text{Increase in length} = \frac{1}{2} F\Delta l$$

$$= \frac{1}{2} \text{Stress} \times \text{Strain} \times \text{Volume of the wire}$$

Elastic potential energy per unit volume

$$U = \frac{1}{2} \times \text{Stress} \times \text{Strain} = \frac{1}{2} (\text{Young's modulus}) \times (\text{Strain})^2$$

Elastic potential energy of a stretched spring $= \frac{1}{2} kx^2$

where, k = Force constant of spring and x = Change in length.

Thermal Stress

When temperature of a rod fixed at its both ends is changed, then the produced stress is called thermal stress.

$$\text{Thermal stress} = \frac{F}{A} = Y\alpha\Delta\theta$$

where, α = coefficient of linear expansion of the material of the rod.

When temperature of a gas enclosed in a vessel is changed, then the thermal stress produced is equal to change in pressure (Δp) of the gas.

$$\text{Thermal stress} = \Delta p = K\gamma\,\Delta\theta$$

where, K = bulk modulus of elasticity and

γ = coefficient of cubical expansion of the gas.

Interatomic force constant

$$K = Yr_0$$

where, r_0 = interatomic distance.

Poisson's Ratio

When a deforming force is applied at the free end of a suspended wire of length l and radius R, then its length increases by dl but its radius decreases by dR. Now two types of strains are produced by a single force.

(i) Longitudinal strain = $\Delta l / l$

(ii) Lateral strain = $-\Delta R / R$

$$\therefore \quad \text{Poisson's ratio } (\sigma) = \frac{\text{Lateral strain}}{\text{Longitudinal strain}} = \frac{-\Delta R / R}{\Delta l / l}$$

The theoritical value of Poisson's ratio lies between – 1 and 0.5.

Its practical value lies between 0 and 0.5.

Relation Between Y, K, η and σ

(i) $Y = 3K(1 - 2\sigma)$

(ii) $Y = 2\eta(1 + \sigma)$

(iii) $\sigma = \dfrac{3K - 2\eta}{2\eta + 6K}$

(iv) $\dfrac{9}{Y} = \dfrac{1}{K} + \dfrac{3}{\eta}$ or $Y = \dfrac{9K\eta}{\eta + 3K}$

Important Points

- Coefficient of elasticity depends upon the material, its temperature and purity but not on stress or strain.
- For the same material, the three coefficients of elasticity γ, η and K have different magnitudes.
- Isothermal elasticity of a gas $E_T = p$ where, p = pressure of the gas.
- Adiabatic elasticity of a gas $E_S = \gamma p$

  where, $\gamma = \frac{C_p}{C_V}$, ratio of specific heats at constant pressure and at constant volume.
- Ratio between isothermal elasticity and adiabatic elasticity $\frac{E_S}{E_T} = \gamma = \frac{C_p}{C_V}$

Cantilever

A beam clamped at one end and loaded at free end is called a cantilever.

Depression at the free end of a cantilever is given by

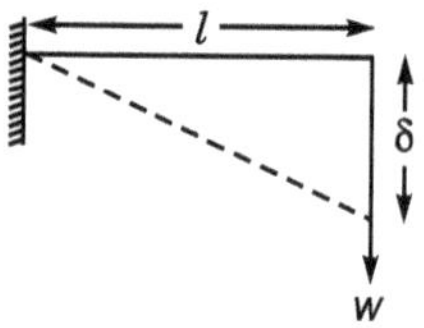

$$\delta = \frac{wl^3}{3YI_G}$$

where, w = load, l = length of the cantilever, Y = Young's modulus of elasticity, and I_G = geometrical moment of inertia.

For a beam of rectangular cross-section having breadth b and thicknes d.

$$I_G = \frac{bd^3}{12}$$

For a beam of circular cross-section area having radius r,

$$I_G = \frac{\pi r^4}{4}$$

Beam Supported at Two Ends and Loaded at the Middle

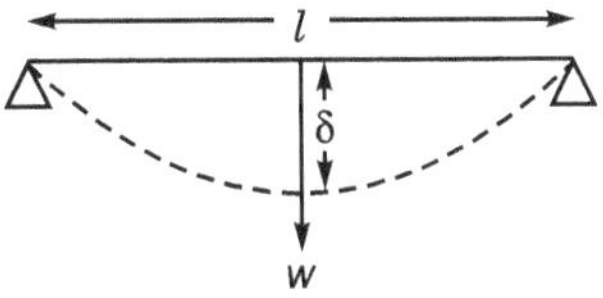

Depression at middle $\delta = \dfrac{wl^3}{48YI_G}$

Torsion of a Cylinder

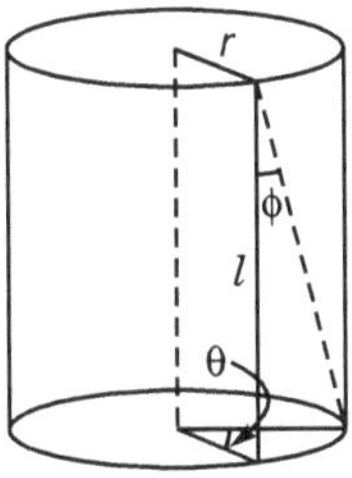

Couple per unit twist $\quad C = \dfrac{\pi\eta r^4}{2l}$

where, η = modulus of rigidity of the material of cylinder,

r = radius of cylinder,

and l = length of cylinder.

Work done in twisting the cylinder through an angle θ

$$W = \frac{1}{2}C\theta^2$$

Relation between angle of twist (θ) and angle of shear (ϕ)

$$r\theta = l\phi \quad \text{or} \quad \phi = \frac{r}{l}\theta$$

10
Hydrostatics

Fluids

Fluids are those substances which can flow when an external force is applied on it.

Liquids and gases are fluids.

Fluids do not have finite shape but takes the shape of the containing vessel.

In fluid mechanics the following properties of fluid would be considered

(i) When the fluid is at rest– **hydrostatics**

(ii) When the fluid is in motion– **hydrodynamics**

Thrust

The total normal force exerted by liquid at rest on a given surface is called **thrust** of liquid.

The SI unit of thrust is newton.

Pressure Exerted by the Liquid

The normal force exerted by a liquid per unit area of the surface in contact is called **pressure of liquid** or **hydrostatic pressure.**

Pressure exerted by a liquid column

$$p = h\rho g$$

where, h = height of liquid column, ρ = density of liquid
and g = acceleration due to gravity

Mean pressure on the walls of a vessel containing liquid upto height h is $\left(\frac{h\rho g}{2}\right)$.

Pascal's Law

The increase in pressure at a point in the enclosed liquid in equilibrium is transmitted equally in all directions in liquid and to the walls of the container.

The working of hydraulic lift, hydraulic pressure and hydraulic brakes are based on Pascal's law.

Variation of Pressure with Depth

Consider a fluid at rest having density ρ (roh) contained in a cylindrical vessel as shown in figure. Let the two points A and B separated by a vertical distance h.

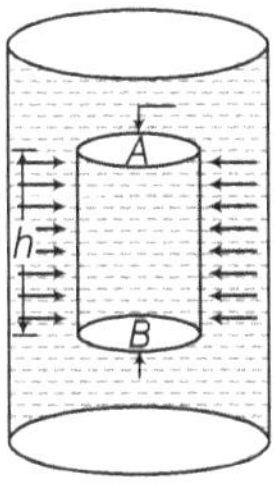

The pressure p at depth below the surface of a liquid open is given by

Pressure, $$p = p_a + h\rho g$$

where, ρ = density of liquid and g = acceleration due to gravity.

Gauge Pressure Gauge pressure at a point in a fluid is the difference of total pressure at that point and atmospheric pressure.

Hydrostatic Paradox The liquid pressure at a point is independent of the quantity of liquid but depends upon the depth of point below the liquid surface. This is known as hydrostatic paradox.

Atmospheric Pressure

The pressure exerted by the atmosphere on earth is called **atmospheric pressure.**

It is about 100000 N/m^2.

It is equivalent to a weight of 10 tones on 1 m^2.

At sea level, atmospheric pressure is equal to 76 cm of mercury column. Then, atmospheric pressure

$$= hdg = 76 \times 13.6 \times 980 \text{ dyne/cm}^2$$

> The atmospheric pressure does not crush our body because the pressure of the blood flowing through our circulatory system balanced this pressure.

Atmospheric pressure is also measured in torr and bar.

$$1 \text{ torr} = 1 \text{ mm of mercury column}$$

$$1 \text{ bar} = 10^5 \text{ Pa}$$

Aneroid barometer is used to measure atmospheric pressure.

Buoyancy

When a body is partially or fully immersed in a fluid an upward force acts on it, which is called buoyant force or simply buoyancy.

The buoyant force acts at the centre of gravity of the liquid displaced by the immersed part of the body and this point is called the centre of buoyancy.

Archimedes' Principle

When a body is partially or fully immersed in a liquid, it loses some of its weight, and it is equal to the weight of the liquid displaced by the immersed part of the body.

If T is the observed weight of a body of density σ when it is fully immersed in a liquid of density ρ, then real weight of the body

$$w = \frac{T}{\left(1 - \frac{\rho}{\sigma}\right)}$$

Laws of Floatation

A body will float in a liquid, if the weight of the body is equal to the weight of the liquid displaced by the immersed part of the body.

If W is the weight of the body and w is the buoyant force, then

(a) If $W > w$, then body will sink to the bottom of the liquid.

(b) If $W < w$, then body will float partially submerged in the liquid.

(c) If $W = w$, then body will float in liquid if its whole volume is just immersed in the liquid.

The floating body will be in stable equilibrium if meta-centre (centre of buoyancy) lies vertically above the centre of gravity of the body.

The floating body will be in unstable equilibrium if meta-centre (centre of buoyancy) lies vertically below the centre of gravity of the body. The floating body will be in neutral equilibrium if meta-centre (centre of buoyancy) coincides with the centre of gravity of the body.

Density and Relative Density

Density of a substance is defined as the ratio of its mass to its volume.

$$\text{Density of a liquid} = \frac{\text{Mass}}{\text{Volume}}$$

$$\text{Density of water} = 1 \text{ g/cm}^3 \text{ or } 10^3 \text{ kg/m}^3$$

It is scalar quantity and its dimensional formula is $[ML^{-3}]$.

Relative density of a substance is defined as the ratio of its density to the density of water at 4°C.

$$\text{Relative density} = \frac{\text{Density of substance}}{\text{Density of water at 4°C}}$$

$$= \frac{\text{Weight of substance in air}}{\text{Loss of weight in water}}$$

Relative density also known as specific gravity has no unit, no dimensions.

For a solid body, density of body = density of substance.

While for a hollow body, density of body is lesser than that of substance.

When immiscible liquids of different densities are poured in a container, the liquid of highest density will be at the bottom while, that of lowest density at the top and interfaces will be plane.

Density of a Mixture of Substances

When two liquids of mass m_1 and m_2 having density ρ_1 and ρ_2 are mixed together, then density of mixture is

$$\rho = \frac{m_1 + m_2}{\left(\frac{m_1}{\rho_1}\right) + \left(\frac{m_2}{\rho_2}\right)} = \frac{\rho_1\rho_2(m_1 + m_2)}{(m_1\rho_2 + m_2\rho_1)}$$

When two liquids of same mass m but of different densities ρ_1 and ρ_2 are mixed together, then density of mixture is $\rho = \frac{2\rho_1\rho_2}{\rho_1 + \rho_2}$

When two liquids of same volume V but of different densities ρ_1 and ρ_2 are mixed together, then density of mixture is $\rho = \frac{\rho_1 + \rho_2}{2}$

Density of a liquid varies with pressure

$$\rho = \rho_0\left[1 + \frac{\Delta p}{K}\right]$$

where, ρ_0 = initial density of the liquid, K = bulk modulus of elasticity of the liquid and Δp = change in pressure.

11 Hydrodynamics

Viscosity

The property of a fluid by virtue of which an internal frictional force acts between its different layers which opposes their relative motion is called **viscosity.**

These internal frictional force is called **viscous force.**

Viscous forces are intermolecular forces acting between the molecules of different layers of liquid moving with different velocities.

$$\text{Viscous force } (F) = -\eta A \frac{dv}{dx}$$

or

$$\eta = -\frac{F}{A\left(\frac{dv}{dx}\right)}$$

where, $\frac{dv}{dx}$ = rate of change of velocity with distance called velocity gradient, A = area of cross-section and η = coefficient of viscosity.

SI unit of η is Nsm^{-2} or pascal-second or decapoise. Its dimensional formula is $[ML^{-1}T^{-1}]$.

The knowledge of the coefficient of viscosity of different oils and its variation with temperature helps us to select a suitable lubricant for a given machine.

Viscosity is due to transport of momentum. The value of viscosity (and compressibility) for ideal liquid is zero.

The viscosity of air and of some liquids is utilised for damping the moving parts of some instruments.

The knowledge of viscosity of some organic liquids is used in determining the molecular weight and shape of large organic moleculars like proteins and cellulose.

Variation of Viscosity

The viscosity of liquids decreases with increase in temperature

$$\eta_t = \frac{\eta_0}{(1 + \alpha t + \beta t^2)}$$

where, η_0 and η_t are coefficient of viscosities at 0°C t°C, α and β are constants.

The viscosity of gases increases with increase in temperatures as

$$\eta \propto \sqrt{T}$$

The viscosity of liquids increases with increase in pressure but the viscosity of water decreases with increase in pressure.

The viscosity of gases do not changes with pressure.

Poiseuille's Formula

The rate of flow (v) of liquid through a horizontal pipe for steady flow is given by

$$v = \frac{\pi}{8} \frac{pr^4}{\eta l}$$

where, p = pressure difference across the two ends of the tube, r = radius of the tube, η = coefficient of viscosity and l = length of the tube.

The Rate of Flow of Liquid

Rate of flow of liquid through a tube is given by

$$v = \frac{p}{R}$$

where, $R = \frac{8\eta l}{\pi r^4}$, called liquid resistance and p = liquid pressure.

(i) **When two tubes are connected in series**

(a) Resultant pressure difference $p = p_1 + p_2$

(b) Rate of flow of liquid (v) is same through both tubes.

(c) Equivalent liquid resistance, $R = R_1 + R_2$

(ii) **When two tubes are connected in parallel**

(a) Pressure difference (p) is same across both tubes.

(b) Rate of flow of liquid $v = v_1 + v_2$

(c) Equivalent liquid resistance $\frac{1}{R} = \frac{1}{R_1} + \frac{1}{R_2}$

Stoke's Law

When a small spherical body falls in a long liquid column, then after sometime it falls with a constant velocity, called **terminal velocity.**

When a small spherical body falls in a liquid column with terminal velocity then viscous force acting on it is

$$F = 6\pi\eta rv$$

where, r = radius of the body, V = terminal velocity and η = coefficient of viscosity.

This is called **Stoke's law.**

Terminal velocity $v = \dfrac{2}{9}\dfrac{r^2(\rho - \sigma)g}{\eta}$

where, ρ = density of body,
σ = density of liquid,
η = coefficient of viscosity of liquid and
g = acceleration due to gravity

(i) If $\rho > \rho_0$, the body falls downwards.

(ii) If $\rho < \rho_0$, the body moves upwards with the constant velocity.

(iii) If $\rho_0 << \rho$, $v = \dfrac{2r^2\rho g}{9\eta}$

Importance of Stoke's Law

(i) This law is used in the determination of electronic charge by Millikan in his oil drop experiment.

(ii) This law helps a man coming down with the help of parachute.

(iii) This law account for the formation of clouds.

Flow of Liquid

(i) **Streamline Flow** The flow of liquid in which each of its particle follows the same path as followed by the proceeding particles, is called streamline flow.

(ii) **Laminar Flow** The steady flow of liquid over a horizontal surface in the form of layers of different velocities, is called laminar flow.

(iii) **Turbulent Flow** The flow of liquid with a velocity greater than its critical velocity is disordered and called turbulent flow.

Critical Velocity

The critical velocity is the velocity of liquid flow, below which its flow is streamlined and above which it becomes turbulent.

Critical velocity $v_c = \frac{K\eta}{r\rho}$

where, K = Reynold's number,

η = coefficient of viscosity of liquid

r = radius of capillary tube and ρ = density of the liquid.

Reynold's Number

Reynold's number is a pure number and it is equal to the ratio of the inertial force per unit area to the viscous force per unit area for a flowing fluid.

$$\text{Reynold number } K = \frac{v_c \rho r}{\eta}$$

where, ρ = density of the liquid and v_c = critical velocity.

For pure water flowing in a cylindrical pipe, K is about 1000.

When $0 < K < 2000$, the flow of liquid is streamlined.

When $2000 < K < 3000$, the flow of liquid is variable between streamlined and turbulant.

When $K > 3000$, the flow of liquid is turbulent.

It has no unit and dimension.

Equation of Continuity

If a liquid is flowing in streamline flow in a pipe of non-uniform cross-section area, then rate of flow of liquid across any cross-section remains constant.

$$a_1 v_1 = a_2 v_2 \Rightarrow av = \text{constant}$$

The velocity of liquid is slower where area of cross-section is larger and faster where area of cross-section is smaller.

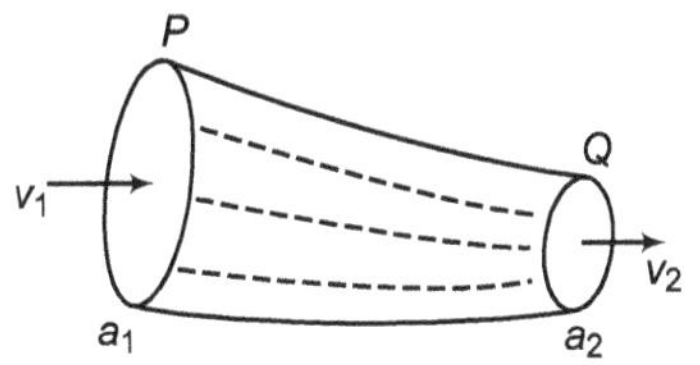

The falling stream of water becomes narrower, as the velocity of falling stream of water increases and therefore its area of cross-section decreases.

Energy of a Liquid

A liquid in motion possess three types of energy

(i) **Pressure Energy**

Pressure energy per unit mass $= \dfrac{p}{\rho}$

where, p = pressure of the liquid
and ρ = density of the liquid.
Pressure energy per unit volume $= p$

(ii) **Kinetic Energy**

Kinetic energy per unit mass $= \dfrac{1}{2} v^2$

Kinetic energy per unit volume $= \dfrac{1}{2} \rho v^2$

(iii) **Potential Energy**

Potential energy per unit mass $= gh$

Potential energy per unit volume $= \rho gh$

Bernoulli's Theorem

If an ideal liquid is flowing in streamlined flow then total energy, *i.e.*, sum of pressure energy, kinetic energy and potential energy per unit volume of the liquid remains constant at every cross-section of the tube.

Mathematically $$p + \frac{1}{2}\rho v^2 + \rho gh = \text{constant}$$

It can be expressed as

$$\frac{p}{\rho g} + \frac{v^2}{2g} + h = \text{constant}$$

where, $\dfrac{p}{\rho g}$ = pressure head,

$\dfrac{v^2}{2g}$ = velocity head and

h = gravitational head.

For horizontal flow of liquid,

$$p + \frac{1}{2}\rho v^2 = \text{constant}$$

where, p is called static pressure and $\dfrac{1}{2}\rho v^2$ is called dynamic pressure.

Therefore in horizontal flow of liquid, if p increases, v decreases and *vice-versa.*

The theorem is applicable to ideal liquid, *i.e.*, a liquid which is non-viscous incompressible and irrotational.

Applications of Bernoulli's Theorem

(i) The action of carburetor, paintgun, scent sprayer atomiser and insect sprayer is based on Bernoulli's theorem.

(ii) The action of Bunsen's burner, gas burner, oil stove and exhaust pump is also based on Bernoulli's theorem.

(iii) Motion of a spinning ball (Magnus effect) is based on Bernoulli's theorem.

(iv) Blowing of roofs by wind storms, attraction between two closely parallel moving boats, fluttering of a flag etc are also based on Bernoulli's theorem.

(v) Bernoulli's theorem helps in explaining blood flow in artery.

Venturimeter

It is a device used for measuring the rate of flow of liquid through pipes. Its working is based on Bernoulli's theorem.

Rate of flow of liquid,

$$v = a_1 a_2 \sqrt{\frac{2gh}{a_1^2 - a_2^2}}$$

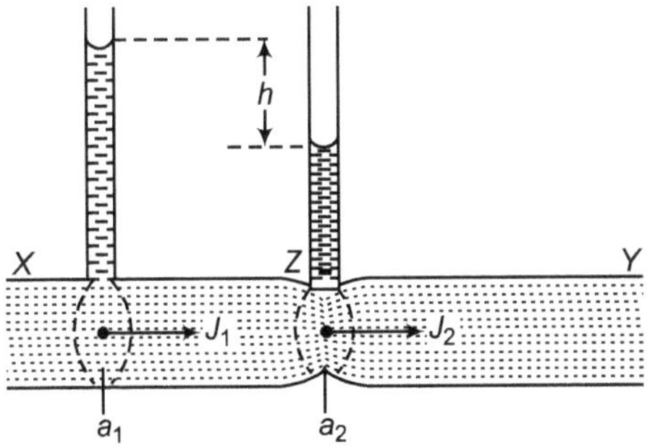

where, a_1 and a_2 are area of cross-sections of tube at broader and narrower part and h is difference of liquid columns in vertical tubes.

Torricelli's Theorem

Velocity of efflux (the velocity with which the liquid flows out of a orifice or narrow hole) is equal to the velocity acquired by a freely falling body through the same vertical distance equal to the depth of orifice below the free surface of liquid.

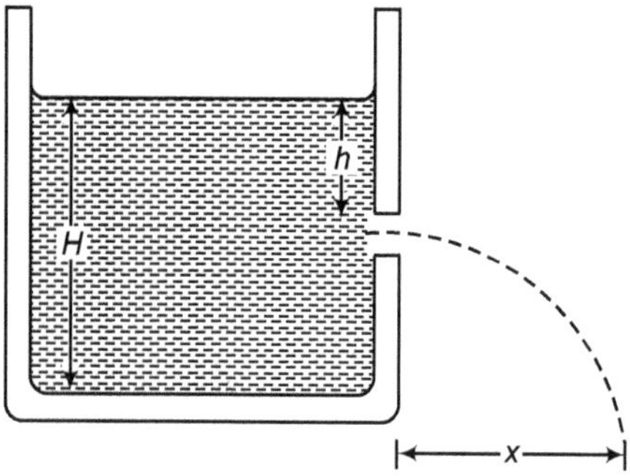

Velocity of efflux, $v = \sqrt{2gh}$

where, h = depth of orifice below the free surface of liquid.

Horizontal range, $S = \sqrt{4h(H - h)}$

where, H = height of liquid column.

Horizontal range is maximum, equal to height of the liquid column H, when orifice is at half of the height of liquid column.

Important Points

- In a pipe the inner layer (central layer) have maximum velocity and the layer in contact with pipe have least velocity.
- Velocity profile of liquid flow in a pipe is parabolic.

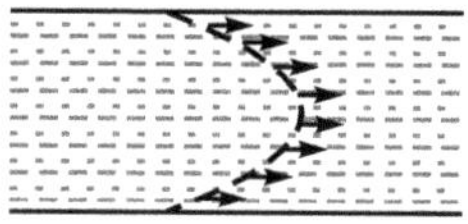

- Solid friction is independent of area of surfaces in contact while viscous force depends on area of liquid layers.
- A lubricant is chosen according to the nature of machinary. In heavy machines lubricating oils of high viscosity are used and in light machines low viscosity oils are used.
- The cause of viscosity in liquids is the cohesive forces among their molecules while cause of viscosity in gases is diffusion.

12

Surface Tension

Surface tension is the property of any liquid by virtue of which it tries to minimize its free surface area.

Surface tension of a liquid is measured as the force acting per unit length on an imaginary line drawn tangentially on the free surface of the liquid.

$$\text{Surface tension } S = \frac{\text{Force}}{\text{Length}} = \frac{F}{l} = \frac{\text{Work done}}{\text{Change in area}}$$

Its SI unit is Nm^{-1} or Jm^{-2} and its dimensional fourmula is $[MT^{-2}]$.

It is a scalar quantity. Surface tension is a molecular phenomenon which is due to cohesive force and root cause of the **force is electrical in nature.**

Surface tension of a liquid depends only on the nature of liquid and is independent of the surface area of film or length of the line considered.

Small liquid drops are spherical due to the property of surface tension.

Adhesive Force

The force of attraction acting between the molecules of different substances is called **adhesive force.** e.g. the force of attraction acting between the molecules of paper and ink, water and glass, etc.

Cohesive Force

The force of attraction acting between the molecules of same substance is called cohesive force. e.g. the force of attraction acting between the molecules of water, glass, etc.

Cohesive forces and adhesive forces are van der Waals' forces.

These forces varies inversely as the seventh power of distance between the molecules.

Molecular Range

The maximum distance upto which a molecule can exert a force of attraction on other molecules is called molecular range.

Molecular range is different for different substances.

In solids and liquids it is of the order of 10^{-9} m.

If the distance between the molecules is greater than 10^{-9} m, the force of attraction between them is negligible.

Surface Energy

If we increase the free surface area of a liquid then work has to be done against the force of surface tension. This work done is stored in liquid surface as potential energy.

This additional potential energy per unit area of free surface of liquid is called surface energy.

$$\text{Surface energy } (E) = S \times \Delta A$$

where, S = surface tension and ΔA = increase in surface area.

(i) **Work Done in Blowing a Liquid Drop** If a liquid drop is blown up from a radius r_1 to r_2, then work done for that is

$$W = S \cdot 4\pi (r_2^2 - r_1^2)$$

(ii) **Work Done in Blowing a Soap Bubble** As a soap bubble has two free surfaces, hence work done in blowing a soap bubble so as to increase its radius from r_1 to r_2 is given by

$$W = S \cdot 8\pi (r_2^2 - r_1^2)$$

(iii) **Work Done in Splitting a Bigger Drop into *n* Smaller Droplets**

If a liquid drop of radius R is split up into n smaller droplets, all of same size, then radius of each droplet

$$r = R \cdot (n)^{-1/3}$$

$$\text{Work done, } W = 4\pi S(nr^2 - R^2)$$

$$= 4\pi SR^2 (n^{1/3} - 1)$$

(iv) **Coalescance of Drops** If n small liquid drops of radius r each combine together so as to form a single bigger drop of radius $R = n^{1/3} \cdot r$, then in the process energy is released. Release of energy is given by

$$\Delta U = S \cdot 4\pi (nr^2 - R^2)$$

$$= 4\pi S\pi^2 n (1 - n^{1/3})$$

Angle of Contact

The angle subtended between the tangents drawn at liquid surface and at solid surface inside the liquid at the point of contact, is called angle of contact (θ).

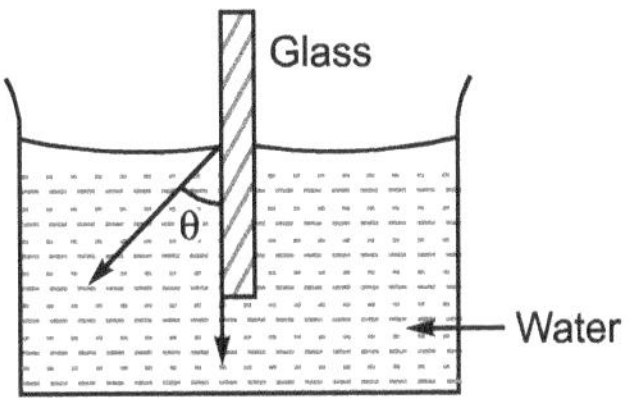

Angle of contact depends upon the nature of the liquid and solid in contact and the medium which exists above the free surface of the liquid.

When wax is coated on a glass capillary tube, it becomes water-proof. The angle of contact increases and becomes obtuse. Water does not rise in it. Rather it falls in the tube by virtue of obtuse angle of contact.

- If θ is acute angle, i.e. $\theta < 90°$, then liquid meniscus will be concave upwards.
- If θ is 90°, then liquid meniscus will be plane.
- If θ is obtuse, i.e. $\theta > 90°$, then liquid meniscus will be convex upwards.
- If angle of contact is acute angle, i.e. $\theta < 90°$, then liquid will wet the solid surface.
- If angle of contact is obtuse angle, i.e. $\theta > 90°$, then liquid will not wet the solid surface.

Angle of contact increases with increase in temperature of liquid. Angle of contact decreases on adding soluble impurity to a liquid.

Angle of contact for pure water and glass is zero. For ordinary water and glass is 8°. For mercury and glass is 140°. For pure water and silver is 90°. For alcohol and clean glass $\theta = 0°$.

Angle of contact, meniscus, shape of liquid surface

Property	Angle of Contact $< 90°$	Angle of Contact $= 90°$	Angle of Contact $> 90°$
Substances	Water and glass	Water and silver	Mercury and glass
Angle of contact	Almost zero, acute angle	Right angle = 90°	Obtuse angle = 135°
Meniscus shape	Concave	Plane	Convex
Capillary action	Liquid rises	No effect	Liquid falls

Property	Angle of Contact $< 90°$	Angle of Contact $= 90°$	Angle of Contact $> 90°$
Sticking to solid	Stick/wets	Does not wet	Does not wet
Relation between cohesive force (F_c) and adhesive force (F_a)	$F_a > \frac{F_c}{\sqrt{2}}$ $F_a > F_c$	$F_a = \frac{F_c}{\sqrt{2}}$	$F_a < \frac{F_c}{\sqrt{2}}$ $F_c > F_a$
Shape of liquid surface	Almost round	Spreads on surface	Flat

Capillarity

The phenomenon of rise or fall of liquid column in a capillary tube is called capillarity.

Ascent of a liquid column in a capillary tube is given by

$$h = \frac{2S\cos\theta}{r\rho g} - \frac{r}{3}$$

If capillary is very narrow, then

$$h = \frac{2S\cos\theta}{r\rho g}$$

where, r = radius of capillary tube, ρ = density of the liquid,

θ = angle of contact and S = surface tension of liquid.

- If $\theta < 90°$, $\cos\theta$ is positive, so h is positive, i.e. liquid rises in a capillary tube.
- If $\theta > 90°$, $\cos\theta$ is negative, so h is negative, i.e. liquid falls in a capillary tube.
- Rise of liquid in a capillary tube does not violate law of conservation of energy.

Some Practical Examples of Capillarity

(i) The kerosene oil in a lantern and the melted wax in a candle, rise in the capillaries formed in the cotton wick and burns.

(ii) Coffee powder is easily soluble in water because water immediately wets the fine granules of coffee by the action of capillarity.

(iii) The water given to the fields rises in the innumerable capillaries formed in the stems of plants and trees and reaches the leaves.

Zurin's Law

If a capillary tube of insufficient length is placed vertically in a liquid, then liquid never come out from the tube its own, as

$$Rh = \text{constant} \Rightarrow R_1h_1 = R_2h_2$$

where, R = radius of curvature of liquid meniscus and

h = height of liquid column.

When a tube is kept in inclined position in a liquid the vertical height remains unchanged then length of liquid column.

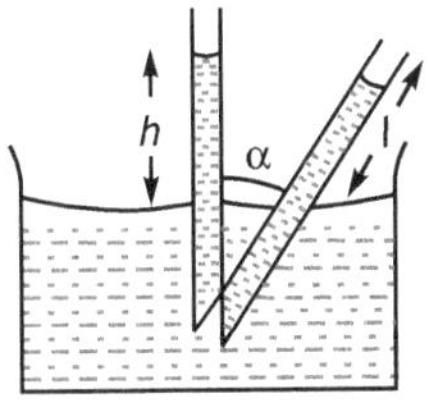

$$\cos\alpha = \frac{h}{l}$$

or
$$l = \frac{h}{\cos\alpha}$$

Liquid rises (water in glass capillary) or falls (mercury in glass capillary) due to property of surface tension

$$T = \frac{R\rho gh}{2\cos\theta}$$

where, R = radius of capillary tube, h = height of liquid,

ρ = density of liquid, θ = angle of contact,

T = surface tension of liquid and g = acceleration due to gravity.

Excess Pressure due to Surface Tension

(i) Excess pressure inside a liquid drop $= \frac{2S}{R}$

(ii) Excess pressure inside an air bubble in a liquid $= \frac{2S}{R}$

(iii) Excess pressure inside a soap bubble $= \frac{4S}{R}$

where, S = surface tension and R = radius of drop/bubble.

(iv) Work done in spraying a liquid drop of radius R into n droplets of radius $r = T \times$ increase in surface area

$$= 4\pi T R^3 \left(\frac{1}{r} - \frac{1}{R}\right)$$

Fall in temperature

$$\Delta\theta = \frac{3T}{J}\left(\frac{1}{r} - \frac{1}{R}\right)$$

where, $J = 4.2$ J/cal.

(v) When n small drops are combined into a bigger drop, then work done is given by

$$W = 4\pi R^2 T\,(n^{1/3} - 1)$$

Temperature increase

$$\Delta\theta = \frac{3T}{J}\left(\frac{1}{r} - \frac{1}{R}\right)$$

(vi) When two bubbles of radii r_1 and r_2 coalesce into a bubble of radius r isothermally, then

$$r^2 = r_1^2 + r_2^2$$

(vii) When two soap bubbles of radii r_1 and r_2 are in contact with each other, then radius (r) of common interface.

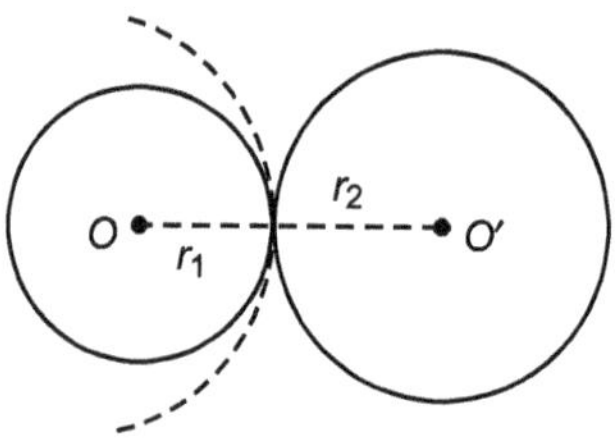

Common interface

$$\frac{1}{r} = \frac{1}{r_1} - \frac{1}{r_2}$$

or $$r = \frac{r_1 r_2}{r_2 - r_1}$$

Factors Affecting Surface Tension

(i) Surface tension of a liquid decreases with increase in temperature and becomes zero at critical temperature.

(ii) At boiling point, surface tension of a liquid becomes zero and becomes maximum at freezing point.

(iii) Surface tension decreases when partially soluble impurities such as soap, detergent, dettol, phenol, etc are added in water.

(iv) Surface tension increases when highly soluble impurities such as salt is added in water.

(v) When dust particles or oil spreads over the surface of water, its surface tension decreases.

When charge is given to a soap bubble, its size increases because surface tension of the liquid decreases due to electrification.

In weightlessness condition liquid does not rise in a capillary tube.

Some Phenomena Based on Surface Tension

(i) Medicines used for washing wounds, as dettol, have a surface tension lower than water.

(ii) Hot soup is more tasteful than the cold one because the surface tension of the hot soup is less than that of the cold and so it spreads over a larger area of the tongue.

(iii) Insects and mosquitoes swim on the surface of water in ponds and lakes due to surface tension. If kerosene oil is sprayed on the water surface, the surface tension of water is lowered and the insects and mosquitoes sink in water and are dead.

(iv) If we deform a liquid drop by pushing it slightly, then due to surface tension it again becomes spherical.

(v) The detergents, used for cleaning the dirty clothes. The molecule of detergent can attached with water and dirt molecules and they take away the dirt with them, when we wash the clothes with detergent.

13 Thermometry and Calorimetry

The branch dealing with measurement of temperature is called **thremometry** and the devices used to measure temperature are called **thermometers.**

Heat

Heat is a form of energy called thermal energy which flows from a higher temperature body to a lower temperature body when they are placed in contact.

Heat or thermal energy of a body is the sum of kinetic energies of all its constituent particles, on account of translational, vibrational and rotational motion.

The SI unit of heat energy is joule (J).

The practical unit of heat energy is calorie.

$$1 \text{ cal} = 4.18 \text{ J}$$

1 calorie is the quantity of heat required to raise the temperature of 1 g of water by 1°C.

Mechanical energy or work (W) can be converted into heat (Q) by

$$W = JQ$$

where J = Joule's mechanical equivalent of heat.

J is a conversion factor (not a physical quantity) and its value is 4.186 J/cal.

Temperature

Temperature of a body is the degree of hotness or coldness of the body.

A device which is used to measure the temperature, is called a **thermometer**.

Highest possible temperature achieved in laboratory is about 10^8 K, while lowest possible temperature attained is 10^{-8} K.

Branch of Physics dealing with production and measurement of temperature close to 0 K is known as **cryagenics**, while that dealing with the measurement of very high temperature is called **pyrometry**.

Temperature of the core of the sun is 10^7 K while that of its surface is 6000 K.

NTP or STP implies 273.15 K (0°C = 32°F).

Different Scale of Temperature

(i) **Celsius Scale** In this scale of temperature, the melting point of ice is taken as 0°C and the boiling point of water as 100°C and the space between these two points is divided into 100 equal parts.

(ii) **Fahrenheit Scale** In this scale of temperature, the melting point of ice is taken as 32°F and the boiling point of water as 212°F and the space between these two points is divided into 180 equal parts.

(iii) **Kelvin Scale** In this scale of temperature, the melting point of ice is taken as 273 K and the boiling point of water as 373 K and the space between these two points is divided into 100 equal parts.

Relation between Different Scales of Temperatures

$$\frac{C}{100} = \frac{F-32}{180} = \frac{K-273}{100} = \frac{R}{80}$$

Absolute Temperature

There is no limit for maximum temperature but there is a sharp point for minimum temperature that nobody can have the temperature lower than this minimum value of temperature, which is known as absolute temperature.

Thermometric Property

The property of an object which changes with temperature, is called a thermometric property. Different thermometric properties and thermometers have been given below

(i) **Pressure of a Gas at Constant Volume**

$$\frac{p_1}{T_1} = \frac{p_2}{T_2} \text{ and } p_t = p_0\left(1 + \frac{t}{273}\right)$$

$$t = \left(\frac{p_t - p_0}{p_{100} - p_0} \times 100\right) {}^\circ\text{C}$$

where p_0, p_{100} and p_t are pressure of a gas at constant volume at 0°C, 100°C and t°C.

A constant volume gas thermometer can measure temperature from – 200°C to 500°C.

(ii) **Electrical Resistance of Metals**

$$R_t = R_0 (1 + \alpha t + \beta t^2)$$

where α and β are constants for a metal.

As β is too small therefore we can take

$$R_t = R_0 (1 + \alpha t)$$

where, α = temperature coefficient of resistance and

R_0 and R_t are electrical resistances at 0°C and t°C.

$$\alpha = \frac{R_2 - R_1}{R_1 t_2 - R_2 t_1}$$

where R_1 and R_2 are electrical resistances at temperatures t_1 and t_2.

or $$t = \frac{R_t - R_0}{R_{100} - R_0} \times 100°C$$

where R_{100} is the resistance at 100°C.

Platinum resistance thermometer can measure temperature from –200°C to 1200°C.

(iii) **Length of Mercury Column in a Capillary Tube**

$$l_t = l_0 (1 + \alpha t)$$

where α = coefficient of linear expansion and l_0, l_t are lengths of mercury column at 0°C and t°C.

Thermo Electro Motive Force

When two junctions of a thermo-couple are kept at different temperatures, then a thermo-emf is produced between the junctions, which changes with temperature difference between the junctions.

Thermo-emf, $$E = at + bt^2$$

where a and b are constants for the pair of metals.

Unknown temperature of hot junction when cold junction is at 0°C.

$$t = \left(\frac{E_t}{E_{100}} \times 100 \right) °C$$

where E_{100} is the thermo-emf when hot junction is at 100°C.

A thermo-couple thermometer can measure temperature from –200°C to 1600°C.

Thermal Expansion

Increase in size on heating is called thermal expansion. There are three types of thermal expansion

1. Expansion of solids
2. Expansion of liquids
3. Expansion of gases

1. Expansion of Solids

Three types of expansion takes place in solid.

(i) **Linear Expansion** Expansion in length on heating is called linear expansion.

Increase in length, $l_2 = l_1 (1 + \alpha \Delta t)$

where, l_1 and l_2 are initial and final lengths, Δt = change in temperature and α = coefficient of linear expansion.

Coefficient of linear expansion

$$\alpha = \frac{\Delta l}{l \times \Delta t}$$

where l = real length and Δl = change in length and

Δt = change in temperature.

(ii) **Superficial Expansion** Expansion in area on heating is called superficial expansion.

Increase in area $A_2 = A_1 (1 + \beta \Delta t)$

where, A_1 and A_2 are initial and final areas and β is a coefficient of superficial expansion.

Coefficient of superficial expansion, $\beta = \dfrac{\Delta A}{A \times \Delta t}$

where, A = area, ΔA = change in area and Δt = change in temperature.

(iii) **Cubical Expansion** Expansion in volume on heating is called cubical expansion.

Increase in volume $V_2 = V_1 (1 + \gamma \Delta t)$

where V_1 and V_2 are initial and final volumes and γ is a coefficient of cubical expansion.

Coefficient of cubical expansion, $\gamma = \dfrac{\Delta V}{V \times \Delta t}$

where V = real volume, ΔV = change in volume and Δt = change in temperature.

Relation between coefficients of linear, superficial and cubical expansions $\beta = 2\alpha$ and $\gamma = 3\alpha$

or $\alpha : \beta : \gamma = 1 : 2 : 3$

2. Expansion of Liquids

In liquids only expansion in volume takes place on heating.

(i) **Apparent Expansion of Liquids** When expansion of the container containing liquid, on heating is not taken into account, then observed expansion is called apparent expansion of liquids.

Coefficient of apparent expansion of a liquid

$$(\gamma_a) = \frac{\text{apparent increase in volume}}{\text{original volume } \times \text{ rise in temperature}}$$

(ii) **Real Expansion of Liquids** When expansion of the container, containing liquid, on heating is also taken into account, then observed expansion is called real expansion of liquids.

Coefficient of real expansion of a liquid

$$(\gamma_r) = \frac{\text{real increase in volume}}{\text{original volume } \times \text{ rise in temperature}}$$

Both, γ_r and γ_a are measured in $^\circ C^{-1}$.

We can show that

$$\gamma_r = \gamma_a + \gamma_g$$

where, γ_r and γ_a are coefficient of real and apparent expansion of liquids and γ_g is coefficient of cubical expansion of the container.

Anamalous Expansion of Water

When temperature of water is increased from 0°C, then its volume decreases upto 4°C, becomes minimum at 4°C and then increases. This behaviour of water around 4°C is called, anamalous expansion of water.

3. Expansion of Gases

There are two types of coefficient of expansion in gases

(i) **Volume Coefficient** (γ_V) At constant pressure, the change in volume per unit volume per degree celsius is called volume coefficient.

$$\gamma_V = \frac{V_2 - V_1}{V_0 (t_2 - t_1)}$$

where V_0, V_1 and V_2 are volumes of the gas at 0°C, t_1°C and t_2°C.

(ii) **Pressure Coefficient** (γ_p) At constant volume, the change in pressure per unit pressure per degree celsius is called pressure coefficient.

$$\gamma_p = \frac{p_2 - p_1}{p_0 (t_2 - t_1)}$$

where p_0, p_1 and p_2 are pressure of the gas at 0°C, t_1°C and t_2°C.

Practical Applications of Expansion

(i) When rails are laid down on the ground, space is left between the end of two rails.

(ii) The transmission cables are not tightly fixed to the poles.

(iii) The iron rim to be put on a cart wheel is always of slightly smaller diameter than that of wheel.

(iv) A glass stopper jammed in the neck of a glass bottle can be taken out by warming the neck of the bottles.

Thermal Equilibrium

When there is no transfer of heat between two bodies in contact, then the bodies are called in thermal equilibrium.

Triple Point of Water

The values of pressure and temperature at which water coexists in equilibrium in all three states of matter, i.e. ice, water and vapour is called triple point of water.

Triple point of water is 273 K temperature and 0.46 cm of mercury pressure.

Specific Heat

The amount of heat required to raise the temperature of unit mass of the substance through 1°C is called its specific heat.

It is denoted by c or s.

Its SI unit is 'joule/kilogram-°C' (J/kg-°C). Its dimensions is $[L^2 T^{-2} \theta^{-1}]$.

The specific heat of water is 4200 J $kg^{-1} °C^{-1}$ or 1 cal $g^{-1} C^{-1}$, which is high compared with most other substances.

Gases have two types of specific heat

(i) The specific heat capacity at constant volume (C_V).

(ii) The specific heat capacity at constant pressure (C_p).

Specific heat at constant pressure (C_p) is greater than specific heat at constant volume (C_V), i.e. $C_p > C_V$.

For molar specific heats $C_p - C_V = R$

where R = gas constant and this relation is called **Mayer's formula.**

The ratio of two principal sepecific heats of a gas is represented by γ.

$$\gamma = \frac{C_p}{C_V}$$

The value of γ depends on atomicity of the gas.

Amount of heat energy required to change the temperature of any substance is given by

$$Q = mc\Delta t$$

where, m = mass of the substance,

c = specific heat of the substance and

Δt = change in temperature.

Thermal (Heat) Capacity

Heat capacity of any body is equal to the amount of heat energy required to increase its temperature through 1°C.

$$\text{Heat capacity} = mc$$

where c = specific heat of the substance of the body and m = mass of the body.

Its SI unit is joule/kelvin (J/K).

Water Equivalent

It is the quantity of water whose thermal capacity is same as the heat capacity of the body. It is denoted by W.

$$W = ms = \text{heat capacity of the body.}$$

Latent Heat

The heat energy absorbed or released at constant temperature per unit mass for change of state is called latent heat.

Heat energy absorbed or released during change of state is given by

$$Q = mL$$

where m = mass of the substance and L = latent heat.

Its unit is cal/g or J/kg and its dimension is $[L^2 T^{-2}]$.

For water at its normal boiling point or condensation temperature (100°C), the latent heat of vaporisation is

$$L = 540 \text{ cal/g} = 40.8 \text{ kJ/ mol} = 2260 \text{ kJ/kg}$$

For water at its normal freezing temperature or melting point (0°C), the latent heat of fusion is

$$L = 80 \text{ cal/ g} = 60 \text{ kJ/mol} = 336 \text{ kJ/kg}$$

It is more painful to get burnt by steam rather than by boiling water at 100°C gets converted to water at 100°C, then it gives out 536 cal of heat. So, it is clear that steam at 100°C has more heat than water at 100°C (i.e. boiling of water).

After snow falls, the temperature of the atmosphere becomes very low. This is because the snow absorbs the heat from the atmosphere to melt down. So, in the mountains, when snow falls, one does not feel too cold, but when ice melts, he feels too cold.

There is more shivering effect of ice cream on teeth as compared to that of water (obtained from ice). This is because when ice cream melts down, it absorbs large amount of heat from teeth.

Melting

Conversion of solid into liquid state at constant temperature is called melting.

Fusion and Freezing Point

The process of change of state from liquid to solid is called fusion. The temperature at which liquid starts to freeze is known as the freezing point of the liquid.

Evaporation

Conversion of liquid into vapour at all temperatures (even below its boiling point) is called evaporation.

Boiling

When a liquid is heated gradually, at a particular temperature the saturated vapour pressure of the liquid becomes equal to the atmospheric pressure, now bubbles of vapour rise to the surface of the liquid. This process is called boiling of the liquid.

The temperature at which a liquid boils, is called **boiling point**.

The boiling point of water increases with increase in pressure and decreases with decrease in pressure.

Sublimation

The conversion of a solid into vapour state is called sublimation.

Hoar Frost

The conversion of vapours into solid state is called hoar frost.

Calorimetry

This is the branch of heat transfer that deals with the measurement of heat. The heat is usually measured in calories or kilo calories.

Principle of Calorimetry

When a hot body is mixed with a cold body, then heat lost by hot body is equal to the heat gained by cold body.

$$\text{Heat lost} = \text{Heat gain}$$

Important Points

- Due to increment in its time period a pendulum clock becomes slow in summer and will lose time.
- Loss of time in a time period $\Delta T = \frac{1}{2}\alpha\, \Delta\theta T$

  $\therefore$ Loss of time in any given time interval t can be given by

  $$\Delta t = \frac{1}{2}\alpha \Delta \theta t$$
- At some higher temperature, a scale will expand and scale reading will be lesser than true values, so that

  true value = scale reading $(1 + \alpha\, \Delta t)$

  Here, Δt is the temperature difference.
- However, at lower temperature scale reading will be more or true value will be less.

14

Kinetic Theory of Gases

Assumptions of Kinetic Theory of Gases

(i) Every gas consists of extremely small particles known as molecules. The molecules of a given gas are all identical but are different from those of another gas.

(ii) The molecules of a gas are identical spherical, rigid and perfectly elastic point masses.

(iii) Their molecular size is negligible in comparison to intermolecular distance (10^{-9} m).

(iv) The speed of gas molecules lies between zero and infinity (very high speed).

(v) The distance covered by the molecules between two successive collisions is known as **free path** and mean of all free path is known as **mean free path**.

(vi) The number of collision per unit volume in a gas remains constant.

(vii) No attractive or repulsive force acts between gas molecules.

(viii) Gravitational attraction among the molecules is ineffective due to extremely small masses and very high speed of molecules.

Gas Laws

Assuming permanent gases to be ideal, through experiments, it was established that gases irrespective of their nature obey the following laws.

Boyle's Law

At constant temperature the volume (V) of given mass of a gas is inversely proportional to its pressure (p), i.e.

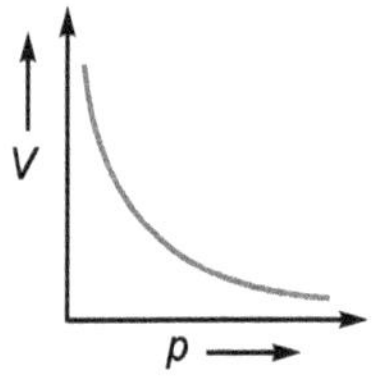

$$V \propto \frac{1}{p} \Rightarrow pV = \text{constant}$$

For a given gas, $p_1V_1 = p_2V_2$

Charles' Law

At constant pressure the volume (V) of a given mass of gas is directly proportional to its absolute temperature (T), i.e.

$$V \propto T \Rightarrow \frac{V}{T} = \text{constant}$$

For a given gas, $$\frac{V_1}{T_1} = \frac{V_2}{T_2}$$

At constant pressure the volume (V) of a given mass of a gas increases or decreases by $\frac{1}{273.15}$ of its volume at 0°C for each 1°C rise or fall in temperature.

Volume of the gas at t°C, $$V_t = V_0\left(1 + \frac{t}{273.15}\right)$$

where V_0 is the volume of gas at 0°C.

Gay Lussac's or Regnault's Law

At constant volume the pressure p of a given mass of gas is directly proportional to its absolute temperature T, i.e.

$$p \propto T \Rightarrow \frac{p}{T} = \text{constant}$$

For a given gas, $$\frac{p_1}{T_1} = \frac{p_2}{T_2}$$

At constant volume, the pressure p of a given mass of a gas increases or decreases by $\frac{1}{273.15}$ of its pressure at 0°C for each 1°C rise or fall in temperature.

Volume of the gas at t°C, $p_t = p_0\left(1 + \frac{t}{273.15}\right)$

where p_0 is the pressure of gas at 0°C.

Avogadro's Law

Avogadro stated that equal volume of all the gases under similar conditions of temperature and pressure contain equal number of molecules. This statement is called Avogadro's hypothesis. According to Avogadro's law

(i) **Avogadro's number** The number of molecules present in 1 g mole of a gas is defined as **Avogadro's number.**

$$N_A = 6.023 \times 10^{23} \text{ per gram mole}$$

(ii) At STP or NTP ($T = 273$ K and $p = 1$ atm 22.4 L of each gas has 6.023×10^{23} molecules.

(iii) One mole of any gas at STP occupies 22.4 L of volume.

Standard or Perfect Gas Equation

Gases which obey all gas laws in all conditions of pressure and temperature are called perfect gases.

Equation of perfect gas

$$pV = nRT$$

where, $p =$ pressure, $V =$ volume, T = absolute temperature, $R =$ universal gas constant and $n =$ number of moles of a gas.

Universal gas constant $R = 8.31 \text{ J mol}^{-1}\text{K}^{-1}$.

Real Gases

Real gases deviate slightly from ideal gas laws because

(i) Real gas molecules attract one another.

(ii) Real gas molecules occupy a finite volume.

Real or van der Waals' Gas Equation

$$\left(p + \frac{a}{V^2}\right)(V - b) = RT$$

where a and b are called van der Waals' constants.

Pressure due to an ideal gas is given by

$$p = \frac{1}{3}\frac{mn}{V}c^2 = \frac{1}{3}\rho c^2$$

For one mole of an ideal gas

$$p = \frac{1}{3}\frac{M}{V}c^2$$

where, m = mass of one molecule, n = number of molecules, V = volume of gas, $c = \frac{c_1^2 + c_2^2 + \ldots + c_n^2}{n}$, is called root mean square (rms) velocity of the gas molecules and M = molecular weight of the gas. If p is the pressure of the gas and E is the kinetic energy per unit volume is E, then

$$p = \frac{2}{3} E$$

Kinetic Energy of a Gas

(i) Average kinetic energy of translation per molecule of a gas is given by

$$E = \frac{3}{2} kT$$

where k = Boltzmann's constant.

(ii) Average kinetic energy of translation per mole of a gas is given by

$$E = \frac{3}{2} RT$$

where R = universal gas constant.

(iii) For a given gas kinetic energy

$$E \propto T \quad \Rightarrow \quad \frac{E_1}{E_2} = \frac{T_1}{T_2}$$

(iv) Root mean square (rms) velocity of the gas molecules is given by

$$c = \sqrt{\frac{3RT}{M}} = \sqrt{\frac{3p}{\rho}}$$

(v) For a given gas, $c \propto \sqrt{T}$

(vi) For different gases, $c \propto \frac{1}{\sqrt{M}}$

(vii) Boltzmann's constant $k = \frac{R}{N}$

where R is ideal gas constant and N = Avogadro number.

Value of Boltzmann's constant is 1.38×10^{-28} J/K.

(viii) The average speed of molecules of a gas is given by

$$\overline{v} = \sqrt{\frac{8kT}{\pi m}} = \sqrt{\frac{8RT}{\pi M}}$$

(ix) The most probable speed of molecules of a gas is given by

$$v_{\text{mp}} = \sqrt{\frac{2kT}{m}} = \sqrt{\frac{2RT}{M}}$$

$$\Rightarrow \qquad v_{\text{rms}} > \bar{v} > v_{\text{mp}}$$

Important Points

(i) With rise in temperature rms speed of gas molecules increases as

$$v_{\text{rms}} \propto \sqrt{T}$$

(ii) With the increase in molecular weight rms speed of gas molecule decrease as

$$v_{\text{rms}} \propto \frac{1}{\sqrt{M}}$$

(iii) Rms speed of gas molecules is of the order of km/s, e.g. at NTP for hydrogen gas

$$v_{\text{rms}} = \sqrt{\frac{3RT}{M}} = \sqrt{\frac{3 \times 8.31 \times 273}{2 \times 10^3}}$$

(iv) Rms speed of gas molecules does not depend on the pressure of gas (if temperature remains constant) because $p \propto \rho$ (Boyle's law). If pressure is increased n times, then density will also increase by n times but v_{rms} remains constant.

Degree of Freedom

The degree of freedom for a dynamic system is the number of directions in which it can move freely or the number of coordinates required to describe completely the position and configuration of the system.

It is denoted by f or N.

Degree of freedom of a system is given by

$$f \text{ or } N = 3A - R$$

where A = number of particles in the system and R = number of independent relations.

Degree of Freedom

(i) For monoatomic gas = 3

(ii) For diatomic gas = 5

(iii) For non-linear triatomic gas = 6

(iv) For linear triatomic gas = 7

Specific heat of a gas

(a) At constant volume, $C_V = \frac{f}{2} R$

(b) At constant pressure, $C_p = \left(\frac{f}{2} + 1\right) R$

(c) Ratio of specific heats of a gas at constant pressure and at constant volume is given by, $\gamma = 1 + \frac{2}{f}$

Specific heat of solids, $C = 3R \Rightarrow C = 24 \cdot 93 \text{ Jmol}^{-1} \text{ K}^{-1}$

Specific heat of water, $C = 9R \Rightarrow C = 74 \cdot 97 \text{ Jmol}^{-1} \text{ K}^{-1}$

Law of Equipartition of Energy

This law states that, for a dynamic system in thermal equilibrium, the total energy is distributed equally amongst all the degree of freedom and the energy associated with each molecule per degree of freedom is $\frac{1}{2} k_B T$.

where, $k_B = 1 \cdot 38 \times 10^{-23} \text{ JK}^{-1}$ is Boltzmann constant and T is absolute temperature of system on the kelvin scale.

Mean Free Path

The average distance travelled by a molecule between two successive collisions is called mean free path (λ).

Mean free path is given by, $\lambda = \frac{kT}{\sqrt{2} \pi \sigma^2 p}$

where σ = diameter of the molecule, p = pressure of the gas,
T = temperature and k = Botlzmann's constant.

Mean free path, $\lambda \propto T$ and $\lambda \propto \frac{1}{p}$

Brownian Motion

The continuous random motion of the particles of microscopic size suspended in air or any liquid, is called Brownian motion.

Brownian motion is observed with many kind of small particles suspended in both liquids and gases.

Brownian motion is due to the unequal bombardment of the suspended particles by the molecules of the surrounding medium.

15

Thermodynamics

The branch of physics which deals with the study of transformation of heat energy into other forms of energy and *vice-versa.*

A thermodynamical system is said to be in thermal equilibrium when macroscopic variables (like pressure, volume, temperature, mass, composition etc) that characterise the system do not change with time.

Adiabatic Wall

It is an insulating wall (can be movable) between two thermodynamic systems that does not allow flow of energy (or heat) from one system to another system.

Diathermic Wall

It is a conducting wall between two thermodynamic systems that allows energy flow (or heat) from one system to another system.

Thermodynamical System

An assembly of an extremely large number of particles whose state can be expressed in terms of pressure, volume and temperature, is called thermodynamic system.

Thermodynamic system is classified into the following three systems

(i) **Open System** It exchanges both energy and matter with surroundings.

(ii) **Closed System** It exchanges only energy (not matter) with surroundings.

(iii) **Isolated System** It exchanges neither energy nor matter with the surroundings.

A thermodynamic system is not always in equilibrium. For example, a gas allowed to expand freely against vacuum. Similary, a mixture of petrol, vapour and air, when ignited by a spark is not an equilibrium state. Equilibrium is acquired eventually with time.

Thermodynamic Parameters or Coordinates or Variables

The state of thermodynamic system can be described by specifying pressure, volume, temperature, internal energy and number of moles, etc. These are called thermodynamic parameters or coordinates or variables.

The state variables are of two such as extensive and intensive.

Extensive State Variables Extensive state variables indicate the 'size' of the system, e.g. volume, total mass, internal energy etc.

Intensive State Variables These are state variables that do not depend on the size of the system, e.g. pressure, temperature and density.

Work Done

Work done by a thermodynamic system is given by

$$W = p \times \Delta V$$

where p = pressure and ΔV = change in volume.

Work done by a thermodynamic system is equal to the area enclosed between the p-V curve and the volume axis.

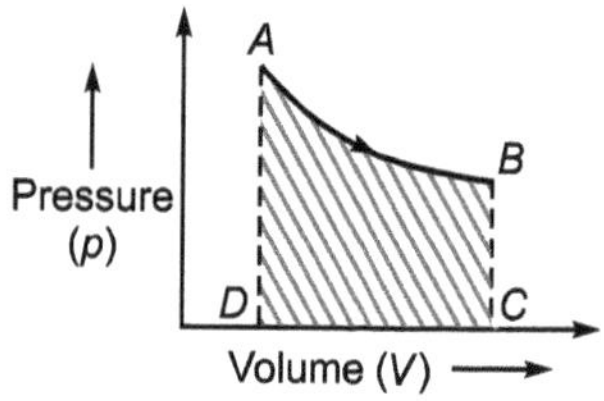

Work done in process A-B = Area $ABCDA$

Work done by a thermodynamic system depends not only upon the initial and final states of the system but also depend upon the path followed in the process.

Work done by the Thermodynamic System is taken as

Positive → as volume increases.

Negative → as volume decreases.

Internal Energy (U)

The total energy possessed by any system due to molecular motion and molecular configuration, is called its internal energy.

Internal energy of a thermodynamic system depends on temperature.

It is the characteristic property of the state of the system.

Zeroth Law of Thermodynamics

According to this law, two systems in thermal equilibrium with a third system separately are in thermal equilibrium with each other. Thus, if A and B are separately in equilibrium with C, that is if $T_A = T_C$ and $T_B = T_C$, then this implies that $T_A = T_B$, i.e. the systems A and B are also in thermal equilibrium.

First Law of Thermodynamics

Heat given to a thermodynamic system (ΔQ) is partially utilized in doing work (ΔW) against the surrounding and the remaining part increases the internal energy (ΔU) of the system.

Therefore, $$\Delta Q = \Delta U + \Delta W$$

First law of thermodynamics is a restatement of the principle of conservation of energy.

In isothermal process, change in internal energy is zero ($\Delta U = 0$).

Therefore, $$\Delta Q = \Delta W$$

In adiabatic process, no exchange of heat takes place, i.e. $\Delta\theta = 0$.

Therefore, $$\Delta U = -\Delta W$$

In adiabatic process, if gas expands, its internal energy and hence, its temperature decreases and *vice-versa*.

In isochoric process, work done is zero, i.e. $\Delta W = 0$, therefore

$$\Delta Q = \Delta U$$

Thermodynamic Processes

A thermodynamical process is said to take place when some changes occur in the state of a thermodynamic system i.e. the thermodynamic parameters of the system change with time.

(i) **Quasi-static Process** Quasi-static is a thermodynamic process which proceeds extremely slowly such that at every instant of time, the temperature and pressure are the same in all parts of the system.

(ii) **Isothermal Process** A process taking place in a thermodynamic system at constant temperature is called an isothermal process.

Isothermal processes are very slow processes.

These process follows **Boyle's law**, according to which

$$pV = \text{constant}$$

From $dU = nC_V dT$ as $dT = 0$ so $dU = 0$, i.e. internal energy is constant.

From first law of thermodynamic $dQ = dW$, i.e. heat given to the system is equal to the work done by system surroundings.

Work done $W = 2.3026\mu RT \log_{10}\left(\frac{V_f}{V_i}\right)$

$$= 2.3026\,\mu RT \log_{10}\left(\frac{p_i}{p_f}\right)$$

where, μ = number of moles, R = ideal gas constant, T = absolute temperature and V_i, V_f and p_i, p_f are initial volumes and pressures.

After differentiating pV = constant, we have

$$\frac{dp}{dV} = -\frac{p}{V} \text{ and } -\frac{dp}{\frac{dV}{V}} = p$$

i.e. bulk modulus of gas in isothermal process, $\beta = p$.

p-V curve for this process is a rectangular hyperbola.

Examples

(a) Melting process is an isothermal change, because temperature of a substance remains constant during melting.

(b) Boiling process is also an isothermal operation.

(iii) **Adiabatic Process** A process taking place in a thermodynamic system for which there is no exchange of heat between the system and its surroundings.

Adiabatic processes are very fast processes.

These process follows **Poisson's law,** according to which

$$pV^{\gamma} = TV^{\gamma - 1} = \frac{T^{\gamma}}{p^{\gamma - 1}} = \text{constant}$$

From $dQ = nCdT, C_{adi} = 0$ as $dQ = 0$, i.e. molar heat capacity for adiabatic process is zero.

From first law, $dU = -dW$, i.e. work done by the system is equal to decrease in internal energy. When a system expands adiabatically, work done is positive and hence internal energy decrease, i.e. the system cools down and *vice-versa.*

Work done in an adiabatic process is

$$W = \frac{nR(T_i - T_f)}{\gamma - 1} = \frac{p_i V_i - p_f V_f}{\gamma - 1}$$

where T_i and T_f are initial and final temperatures.

Examples

(a) Sudden compression or expansion of a gas in a container with perfectly non-conducting wall.

(b) Sudden bursting of the tube of a bicycle tyre.

(c) Propagation of sound waves in air and other gases.

(iv) **Isobaric Process** A process taking place in a thermodynamic system at constant pressure is called an **isobaric process.**

Molar heat capacity of the process is C_p and $dQ = nC_p dT$.

Internal energy $dU = nC_V\ dT$

From the first law of thermodynamics

$$dQ = dU + dW$$

$$dW = pdV = nRdT$$

Process equation is $\frac{V}{T}$ = constant.

p-V curve is a straight line parallel to volume axis.

(v) **Isochoric Process** A process taking place in a thermodynamic system at constant volume is called an **isochoric process.**

$dQ = nC_V dT$, molar heat capacity for isochoric process is C_V.

Volume is constant, so $dW = 0$.

Process equation is $\frac{p}{T}$ = constant.

p-V curve is a straight line parallel to pressure axis.

(vi) **Cyclic Process** When a thermodynamic system returns to its initial state after passing through several states, then it is called a cyclic process.

Efficiency of the cycle is given by

$$\eta = \frac{\text{Work done}}{\text{Heat supplied}}$$

Work done by the cycle can be computed from area enclosed by cycle on p-V curve.

Isothermal and Adiabatic Curves

The graph drawn between the pressure p and the volume V of a given mass of a gas for an isothermal process is called **isothermal curve** and for an adiabatic process it is called **adiabatic curve.**

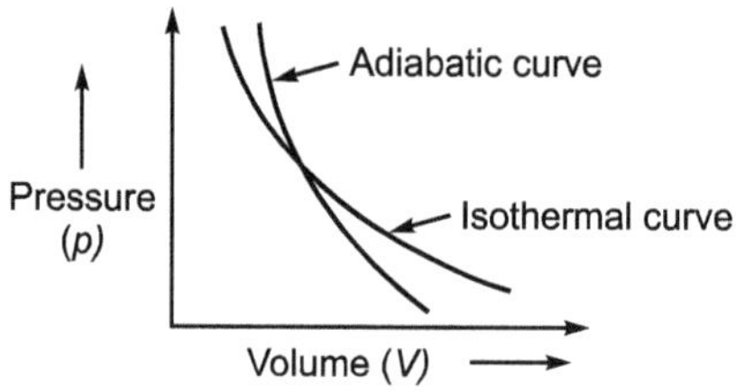

The slope of the adiabatic curve

$$= \gamma \times \text{the slope of the isothermal curve}$$

Volume Elasticities of Gases

There are two types of volume elasticities of gases

(i) Isothermal modulus of elasticity $E_S = p$

(ii) Adiabatic modulus of elasticity $E_T = \gamma p$

Ratio between isothermal and adiabatic modulus

$$\frac{E_S}{E_T} = \gamma = \frac{C_p}{C_V}$$

where C_p and C_V are specific heats of gas at constant pressure and at constant volume.

For an isothermal process $\Delta t = 0$, therefore specific heat,

$$c = \frac{\Delta\theta}{m\,\Delta t} = \infty$$

For an adiabatic process $\Delta\theta = 0$, therefore specific heat,

$$c = \frac{0}{m\,\Delta t} = 0$$

Second Law of Thermodynamics

The second law of thermodynamics gives a fundamental limitation to the efficiency of a heat engine and the coefficient of performance of a refrigerator. It says that efficiency of a heat engine can never be unity (or 100%). This implies that heat released to the cold reservoir can never be made zero.

Kelvin's Statement

It is impossible to obtain a continuous supply of work from a body by cooling it to a temperature below the coldest of its surroundings.

Clausius' Statement

It is impossible to transfer heat from a lower temperature body to a higher temperature body without use of an external agency.

Planck's Statement

It is impossible to construct a heat engine that will convert heat completely into work.

All these statements are equivalent as one can be obtained from the other.

Reversible and Irreversible Processes

Reversible Process A process which could be reserved in such a way that the system and its surrounding returns exactly to their initial states with no other changes in the universe is known as reversible process.

Irreversible Process Any process which is not reversible exactly is an irreversible process.

Entropy

Entropy is a physical quantity that remains constant during a reversible adiabatic change.

Change in entropy is given by $dS = \dfrac{\delta Q}{T}$

where, δQ = heat supplied to the system

and T = absolute temperature.

Entropy of a system never decreases, *i.e.*, $dS \geq 0$.

Entropy of a system increases in an irreversible process.

Heat Engine

A heat engine is a device which converts heat energy into mechanical energy.

A heat engine consists of three parts

(i) Source of heat at higher temperature

(ii) Working substance

(iii) Sink of heat at lower temperature

Thermal efficiency of a heat engine is given by

$$\eta = \frac{\text{Work done/ cycle}}{\text{Total amount of heat absorbed / cycle}}$$

$$\eta = 1 - \frac{Q_2}{Q_1} = 1 - \frac{T_2}{T_1}$$

where Q_1 is heat absorbed from the source,
Q_2 is heat rejected to the sink and T_1 and T_2 are temperatures of source and sink.

Heat engine are of two types

(i) **External Combustion Engine** In this engine fuel is burnt in a chamber outside the main body of the engine. *e.g.*, steam engine. In practical life thermal efficiency of a steam engine varies from 12% to 16%.

(ii) **Internal Combustion Engine** In this engine fuel is burnt in inside the main body of the engine. *e.g.*, petrol and diesel engine. In practical life thermal efficiency of a petrol engine is 26% and for a diesel engine is 40%.

Carnot's Cycle

Carnot devised an ideal cycle of operation for a heat engine, called Carnot's cycle.

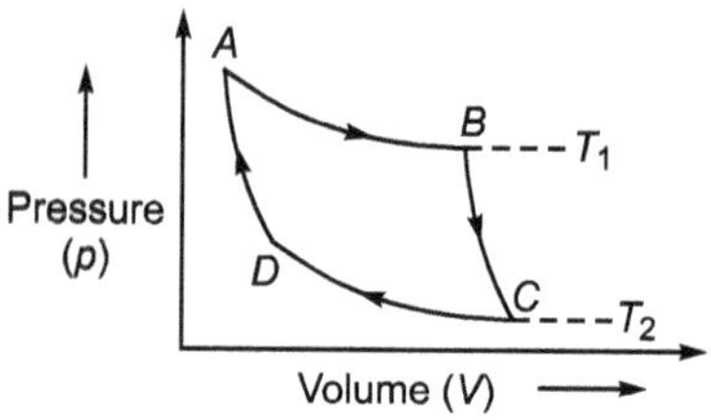

A Carnot's cycle contains the following four processes

(i) Isothermal expansion (*AB*)

(ii) Adiabatic expansion (*BC*)

(iii) Isothermal compression (*CD*)

(iv) Adiabatic compression (*DA*)

The net work done per cycle by the engine is numerically equal to the area of the loop representing the Carnot's cycle.

After doing the calculations for different processes we can show that

$$\frac{\theta_2}{\theta_1} = \frac{T_2}{T_1}$$

Therefore, efficiency of the cycle is

$$\eta = 1 - \frac{T_2}{T_1}$$

Efficiency of Carnot engine is maximum (not 100%) for given temperatures T_1 and T_2. But still Carnot engine is not a practical engine because many ideal situations have been assumed while designing this engine which can practically not be obtained.

Refrigerator or Heat Pump

A refrigerator or heat pump is a device used for cooling things. It absorb heat from sink at lower temperature and reject a large amount of heat to source at higher temperature.

Coefficient of performance of refrigerator is given by

$$\beta = \frac{Q_2}{W} = \frac{Q_2}{Q_1 - Q_2} = \frac{T_2}{T_1 - T_2}$$

where Q_2 is heat absorbed from the sink, Q_1 is heat rejected to source and T_1 and T_2 are temperatures of source and sink.

Relation between efficiency (η) and coefficient of performance (β)

$$\beta = \frac{1 - \eta}{\eta}$$

16

Transmission of Heat

Heat Transmission

Heat can be transferred from one part of system to another. It is called Heat transmission.

There are three methods of heat transmission

(i) **Conduction** In solids, heat is transmitted from higher temperature to lower temperature without actual movements of the particles. This mode of transmission of heat, is called conduction.

(ii) **Convection** The process of heat-transmission in which the particles of the fluid (liquid or gas) move, is called convection.

Land breeze, see-breeze and trade wind are formed due to convection.

(iii) **Radiation** The process of heat transmission in the form of electromagnetic waves, is called radiation.

Radiation do not require any medium for propagation.

It propagates without heating the intervening medium.

The heat energy transferred by radiation, is called radiant energy.

Heat from the sun reaches the earth by radiation.

Thermal Conductivity

In solids, heat is transferred through conduction. We will study conduction of heat through a solid bar.

Conduction of Heat in a Conducting Rod

Steady State

The state of a conducting rod in which no part of the rod absorbs heat, is called the steady state.

Isothermal Surface

A surface of a material whose all points are at the same temperature, is called an isothermal surface.

Temperature Gradient

The rate of change of temperature with distance between two isothermal surfaces is called temperature gradient.

$$\text{Temperature gradient} = \frac{\text{Change in temperature}}{\text{Perpendicular distance}} = -\frac{\Delta\theta}{\Delta x}$$

Its SI unit is "°C per metre' and dimension is $[L^{-1}\theta]$.

The amount of heat flow in a conducting rod

$$Q = \frac{KA\,\Delta\theta t}{l}$$

where K = coefficient of thermal conductivity,
A = area of cross-section,
l = length of rod,
$\Delta\theta$ = temperature difference between the ends of the rod and
t = time.

The SI unit of K is '$Wm^{-1}\,K^{-1}$' and its dimension is $[MLT^{-3}\theta^{-1}]$.

The value of K is large for good conductors and very small for insulators.

Thermal resistance is given by $R = \frac{\Delta\theta}{H} = \frac{l}{KA}$

where, $\Delta\theta$ is temperature difference at the ends of the rod and H is rate of flow of heat.

Its SI unit is K/W and its dimensional formula is $[ML^2T^{-3}\theta^{-1}]$.

When Conducting Rods are Connected in Series

The amount of heat flow, $Q = \dfrac{A(\theta_1 - \theta_2)\,t}{\dfrac{l_1}{K_1} + \dfrac{l_2}{K_2} + \dfrac{l_3}{K_3}}$

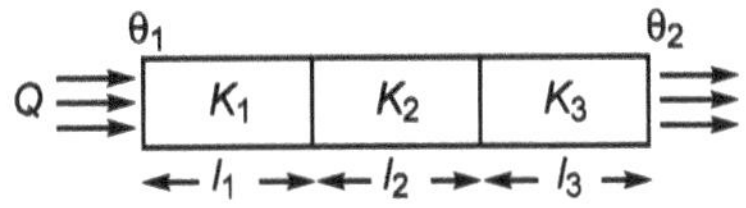

Equivalent thermal conductivity

$$H = \frac{l_1 + l_2 + l_3}{\frac{l_1}{K_1} + \frac{l_2}{K_2} + \frac{l_3}{K_3}}$$

When Two Conducting Rods are Connected in Series

Rate of heat flow

$$H = \frac{Q}{t} = \frac{K_1 A(\theta_1 - \theta)}{l_1} = \frac{K_2 A(\theta - \theta_2)}{l_2}$$

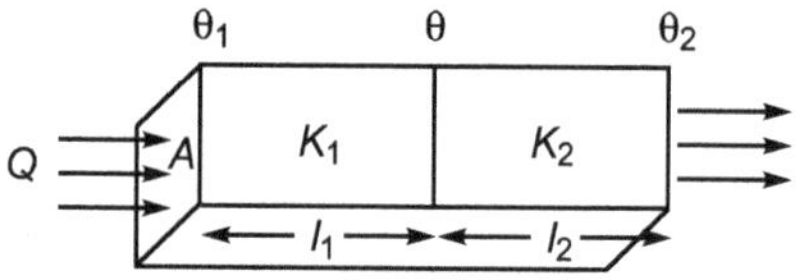

Temperature of contact surface

$$\theta = \frac{\frac{K_1\theta_1}{l_1} + \frac{K_2\theta_2}{l_2}}{\frac{K_1}{l_1} + \frac{K_2}{l_2}} = \frac{K_1\theta_1 l_2 + K_2\theta_2 l_1}{K_1 l_2 + K_2 l_1}$$

Equivalent thermal conductivity $H = \dfrac{l_1 + l_2}{\frac{l_1}{K_1} + \frac{l_2}{K_2}}$

Equivalent thermal resistance $R = R_1 + R_2$

When Conducting Rods are Connected in Parallel

Rate of heat flow

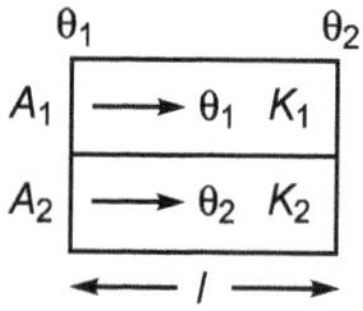

$$H = \frac{Q}{t} = \left(\frac{K_1 A_1}{l} + \frac{K_2 A_2}{l}\right)(\theta_1 - \theta_2)$$

Equivalent thermal conductivity

$$K = \frac{K_1 A_1 + K_2 A_2}{A_1 + A_2}$$

Equivalent thermal resistance

$$\frac{1}{R} = \frac{1}{R_1} + \frac{1}{R_2}$$

Ingen-Hausz Experiment

The thermal conductivities of different materials are proportional to the square of the lengths of the melted wax on the rods of these materials in the steady state.

If $l_1, l_2, l_3, \ldots$ are the lengths of the melted wax on the rods of different materials having coefficient of thermal conductivities $K_1, K_2, K_3, \ldots$, then

$$K_1 : K_2 : K_3 : \ldots = l_1^2 : l_2^2 : l_3^2 : \ldots$$

or

$$\frac{K_1}{K_2} = \frac{l_1^2}{l_2^2}$$

Formation of Ice on Lakes

Time taken to form x thickness of ice on lake

$$t = \frac{\rho L}{2K\theta} x^2$$

where ρ = density of ice, L = latent heat of freezing of ice,

K = coefficient of thermal conductivity of ice and

θ = temperature above lake.

Time taken to increase the thickness of ice from x_1 to x_2

$$t = \frac{\rho L}{2K\theta}(x_2^2 - x_1^2)$$

Reflectance or Reflecting Power

The ratio of the amount of thermal radiations reflected by a body in a given time to the total amount of thermal radiations incident on the body in that time is called reflectance or reflecting power of the body.

It is denoted by r.

Absorptance or Absorbing Power

The ratio of the amount of thermal radiations absorbed by a body in a given time to the total amount of thermal radiations incident on the body in that time, is called absorptance or absorbing power of the body.

It is denoted by a.

Transmittance or Transmitting Power

The ratio of the amount of thermal radiations transmitted by the body in a given time to the total amount of thermal radiations incident on the body in that time, is called transmittance or transmitting power of the body.

It is denoted by t.

Relation among reflecting power, absorbing power and transmitting power,

$$r + a + t = 1$$

If body does not transmit any heat radiations, then $t = 0$

$$\therefore \qquad r + a = 1$$

(i) r, a and t all are the pure ratio, so they have no unit and dimension.

(ii) For perfect reflector, $r = 1, a = 0$ and $t = 0$.

(iii) For perfect absorber, $a = 1, r = 0$ and $t = 0$ (perfect black body).

(iv) For perfect transmitter, $t = 1, a = 0$ and $r = 0$.

Emissive Power

Emissive power of a body at a particular temperature is the total amount of thermal energy emitted per unit time per unit area of the body for all possible wavelengths.

It is denoted by e_λ.

Its SI unit is 'joule sec^{-1} $metre^{-2}$ or 'watt-$metre^{-2}$'.

Its dimensional formula is $[MT^{-3}]$.

Emissivity

Emissivity of a body at a given temperature is equal to the ratio of the total emissive power of the body (e_λ) to the total emissive power of a perfectly black body (E_λ) at that temperature.

Emissivity $$\varepsilon = \frac{e_\lambda}{E_\lambda}$$

Perfectly Black Body

A body which absorbs completely the radiations of all wavelengths incident on it, is called a perfectly black body.

For a perfectly black body, emissive power $(E_\lambda) = 1$.

An ideal black body need not be black in colour.

The radiation from a black body depend upon its temperature only. These heat radiations do not depend on density mass, size or the nature of the body.

Lamp black is 96% black and platinum black is about 98% black.

A perfectly black body cannot be realised in practice. The nearest example of an ideal black body is the Ferry's black body.

Kirchhoff's Law

The ratio of emissive power (e_λ) to the absorptive power (a_λ) corresponding to a particular wavelength and at any given temperature is always a constant for all bodies and it is equal to the emissive power (E_λ) of a perfectly black body at the same temperature and corresponding to the same wavelength.

Mathematically $\frac{e_\lambda}{a_\lambda} = \text{constant } (E_\lambda)$

Fraunhofer Lines

These are typical spectral absorption lines. These are dark but not perfectly black due to their lower intensity as compared to the remaining part of spectrum.

These lines are produced when a cold gas is in between a broad spectrum photon source and detector.

A set of several hundred dark lines appearing against the bright back ground of the continuous solar spectrum and produced by absorption of light by the cooler gases in the sun's outer atmosphere at frequencies corresponding to the atomic transition frequencies of these gases.

About 20000 such dark lines have been detected so far. These dark lines belong to hydrogen, helium, sodium, iron, calcium etc.

Kirchhoff's law explains this phenomenon.

Moon covers the photosphere the central part of sun at the time of total solar eclipse. Thus, the elements present in the chromosphere emit certain wavelength which they had absorbed. Because of this reason the Fraunhofer lines appears as bright lines at the time of total solar eclipse.

Variation of Colour with Temperature

(i) At temperature 525°C, the colour of object is red.

(ii) At temperature 900°C, the colour of object is cherry red.

(iii) At temperature 1100°C, the colour of object is orange red.

(iv) At temperature 1200°C, the colour of object is yellow.

(v) At temperature 1600°C, the colour of object is white.

Stefan's Law

Heat energy emitted per second per unit area of a perfectly black body

$$E \propto T^4 \quad \Rightarrow \quad E = \sigma T^4$$

where σ is Stefan's constant and its value is 5.67×10^{-8} Wm^{-2} K^{-4}.

If T_0 is the temperature of the surroundings, then

$$E = \sigma (T^4 - T_0^4)$$

If ε is the emissivity of the body, then

$$E = \varepsilon \sigma T^4$$

Energy radiated by whole body in t time,

$$E = \sigma A t T^4$$

Newton's Law of Cooling

The rate of loss of heat of a liquid is directly proportional to the difference in temperature of the liquid and its surroundings, i.e.

$$\frac{dT}{dt} = E \propto (T - T_0)$$

where T and T_0 are the temperatures of the liquid and its surroundings.

Wien's Displacement Law

For black body radiation the rate of energy radiation per unit area per unit wavelength range at constant kelvin temperature can be plotted against wavelength.

$$\lambda_m T = \text{constant } (b)$$

where λ_m = wavelength corresponding to which maximum energy is radiated,

T = absolute temperature and

b = Wien's constant = $2.898 \times 10^{-3} \approx 3 \times 10^{-3}$ mK

When temperature of a black body increases, colour changes towards higher frequency, i.e. from red → orange → yellow → green → blue → violet.

Solar Constant

The amount of heat received from the sun by one square centimetre area of a surface placed normally to the sun rays at mean distance of the earth from the sun is known as solar constant. It is denoted by S.

$$S = \left(\frac{r}{R}\right)^2 \sigma T^4$$

Here, r is the radius of sun and R is the mean distance of earth from the centre of sun. Value of solar constant is 1.937 cal cm^{-2} min^{-1}.

Temperature of violet star is maximum while temperature of red star is minimum.

17

Oscillations

Periodic Motion

A motion which repeats itself identically after a fixed interval of time is called periodic motion, e.g. orbital motion of the earth around the sun, motion of arms of a clock, motion of a simple pendulum etc.

Oscillatory Motion

A periodic motion taking place to and fro or back and forth about a fixed point, is called oscillatory motion, e.g. motion of a simple pendulum, motion of a loaded spring etc.

Note *Every oscillatory motion is periodic motion but every periodic motion is not oscillatory motion.*

Harmonic Oscillation

The oscillation which can be expressed in terms of single harmonic function, i.e. sine or cosine function, is called harmonic oscillation.

Simple Harmonic Motion

A harmonic oscillation of constant amplitude and of single frequency under a restoring force whose magnitude is proportional to the displacement and always acts towards mean position is called Simple Harmonic Motion (SHM).

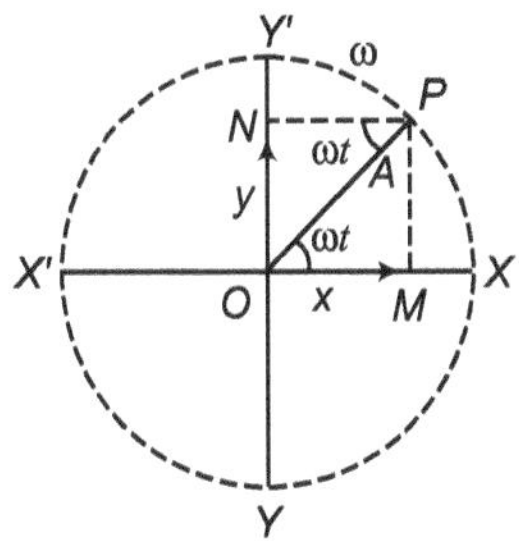

A simple harmonic oscillation can be expressed as

$$y = a \sin \omega t$$

or

$$y = a \cos \omega t$$

where a = amplitude of oscillation.

Non-harmonic Oscillation

A non-harmonic oscillation is a combination of two or more than two harmonic oscillations.

It can be expressed as $y = a \sin \omega t + b \sin 2\omega t$

Some Terms Related to SHM

(i) **Time Period** Time taken by the body to complete one oscillation is known as time period. It is denoted by T.

(ii) **Frequency** The number of oscillations completed by the body in one second is called frequency. It is denoted by ν.

Its SI unit is 'hertz' or 'second^{-1}'.

$$\text{Frequency} = \frac{1}{\text{Time period}}$$

(iii) **Angular Frequency** The product of frequency with factor 2π, is called angular frequency. It is denoted by ω.

$$\text{Angular frequency } (\omega) = 2\pi\nu.$$

Its SI unit is 'hertz' or 'second^{-1}'.

(iv) **Displacement** A physical quantity which changes uniformely with time in a periodic motion, is called displacement. It is denoted by y.

(v) **Amplitude** The maximum displacement in any direction from mean position is called amplitude. It is denoted by a.

(vi) **Phase** A physical quantity which express the position and direction of motion of an oscillating particle, is called phase. It is denoted by ϕ.

Simple harmonic motion is defined as the projection of a uniform circular motion on any diameter of a circle of reference.

Some Important Formulae of SHM

(i) Displacement in SHM at any instant is given by

$$y = a \sin \omega t$$

or

$$y = a \cos \omega t$$

where a = amplitude and

ω = angular frequency.

(ii) Velocity of a particle executing SHM at any instant is given by

$$v = \omega\sqrt{(a^2 - y^2)}$$

At mean position $y = 0$ and v is maximum

$$v_{max} = a\omega$$

At extreme position $y = a$ and v is zero.

(iii) Acceleration of a particle executing SHM at any instant is given by

$$A \text{ or } \alpha = -\omega^2 y$$

Negative sign indicates that the direction of acceleration is opposite to the direction in which displacement increases, i.e. towards mean position.

At mean position $y = 0$ and acceleration is also zero.

At extreme position $y = a$ and acceleration is maximum

$$A_{max} = -a\omega^2$$

(iv) Time period in SHM is given by

$$T = 2\pi \sqrt{\frac{\text{Displacement}}{\text{Acceleration}}}$$

Graphical Representation

(i) **Displacement-Time Graph**

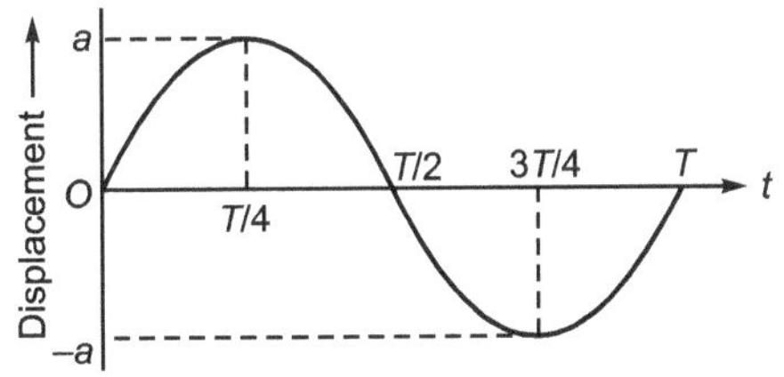

(ii) **Velocity-Time Graph**

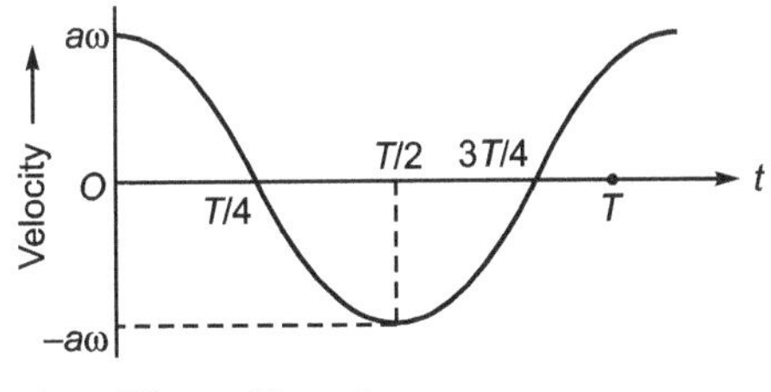

(iii) **Acceleration-Time Graph**

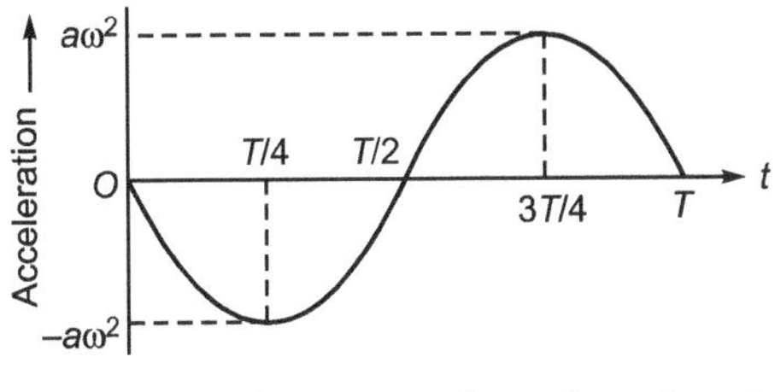

Note *The acceleration is maximum at a place where the velocity is minimum and vice-versa.*

For a particle executing SHM, the phase difference between

(i) Instantaneous displacement and instantaneous velocity

$$= \left(\frac{\pi}{2}\right) \text{rad}$$

(ii) Instantaneous velocity and instantaneous acceleration

$$= \left(\frac{\pi}{2}\right) \text{rad}$$

(iii) Instantaneous acceleration and instantaneous displacement

$$= \pi \text{ rad}$$

The graph between velocity and displacement for a particle executing SHM is elliptical.

Force in SHM

We know that, the acceleration of body in SHM is $a = -\omega^2 x$.

Applying the equation of motion $\mathbf{F} = m\mathbf{a}$,

We have, $F = -m\omega^2 x = -kx$

where, $\omega = \sqrt{\frac{k}{m}}$ and $k = m\omega^2$ is a constant and sometimes it is called the elastic constant.

In SHM, the force is proportional and opposite to the displacement.

Energy in SHM

The kinetic energy of the particle is $K = \frac{1}{2} m\omega^2 (A^2 - x^2)$

From this expression, we can see that, the kinetic energy is maximum at the centre ($x = 0$) and zero at the extremes of oscillation ($x \pm A$).

The potential energy of the particle is $U = \frac{1}{2} m\omega^2 x^2$

From this expression, we can see that, the potential energy has a minimum value at the centre ($x = 0$) and increases as the particle approaches either extreme of the oscillation ($x \pm A$).

Total energy can be obtained by adding potential and kinetic energies. Therefore,

$$E = K + U$$

$$= \frac{1}{2} m(A^2 - x^2)\omega^2 + \frac{1}{2} m\omega^2 x^2 = \frac{1}{2} m\omega^2 A^2$$

where A = amplitude

m = mass of particle executing SHM,

ω = angular frequency and

ν = frequency

Changes of kinetic and potential energies during oscillations.

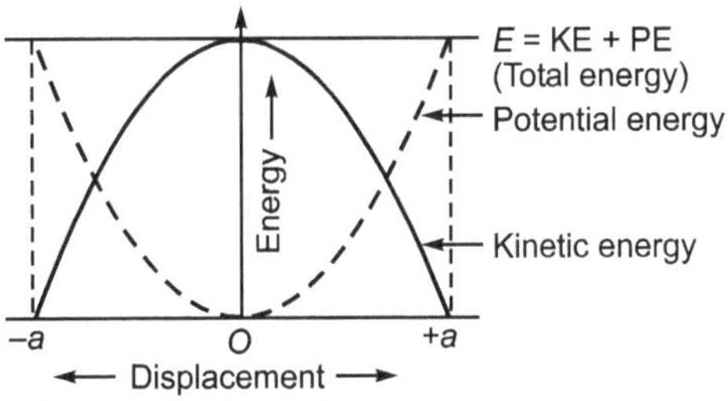

The frequency of kinetic energy or potential energy of a particle executing SHM is double than that of the frequency in SHM.

The frequency of total energy of particles executing SHM is zero as total energy in SHM remains constant at all positions.

When a particle of mass m executes SHM with a constant angular frequency ω, then time period of oscillation

$$T = 2\pi\sqrt{\frac{\text{Inertia factor}}{\text{Spring factor}}}$$

In general, inertia factor = m, (mass of the particle)

spring factor = k (force constant)

How the different physical quantities (e.g. displacement, velocity, acceleration, kinetic energy etc) vary with time or displacement are listed ahead in tabular form.

S.No.	Name of the equation	Expression of the equation	Remarks
1.	Displacement-time	$x = A\cos(\omega t + \phi)$	x varies between $+A$ and $-A$
2.	Velocity-time $\left(v = \frac{dx}{dt}\right)$	$v = -A\omega\sin(\omega t + \phi)$	v varies between $+A\omega$ and $-A\omega$
3.	Acceleration-time $\left(a = \frac{dv}{dt}\right)$	$a = -A\omega^2\cos(\omega t + \phi)$	a varies between $+A\omega^2$ and $-A\omega^2$

S.No.	Name of the equation	Expression of the equation	Remarks
4.	Kinetic energy-time $\left(K = \frac{1}{2}mv^2\right)$	$K = \frac{1}{2}mA^2\omega^2 \sin^2(\omega t + \phi)$	K varies between $-\frac{1}{2}mA^2\omega^2$ and $\frac{1}{2}mA^2\omega^2$
5.	Potential energy-time $\left(U = \frac{1}{2}m\omega^2x^2\right)$	$U = -\frac{1}{2}m\omega^2A^2\cos^2(\omega t + \phi)$	U varies between $\frac{1}{2}mA^2\omega^2$ and 0
6.	Total energy-time $(E = K + U)$	$E = \frac{1}{2}m\omega^2A^2$	E is constant
7.	Velocity-displacement	$v = \omega\sqrt{A^2 - x^2}$	$v = 0$ at $x = \pm A$ and at $x = 0$, $v = \pm A\omega$
8.	Acceleration-displacement	$a = -\omega^2x$	$a = 0$ at $x = o$ and $a = \pm\omega^2A$ at $x = \mp Ax$
9.	Kinetic energy-displacement	$K = \frac{1}{2}m\omega^2(A^2 - x^2)$	$K = 0$ at $x = \mp A$ and $K = \frac{1}{2}m\omega^2A^2$ at $x = 0$
10.	Potential energy-displacement	$U = \frac{1}{2}m\omega^2x^2$	$U = 0$ at $x = 0$ and $U = \frac{1}{2}m\omega^2A^2$ at $x = \pm A$
11.	Total energy-displacement	$E = \frac{1}{2}m\omega^2A^2$	E is constant

Simple Pendulum

A simple pendulum consists of a heavy point mass suspended from a rigid support by means of an elastic inextensible string.

The time period of the simple pendulum is given by

$$T = 2\pi\sqrt{\frac{l}{g}}$$

where l = effective length of the pendulum and g = acceleration due to gravity.

If the effective length l of simple pendulum is very large and comparable with the radius of earth (R), then its time period is given by

$$T = 2\pi\sqrt{\frac{Rl}{(l + R)g}}$$

For a simple pendulum of infinite length ($l >> R$)

$$T = 2\pi\sqrt{\frac{R}{g}} = 84.6 \text{ min}$$

For a simple pendulum of length equal to radius of earth,

$$T = 2\pi\sqrt{\frac{R}{2g}} = 60 \text{ min}$$

If the bob of the simple pendulum is suspended by a metallic wire of length l, having coefficient of linear expansion α, then due to increase in temperature by $d\theta$, then

$$\text{Effective length } l' = l\,(1 + \alpha\, d\theta)$$

Percentage increase in time period

$$\left(\frac{T'}{T} - 1\right) \times 100 = 50\,\alpha\, d\theta$$

When a bob of simple pendulum of density ρ oscillates in a fluid of density ρ_0 ($\rho_0 < \rho$), then time period get increased.

Increased time period,

$$T' = T\sqrt{\frac{\rho}{\rho - \rho_0}}$$

When simple pendulum is in a horizontally accelerated vehicle, then its time period is given by

$$T = 2\pi\sqrt{\frac{l}{\sqrt{(a^2 + g^2)}}}$$

where a = horizontal acceleration of the vehicle.

When simple pendulum is in a vehicle sliding down an inclined plane, then its time period is given by

$$T = 2\pi\sqrt{\frac{l}{g\cos\theta}}$$

where θ = inclination of plane.

Second's Pendulum

A simple pendulum having time period of 2 second is called second's pendulum.

The effective length of a second's pendulum is 99.992 cm of approximately 1 metre on earth.

Conical Pendulum

If a simple pendulum is fixed at one end and the bob is rotating in a horizontal circle, then it is called a conical pendulum.

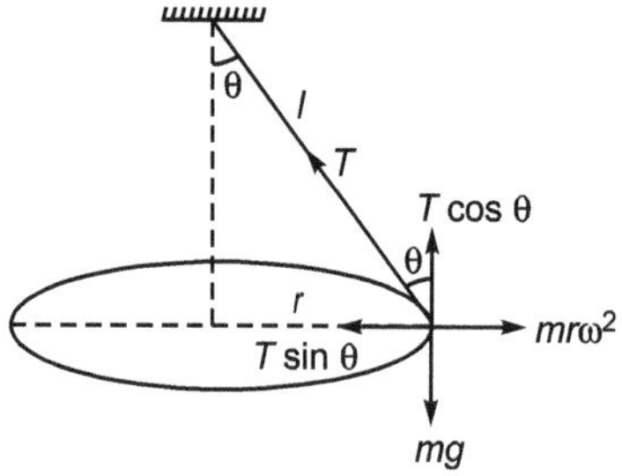

In equilibrium $T \sin\theta = mr\omega^2$

Its time period $T = 2\pi\sqrt{\dfrac{mr}{T\sin\theta}}$

Compound Pendulum

Any rigid body mounted, so that it is capable of swinging in a vertical plane about some axis passing through it is called a physical or compound pendulum.

Its time period is given by

$$T = 2\pi\sqrt{\frac{I}{mgl}}$$

where, I = moment of inertia of the body about an axis passing through the centre of suspension,

m = mass of the body and

l = distance of centre of gravity from the centre of suspension.

Torsional Pendulum

Time period of torsional pendulum is given by

$$T = 2\pi\sqrt{\frac{I}{C}}$$

where, I = moment of inertia of the body about the axis of rotation and

C = restoring couple per unit twist.

Physical Pendulum

When a rigid body of any shape is capable of oscillating about an axis (may or may not be passing through it), it constitutes a physical pendulum.

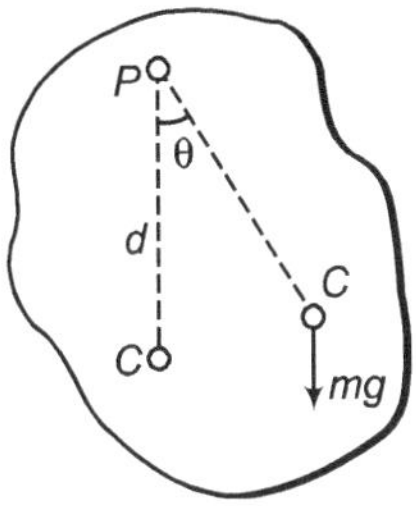

$$T = 2\pi\sqrt{\frac{I}{mgd}}$$

- The simple pendulum whose time period is same as that of a physical pendulum is termed as an equivalent simple pendulum.

$$T = 2\pi\sqrt{\frac{I}{mgd}} = 2\pi\sqrt{\frac{l}{g}}$$

- The length of an equivalent simple pendulum is given by $l = \frac{I}{md}$

Spring Pendulum

A point mass suspended from a massless (or light) spring constitutes a spring pendulum. If the mass is once pulled downwards so as to stretch the spring and then released, the system oscillated up and down about its mean position simple harmonically. Time period and frequency of oscillations are given by

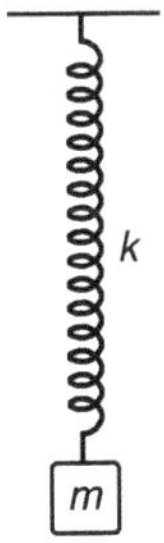

$$T = 2\pi\sqrt{\frac{m}{k}} \text{ or } \nu = \frac{1}{2\pi}\sqrt{\frac{k}{m}}$$

If the spring is not light but has a definite mass m_s, then it can be easily shown that period of oscillation will be

$$T = 2\pi\sqrt{\frac{m + \frac{m_s}{3}}{k}}$$

Oscillations of Liquid in a U-tube

If a liquid is filled up to height h in both limbs of a U-tube and now liquid is depressed upto a small distance y in one limb and then released, then liquid column in U-tube start executing SHM.

The time period of oscillation is given by

$$T = 2\pi\sqrt{\frac{h}{g}}$$

Oscillations of a floating cylinder in liquid is given by

$$T = 2\pi\sqrt{\frac{l}{g}}$$

where l = length of the cylinder submerged in liquid in equilibrium.

Vibrations of a Loaded Spring

When a spring is compressed or stretched through a small distance y from mean position, a restoring force acts on it.

$$\text{Restoring force } (F) = -ky$$

where k = force constant of spring.

If a mass m is suspended from a spring then in equilibrium,

$$mg = kl$$

This is also called **Hooke's law.**

Time period of a loaded spring is given by

$$T = 2\pi\sqrt{\frac{m}{k}}$$

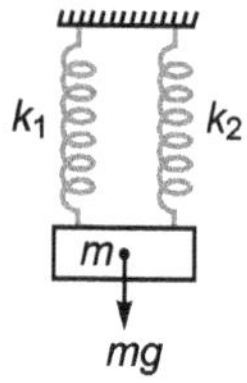

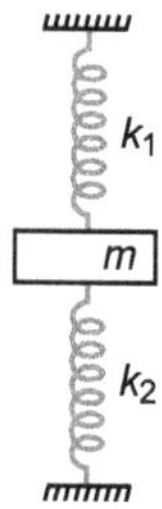

When two springs of force constants k_1 and k_2 are connected in parallel to mass m as shown in figure, then

(i) Effective force constant of the spring combination

$$k = k_1 + k_2$$

(ii) Time period $\quad T = 2\pi\sqrt{\frac{m}{(k_1 + k_2)}}$

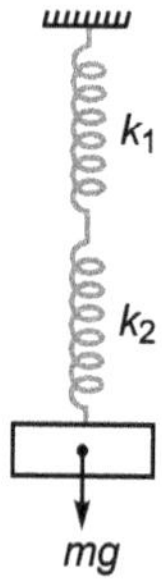

When two springs of force constant k_1 and k_2 are connected in series to mass m as shown in figure, then

(i) Effective force constant of the spring combination

$$\frac{1}{k} = \frac{1}{k_1} + \frac{1}{k_2}$$

(ii) Time period $\quad T = 2\pi\sqrt{\frac{m(k_1 + k_2)}{k_1 k_2}}$

Free Oscillations

When a body which can oscillate about its mean position is displaced from mean position and then released, it oscillates about its mean position. These oscillations are called free oscillations and the frequency of oscillations is called **natural frequency.**

Damped Oscillations

Oscillations with a decreasing amplitude with time are called damped oscillations.

The displacement of the damped oscillator at an instant t is given by

$$x = x_0 e^{-bt/2m} \cos(\omega' t + \phi)$$

where $x_0 e^{-bt/2m}$ is the amplitude of oscillator which decreases continuously with time t and ω'.

The mechanical energy E of the damped oscillator at an instant t is given by

$$E = \frac{1}{2} k x_0^2 e^{-bt/m}$$

Un-damped Oscillations

Oscillations with a constant amplitude with time are called un-damped oscillations.

Forced Oscillations

Oscillations of any object with a frequency different from its natural frequency under a periodic external force are called forced oscillations.

Resonant Oscillations

When an external force is applied on a body whose frequency is an integer multiple of the natural frequency of the body, then its amplitude of oscillation increases and these oscillations are called resonant oscillations.

Lissajous' Figures

If two SHMs are acting in mutually perpendicular directions, then due to their superpositions the resultant motion, in general, is a curve/loop. The shape of the curve depends on the frequency ratio of two SHMs and initial phase difference between them. Such figures are called Lissajous' figures.

1. Let two SHMs be of same frequency (*e.g.*, $x = a_1 \sin \omega t$ and $y = a_2 \sin(\omega t + \phi)$, then the general equation of resultant motion is found to be

$$\frac{x^2}{a_1^2} + \frac{y^2}{a_2^2} - \frac{2xy}{a_1 a_2} \cos \phi = \sin^2 \phi$$

   The equation represents an ellipse. However, if $\phi = 0°$ or π or $n\pi$, then the resultant curve is a straight inclined line.

2. Let two SHMs be having frequencies in the ratio 1:2, then in general, the Lissajous figure is a figure of eight (8).

18

Waves and Sound

Wave

A wave is a vibratory disturbance in a medium which carries energy from one point to another point without any actual movement of the medium. There are three types of waves

(i) **Mechanical Waves** Those waves which require a material medium for their propagation, are called mechanical waves, e.g. sound waves, water waves etc.

(ii) **Electromagnetic Waves** Those waves which do not require a material medium for their propagation, are called electromagnetic waves, e.g. light waves, radio waves etc.

(iii) **Matter Waves** These waves are commonly used in modern technology but they are unfamiliar to us. These waves are associated with electrons, protons and other fundamental particles.

Nature of Waves

(i) **Transverse Waves** A wave in which the particles of the medium vibrate at right angles to the direction of propagation of wave, is called a transverse wave.

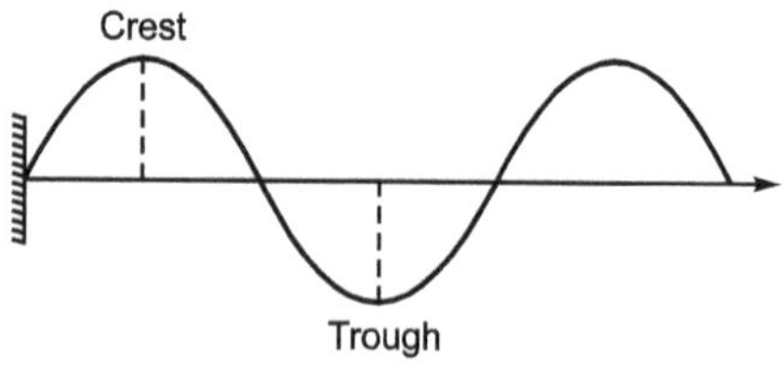

These waves travel in the form of crests and troughs.

(ii) **Longitudinal Waves** A wave in which the particles of the medium vibrate in the same direction in which wave is propagating, is called a longitudinal wave.

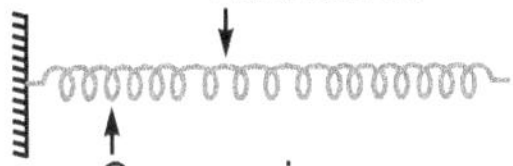

These waves travel in the form of compressions and rarefactions.

Some Important Terms of Wave Motion

(i) **Amplitude** The amplitude of a wave is the magnitude of maximum displacement of the particles from their equilibrium position, as the wave passes through them.

(ii) **Wavelength** The distance between two nearest points in a wave which are in the same phase of vibration is called the wavelength (λ).

(iii) **Time Period** Time taken to complete one vibration is called time period (T).

(iv) **Frequency** The number of vibrations completed in one second is called frequency of the wave.

$$\text{Frequency, } f = \frac{1}{\text{Time period } (T)}$$

Its SI unit is hertz.

$$\text{Angular frequency, } \omega = \frac{2\pi}{T}$$

Its SI unit is rad s^{-1}.

(v) **Velocity of Wave or Wave Velocity** The distance travelled by a wave in one second is called velocity of the wave (v).

Relation among velocity, frequency and wavelength of a wave is given by

$$v = f\lambda$$

(vi) **Particle Velocity** The velocity of the particles executing SHM is called particle velocity.

$$\text{Particle velocity} = \frac{dy}{dt}$$

where, y = displacement

Sound Waves

Sound waves of all the mechanical waves that occur in nature, the most important in our everyday lives are longitudinal waves in a medium, usually air, called sound waves.

Sound waves are of three types

(i) **Infrasonic Waves** The sound waves of frequency lies between 0 to 20 Hz are called infrasonic waves.

(ii) **Audiable Waves** The sound waves of frequency lies between 20 Hz to 20000 Hz are called audiable waves.

(iii) **Ultrasonic Waves** The sound waves of frequency greater than 20000 Hz are called ultrasonic waves.

Sound waves are mechanical longitudinal waves and require medium for their propagation. Sound waves can travel through any material medium (i.e. solids, liquid and gases) with speed that depends on the properties of the medium.

Sound waves cannot propagate through vacuum.
If v_s, v_l and v_g are speed of sound waves in solid, liquid and gases, then

$$v_s > v_l > v_g$$

Sound waves (longitudinal waves) can reflect, refract, interfere and diffract but cannot be polarised as only transverse waves can polarised.

Velocity of Longitudinal (Sound) **Waves**

Velocity of longitudinal (sound) wave in any medium is given by

$$v = \sqrt{\frac{E}{\rho}}$$

where, E is coefficient of elasticity of the medium and ρ is density of the medium.

Newton's Formula

According to Newton, the propagation of longitudinal waves in a gas is an **isothermal process**. Therefore, velocity of longitudinal (sound) waves in gas

$$v = \sqrt{\frac{E_T}{\rho}} = \sqrt{\frac{p}{\rho}}$$

where, E_T is the isothermal coefficient of volume elasticity and it is equal to the pressure of the gas.

Laplace's Correction

According to Laplace, the propagation of longitudinal wave is an **adiabatic process**. Therefore, velocity of longitudinal (sound) wave in gas should be

$$v = \sqrt{\frac{E_s}{\rho}} = \sqrt{\frac{\gamma p}{\rho}}$$

where, E_s is the adiabatic coefficient of volume elasticity and it is equal to γp.

Factors Affecting Velocity of Longitudinal (Sound) Wave

(i) **Effect of Pressure** The formula for velocity of sound in a gas.

$$v = \sqrt{\frac{p}{\rho}} = \sqrt{\frac{RT}{m}}$$

Therefore, $\left(\frac{p}{\rho}\right)$ remains constant at constant temperature.

Hence, there is no effect of pressure on velocity of longitudinal wave.

(ii) **Effect of Temperature** Velocity of longitudinal wave in a gas

$$v = \sqrt{\frac{\gamma p}{\rho}} = \sqrt{\frac{RT}{M}}$$

$$\therefore \quad v \propto \sqrt{T}$$

$$\Rightarrow \quad \frac{v_1}{v_2} = \sqrt{\frac{T_1}{T_2}}$$

Velocity of sound in a gas is directly proportional to the square root of its absolute temperature.

If v_0 and v_t are velocities of sound in air at 0°C and t°C, then

$$v_t = v_0 \left(1 + \frac{t}{273}\right)^{1/2}$$

or

$$v_t = v_0 + 0.61\, t$$

(iii) **Effect of Density** The velocity of sound in gaseous medium

$$v \propto \frac{1}{\sqrt{\rho}}$$

$$\Rightarrow \qquad \frac{v_1}{v_2} = \sqrt{\frac{\rho_2}{\rho_1}}$$

The velocity of sound in a gas is inversely proportional to the square root of density of the gas.

(iv) **Effect of Humidity** The velocity of sound increases with increase in humidity in air.

Shock Waves

If speed of a body in air is greater than the speed of sound, then it is called **supersonic speed**. Such a body leaves behind it a conical region of disturbance which spreads continuously. Such a disturbance is called a shock wave.

Speed of Transverse Motion

On a stretched string $v = \sqrt{\frac{T}{m}}$

where, T = tension in the string and m = mass per unit length of the string.

Speed of transverse wave in a solid $v = \sqrt{\frac{\eta}{\rho}}$

where, η is modulus of rigidity and ρ is density of solid.

Plane Progressive Simple Harmonic Wave

Equation of a plane progressive simple harmonic wave

$$y = a \sin 2\pi \left(\frac{t}{T} - \frac{x}{\lambda}\right)$$

or

$$y = a \sin \frac{2\pi}{\lambda} (vt - x)$$

where, y = displacement,
a = amplitude of vibration of particle,
λ = wavelength of wave,
T = time period of wave,
x = distance of particle from the origin and
v = velocity of wave.

Important Relation Related to Equation of Progressive Wave

Wave velocity,

$$v = \frac{\text{Coefficient of } t}{\text{Coefficient of } x} = \frac{\omega}{k}$$

where, $k = \frac{2\pi}{\lambda}$ is called propagation constant of wave motion. It is also called angular wave number.

Wavelength, $\lambda = \frac{2\pi}{\text{Coefficient of } x} = \frac{2\pi}{k}$

Angular wave number, $k = \frac{2\pi}{\lambda}$ rad m^{-1}

Time period, $T = \frac{2\pi}{\text{Coefficient of } t} = \frac{2\pi}{\omega}$

Frequency, $\eta = \frac{\text{Coefficient of } t}{2\pi} = \frac{\omega}{2\pi}$

Phase of the vibration is the angle of sine in equation of plane progressive wave. It is denoted by ϕ.

$$\phi = 2\pi\left(\frac{t}{T} - \frac{x}{\lambda}\right)$$

Relation between phase difference and path difference and time difference

$$\Delta\phi = \frac{2\pi}{\lambda}\Delta x$$

or

$$\Delta\phi = \frac{2\pi}{T}\Delta t$$

Echo

The repetition of sound caused by the reflection of sound waves is called an echo.

Sound persists on ear for 0.1 s.

The minimum distance from a sound reflecting surface to hear an echo is 16.5 m.

If first echo be heard after t_1 second, second echo after t_2 second, then third echo will be heard after $(t_1 + t_2)$ s.

Superposition of Waves

Two or more progressive waves can travel simultaneously in the medium without effecting the motion of one another. Therefore, resultant displacement of each particle of the medium at any instant is equal to vector sum of the displacements produced by two waves separately. This principle is called principle of superposition.

Interference

When two waves of same frequency travel in a medium simultaneously in the same direction, then due to their superposition, the resultant intensity at any point of the medium is different from the sum intensities of the two waves. At some points the intensity of the resultant wave is very large while at some other points it is very small or zero. This phenomenon is called **interference of waves.**

Constructive Interference

Phase difference between two waves $= 0, 2\pi, 4\pi$

Maximum amplitude $= (a + b)$

Intensity $\propto (\text{Amplitude})^2 \propto (a + b)^2$

In general, amplitude $= \sqrt{a^2 + b^2 + 2ab\cos\phi}$

Destructive Interference

Phase difference between two waves $= \pi, 3\pi, 5\pi$

Minimum amplitude $= (a \sim b) =$ Difference of component amplitudes.

$$\text{Intensity} \propto (\text{Amplitude})^2$$
$$\propto (a - b)^2$$

A vibrating tuning fork, when rotated near ear, produced loud sound and silence due to constructive and destructive interference.

Reflection of Wave

The rebouncing back of waves when it strikes a hard surface is called reflection of wave.

Refracted Wave

If a wave is incident obliquely on the boundary between two different media, the transmitted wave is called the refracted wave.

Beats

When two sound waves of nearly equal frequencies are produced simultaneously, then intensity of the resultant sound produced by their superposition increases and decreases alternately with time. This rise and fall intensity of sound is called beats.

The number of maxima or minima heard in one second is called beats frequency.

> The difference of frequencies should not be more than 10. Sound persists on human ear drums for 0.1 second. Hence, beats will not be heard if the frequency difference exceeds 10.

Number of beats heard per second = $n_1 - n_2$
= difference of frequencies of two waves.

Maximum amplitude = $(a_1 + a_2)$

Maximum intensity = (Maximum amplitude)$^2 = (a_1 + a_2)^2$

For loudness, time intervals are $\frac{1}{n_1 - n_2}, \frac{2}{n_1 - n_2}, \ldots$

Stationary or Standing Waves

When two similar waves propagate in a bounded medium in opposite directions, then due to their superposition a new type of wave is obtained, which appears stationary in the medium. This wave is called stationary or standing waves.

Equation of a stationary wave

$$y = 2a \sin \frac{2\pi t}{T} \cos \frac{2\pi x}{\lambda}$$

Nodes and antinodes are obtained alternatively in a stationary waves.

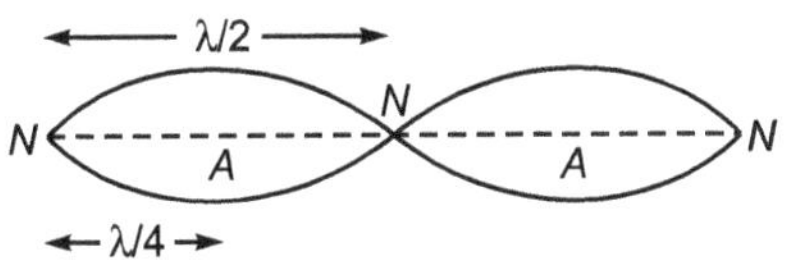

At nodes, the displacement of the particles remains minimum, strain is maximum, pressure and density variations are maximum.

At antinodes, the displacement of the particles remains maximum, strain is minimum, pressure and density variations are minimum.

The distance between two consecutive nodes or two consecutive antinodes $= \frac{\lambda}{2}$.

The distance between a node and adjoining antinode $= \frac{\lambda}{4}$.

All the particles between two nodes vibrate in same phase.

Particles on two sides of a node vibrate in opposite phase.

n consecutive nodes are separated n by $\frac{(n-1)\lambda}{2}$.

Vibrations in a Stretched String

Velocity of a transverse wave in a stretched string.

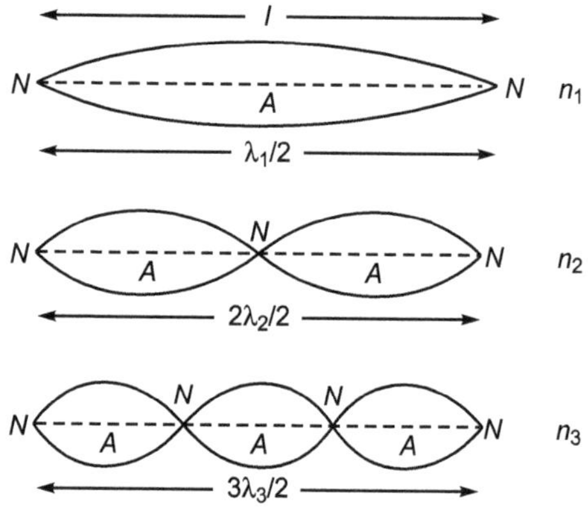

$$v = \sqrt{\frac{T}{m}}$$

where, T is tension in the string and m is mass per unit length of the string.

Fundamental frequency or frequency of first harmonic

$$n_1 = \frac{v}{2l}$$
$$= \frac{1}{2l}\sqrt{\frac{T}{m}}$$

Frequency of first overtone or second harmonic

$$n_2 = 2 \cdot \frac{v}{2l} = 2n_1$$

Frequency of second overtone or third harmonic

$$n_3 = 3 \cdot \frac{v}{2l} = 3n_1$$

$\therefore$ $$n_1 : n_2 : n_3 : \ldots = 1 : 2 : 3 : \ldots$$

Organ Pipes

Organ pipes are those cylindrical pipes which are used for producing musical (longitudial) sounds. Organ pipes are of two types

(i) **Open Organ Pipe** Cylindrical pipes open at both ends.

(ii) **Closed Organ Pipe** Cylindrical pipes open at one end and closed at other end.

Fundamental Note

It is the sound of lowest frequency produced in fundamental note of vibration of a system.

Overtones Tones having frequencies greater than the fundamental note are called overtones.

Harmonics When the frequencies of overtone are integral multiples of the fundamental, then they are known as **harmonics.** Thus the note of lowest frequency n is called fundamental note or **first harmonics.** The note of frequency $2n$ is called **second harmonic** or **first overtone.**

Vibrations in Open Organ Pipe

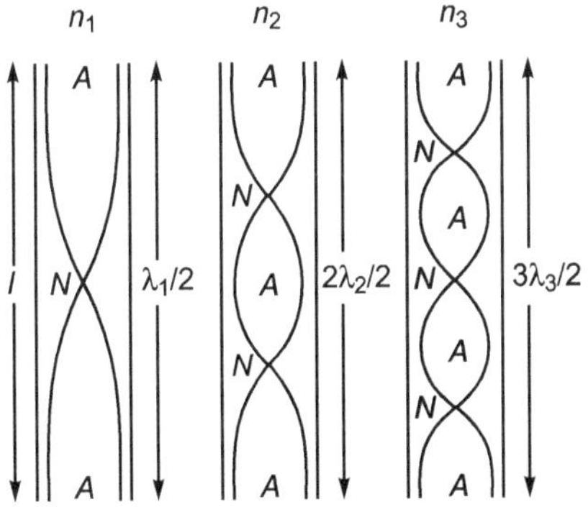

Fundamental frequency or frequency of first harmonic $n_1 = \frac{v}{2l}$

Frequency of first overtone or second harmonic $n_2 = 2\frac{v}{2l} = 2n_1$

Frequency of second overtone or third harmonic $n_3 = 3\frac{v}{2l} = 3n_1$

$\therefore$ $$n_1 : n_2 : n_3 : \ldots = 1 : 2 : 3 : \ldots$$

Therefore, even and odd harmonics are produced by an open organ pipe.

Vibrations in Closed Organ Pipe

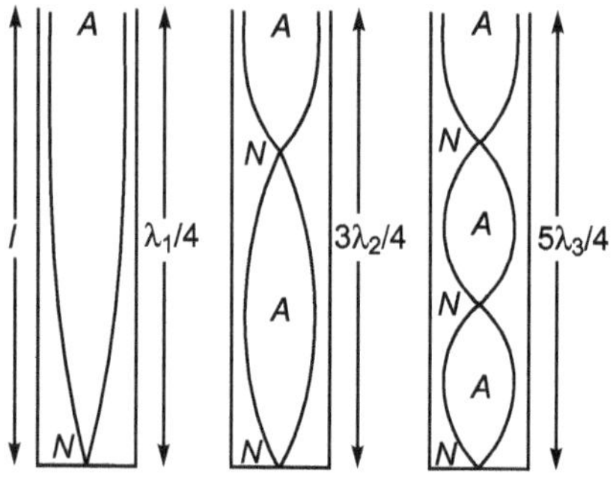

Fundamental frequency or frequency of first harmonic

$$n_1 = \frac{v}{4l}$$

Frequency of first harmonic or third harmonic

$$n_2 = \frac{3v}{4l} = 3n_1$$

Frequency of second harmonic or fifth harmonic

$$n_3 = 5 \cdot \frac{v}{4l} = 5n_1$$

$\therefore$ $$n_1 : n_2 : n_3 : \ldots = 1 : 3 : 5 : \ldots$$

Therefore only even harmonics are produced by a closed organ pipe.

End Correction

Antinode is not obtained at exact open end but slightly above it. The distance between open end and antinode is called end correction.

It is denoted by e.

Effective length of an open organ pipe $= (l + 2e)$

Effective length of a closed organ pipe $= (l + e)$

If r is the radius of organ pipe, then $e = 0.6\,r$

Factors Affecting Frequency of Pipe

(i) Length of air column, $n \propto \frac{1}{l}$

(ii) Radius of air column, $n \propto \frac{1}{r}$

(iii) Temperature of air column, $n \propto \sqrt{T}$

(iv) Pressure of air inside air column, $n \propto \sqrt{p}$

(v) Density of air, $n \propto \frac{1}{\sqrt{\rho}}$

(vi) Velocity of sound in air column, $n \propto v$

Resonance Tube

Resonance tube is a closed organ pipe in which length of air column can be changed by changing height of liquid column in it.

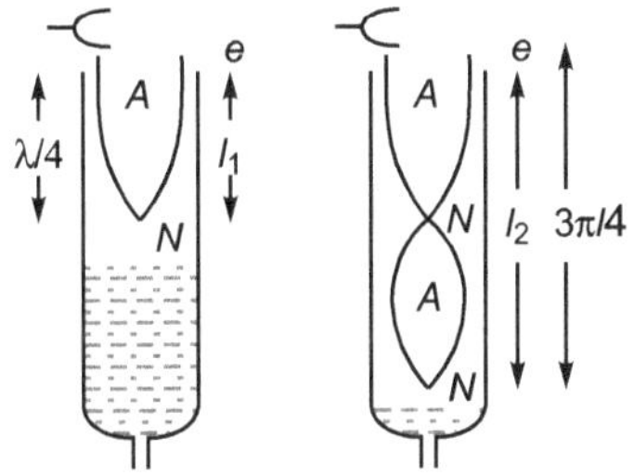

For first resonance $\frac{\lambda}{4} = l_1 + e$

For second resonance $\frac{3\lambda}{4} = l_2 + e$

Velocity of sound $v = n\lambda = 2n\,(l_2 - l_1)$

End correction $\quad e = \frac{l_2 - 3l_1}{2}$

Melde's Experiment

In longitudinal mode, vibrations of the prongs of tuning fork are along the length of the string.

Frequency of vibration of string $= \frac{\text{frequency of tuning fork}}{2}$

$$n_L = \frac{p}{l}\sqrt{\frac{T}{m}}$$

In transverse mode, vibrations of tuning fork are at 90° to the length of string.

Frequency of vibration of string = Frequency of tuning fork.

$$n_T = \frac{p}{2l}\sqrt{\frac{T}{m}} = \frac{n_L}{2}$$

In both modes of vibrations, Melde's law

$$p^2T = \text{constant, is obeyed.}$$

Characteristics of Musical Sound

Musical sound has three characteristics

(i) **Intensity or Loudness** Intensity of sound is energy transmitted per second per unit area by sound waves. Its SI unit is watt/metre^2. Intensity is measured in decibel (dB).

(ii) **Pitch or Frequency** Pitch of sound directly depends on frequency.

A shrill and sharp sound has higher pitch and a grave and dull sound has lower pitch.

(iii) **Quality or Timbre** Quality is the characteristic of sound that differentiates between two sounds of same intensity and same frequency.

Quality depends on harmonics and their relative order and intensity.

Doppler's Effect

The phenomena of apparent change in frequency of source due to a relative motion between the source and observer is called Doppler's effect.

(i) **When Source is Moving and Observer is at Rest** When source is moving with velocity v_s, towards an observer at rest, then apparent frequency

$$n' = n\left(\frac{v}{v - v_s}\right)$$

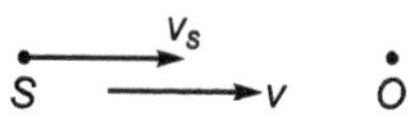

If source is moving away from observer, then

$$n' = n\left(\frac{v}{v + v_s}\right)$$

(ii) **When Source is at Rest and Observer is Moving** When observer is moving with velocity v_o, towards a source at rest, then apparent frequency.

$$n' = n\left(\frac{v + v_o}{v}\right)$$

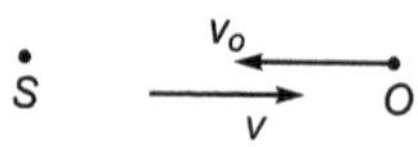

When observer is moving away from source, then

$$n' = n\left(\frac{v - v_o}{v}\right)$$

(iii) **When Source and Observer Both are Moving**

(a) When both are moving in same direction along the direction of propagation of sound, then

$$n' = n\left(\frac{v - v_o}{v - v_s}\right)$$

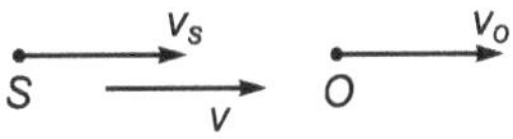

(b) When both are moving in same direction opposite to the direction of propagation of sound, then

$$n' = n\left(\frac{v + v_o}{v + v_s}\right)$$

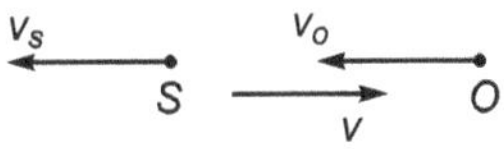

(c) When both are moving towards each other, then

$$n' = n\left(\frac{v + v_o}{v - v_s}\right)$$

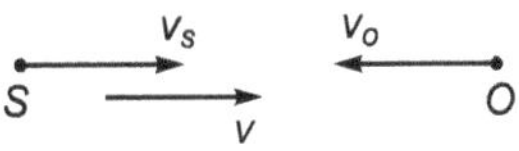

(d) When both are moving in opposite direction, away from each other, then

$$n' = n\left(\frac{v - v_o}{v + v_s}\right)$$

Transverse Doppler's Effect

(i) The Doppler's effect in sound does not take place in the transverse direction.

(ii) As shown in figure, the position of a source is S and of observer is O. The component of velocity of source towards the observer is $v \cos \theta$. For this situation, the approach frequency is

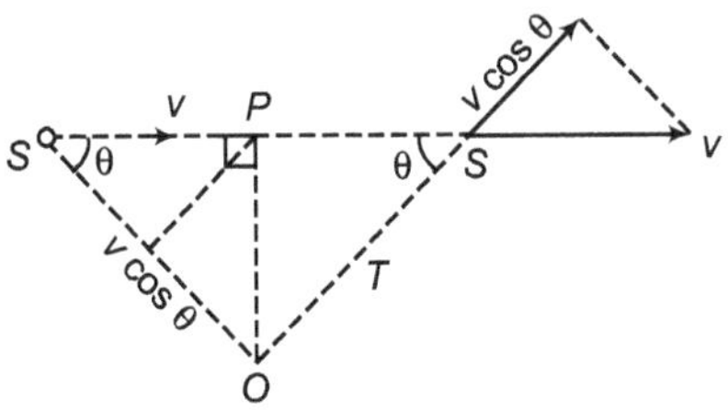

$$f' = \frac{v}{v - v_s \cos\theta} \times f$$

f' which will now be a function of θ so, it will no more be constant.

Similarly, if the source is moving away from the observer as shown above, with velocity component $v_s \cos\theta$, then

$$f' = \frac{v}{v + v_s \cos\theta} \times f$$

(iii) If $\theta = 90°$, the $v_s \cos\theta = 0$ and there is no shift in the frequency. Thus, at point P, Doppler's effect does not occur.

Effect of Wind

If wind is also blowing with a velocity w in the direction of sound, then its velocity is added to the velocity of sound. Hence, in this condition the apparent frequency is given by

$$n' = n\left(\frac{v + w - v_o}{v + w - v_s}\right)$$

Applications of Doppler's Effect

The measurement of Doppler shift has been used

(i) by police to check over speeding of vehicles.

(ii) at airports to guide the aircraft.

(iii) to study heart beats and blood flow in different parts of the body.

(iv) by astrophysicist to measure the velocities of plants and stars.

19

Electrostatics

Charge

Charge is that property of an object by virtue of which it apply electrostatic force of interaction on other objects.

Charges are of two types

(i) Positive charge

(ii) Negative charge

Like charges repel and unlike charges attract each other.

Conductors

Conductors are those substances which can be used to carry or conduct electric charge from one point to other, e.g. silver, copper, aluminium etc.

Insulators

Insulators are those substances which cannot conduct electric charge, e.g. glass, rubber, plastic etc.

Charging by Induction

The process of changing a neutral body nearby it without making contact between two bodies is known as charging by induction.

Additivity of Charge

If a system consists of n charges $q_1, q_2, q_3 \ldots\ldots\ldots q_n$, then the total charge of the system will be $q_1 + q_2 + q_3 + q_4 + \ldots\ldots q_n$.

Quantisation of Charge

Charge on any object can be an integer multiple of a smallest charge (e).

$$Q = \pm ne$$

where, $n = 1, 2, 3, \ldots$ and $e = 1.6 \times 10^{-19}$ C.

Conservation of Charge

Charge can neither be created nor be destroyed, but can be transferred from one object to another object.

Recently a new particle has been discovered called 'Quark'. It contains charge $\pm \frac{e}{3}, \pm \frac{2e}{3}$.

The protons and neutrons are combination of other entities called quarks, which have charges $\frac{1}{3}$e. However, isolated quarks have not been observed, so, quantum of charge is still e.

Coulomb's Law of Electrostatics

Electrostatic force of interaction acting between two stationary point charges is given by

$$F = \frac{1}{4\pi \varepsilon_0} \frac{q_1 q_2}{r^2}$$

where q_1, q_2 are magnitude of point charges, r is the distance between them and ε_0 is permittivity of free space.

Here, $$\frac{1}{4\pi\varepsilon_0} = \left(10^{-7} \frac{\text{N} - \text{s}^2}{\text{C}^2}\right) c^2$$

Substituting value of $c = 2.99792458 \times 10^8$ m/s,

we get $$\frac{1}{4\pi\varepsilon_0} = 8.99 \times 10^9 \text{ N-m}^2/\text{C}^2$$

In examples and problems we will often use the approximate value,

$$\frac{1}{4\pi\varepsilon_0} = 9 \times 10^9 \text{ N-m}^2/\text{C}^2$$

The value of ε_0 is 8.85×10^{-12} C^2/N-m^2.

If there is another medium between the point charges except air or vacuum, then ε_0 is replaced by $\varepsilon_0 K$ or $\varepsilon_0 \varepsilon_r$ or ε.

where K or ε_r is called dielectric constant or relative permittivity of the medium.

$$K = \varepsilon_r = \frac{\varepsilon}{\varepsilon_0}$$

where, ε = permittivity of the medium.

For air or vacuum, $K = 1$

For water, $K = 81$

For metals, $K = \infty$

Coulomb's Law in Vector Form

Let q_1 and q_2 both are positive.

Force on q_2 due to q_1,

$$\mathbf{F}_{21} = \frac{1}{4\pi\varepsilon_0} \cdot \frac{q_1 q_2}{r^2} \hat{\mathbf{r}}_{AB}$$

$$\mathbf{F}_{21} = \frac{1}{4\pi\varepsilon_0} \cdot \frac{q_1 q_2}{r^3} \mathbf{r}_{AB}$$

$$\mathbf{F}_{21} = \frac{q_1 q_2}{4\pi\varepsilon_0} \cdot \frac{\mathbf{r}_2 - \mathbf{r}_1}{|\mathbf{r}_2 - \mathbf{r}_1|^3}$$

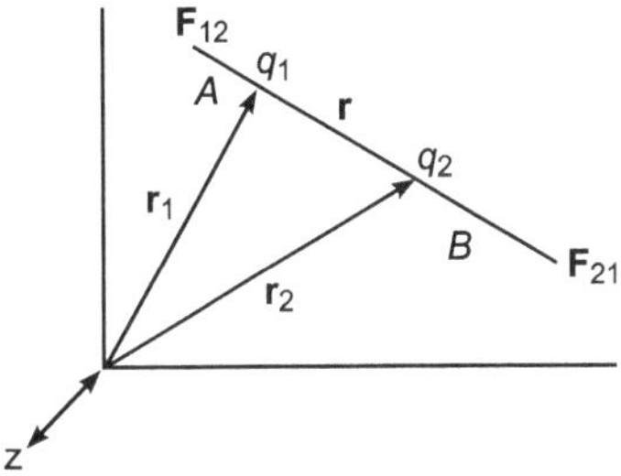

The above equations give the Coulomb's law in vector form.

$\because$ Force on q_1 due to q_2 = – Force on q_2 due to q_1

$\because$
$$\mathbf{F}_{12} = -\mathbf{F}_{21}$$

$$\mathbf{F}_{12} = \frac{q_1 q_2}{4\pi\varepsilon_0} \cdot \frac{\mathbf{r}_1 - \mathbf{r}_2}{|\mathbf{r}_1 - \mathbf{r}_2|^3}$$

The forces due to two point charges are parallel to the line joining the point charges; such forces are called **central forces** and so electrostatic forces are **conservative forces**.

Forces between Multiple Charges

Consider a system of n point charges $q_1, q_2, q_3 \ldots\ldots q_n$ are distributed in space. Let the charges be $q_2, q_3 \ldots q_n$, exert forces $F_{12}, F_{13}, \ldots\ldots F_{1n}$ on charge q_1. The total force on charge q_1 is given by

$$F_1 = \frac{1}{4\pi\varepsilon_0}\left(\frac{q_1 q_2}{r_{12}^2}\hat{\mathbf{r}}_{21} + \frac{q_1 q_2}{r_{13}^2}\hat{\mathbf{r}}_{31} + \ldots\ldots + \frac{q_1 q_2}{r_n^2}\hat{\mathbf{r}}_{n1}\right)$$

Contiuous Charge Distribution

The region in which charges are closely spaced is said to have continuous distribution of charge.

Linear Charge Density (λ)

It is defined as the charge per unit length of linear charge distribution. Its unit is coulomb/metre.

Surface Charge Density (σ)

It is defined as the charge per unit surface area of surface charge distribution. Its unit is coulomb/metre^2.

Volume Charge Density (ρ)

It is defined as the charge per unit volume of volume charge distribution. Its unit is coulomb/metre^3.

Electric Field

The space in the surrounding of any charge in which its influence can be experienced by other charges is called electric field.

Electric Field Lines

"An electric field line is an imaginary line or curve drawn through a region of space so that its tangent at any point is in the direction of the electric field vector at that point.

The relative closeness of the lines at some place give an idea about the intensity of electric field at that point."

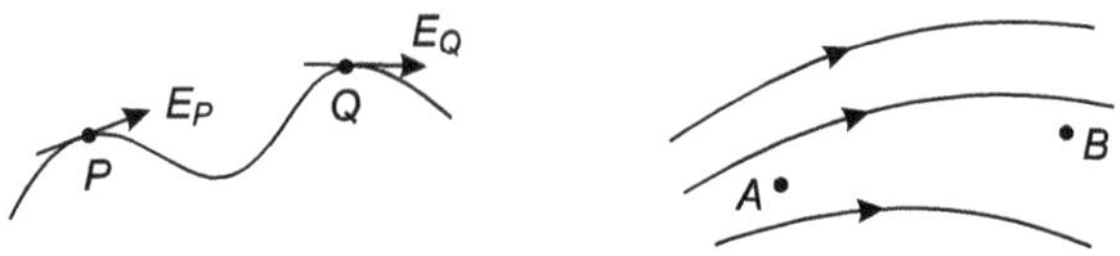

$|E_A| > |E_B|$

Two lines can never intersect.

Electric field lines always begin on a positive charge and end on a negative charge and do not start or stop in mid space.

Electric Field Intensity (E)

The electrostatic force acting per unit positive charge on a point in electric field is called electric field intensity at that point.

Electric field intensity $\mathbf{E} = \lim_{q_0 \to 0} \frac{\mathbf{F}}{q_0}$

where $\mathbf{F}$ = force experienced by the test charge q_0.

Its SI unit is NC^{-1} or V/m and its dimension is $[MLT^{-3}A^{-1}]$.

It is a vector quantity and its direction is in the direction of electrostatic force acting on positive charge.

Electric field intensity due to a point charge $\boldsymbol{q}$ at a distance r is given by

$$E = \frac{1}{4\pi\varepsilon_0} \frac{q}{r^2}$$

Electric Field due to System of Charges

If E is electric field at point P due to the systems of charges is given by

$$E = \frac{1}{4\pi\varepsilon_0} \sum_{i=1}^{n} \frac{q_i}{r_i^2} \hat{\mathbf{r}}_i$$

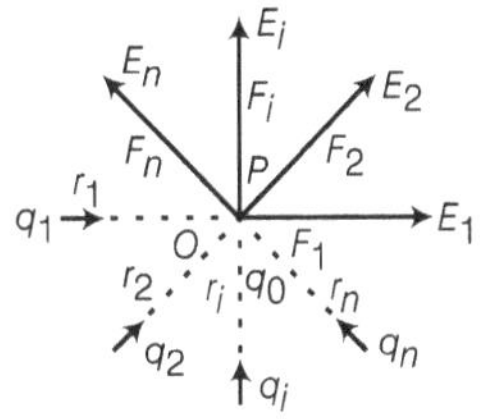

Electric Potential (V)

Electric potential at any point is equal to the work done per unit positive charge in carrying it from infinity to that point in electric field.

Electric potential, $V = \frac{W}{q}$

Its SI unit is J/C or volt and its dimension is $[ML^2T^{-3}A^{-1}]$.

It is a scalar quantity.

Electric potential due to a point charge at a distance r is given by

$$V = \frac{1}{4\pi\,\varepsilon_0} \frac{q}{r}$$

Potential due to System of Charges

Let there be a number of point charges $q_1, q_2, q_3, \ldots, q_n$ at distances $r_1, r_2, r_3 \ldots, r_n$ respectively from the point P, where electric potential is given by

$$V = \frac{1}{4\pi\varepsilon_0} \sum_{i=1}^{n} \frac{q_i}{r_i}$$

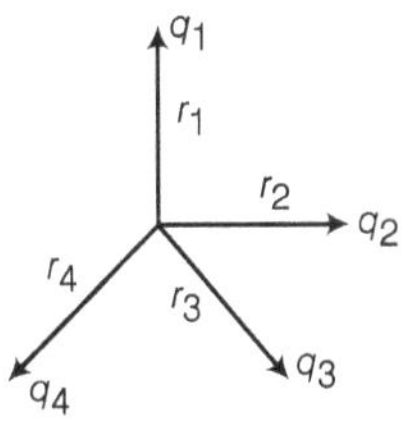

Potential Gradient

The rate of change of potential with distance in electric field is called potential gradient.

Potential gradient $= \frac{dV}{dr}$

Its unit is V/m.

Relation between potential gradient and electric field intensity is given by

$$E = -\left(\frac{dV}{dr}\right)$$

Equipotential Surface

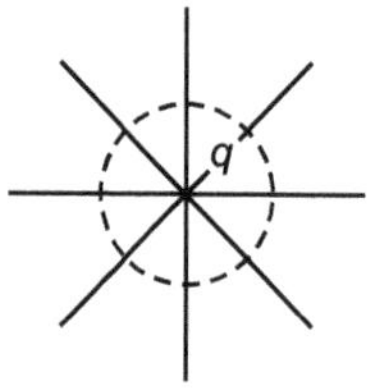

Equipotential surface is an imaginary surface joining the points of same potential in an electric field. So, we can say that the potential difference between any two points on an equipotential surface is zero. The electric lines of force at each point of an equipotential surface are normal to the surface.

(i) Equipotential surface may be planer, solid etc. But equipotential surface can never be point size.

(ii) Electric field is always perpendicular to equipotential surface.

(iii) Equipotential surface due to an isolated point charge is spherical.

(iv) Equipotential surface are planer in an uniform electric field.

(v) Equipotential suface due to a line charge is cylindrical.

Electric Lines of Force

Electric lines of force are the imaginary lines drawn in electric field at which a positive test charge will move if it is free to do so.

Electric lines of force start from positive charge and terminate on negative charge.

A tangent drawn at any point on electric field represents the direction of electric field at that point.

Two electric lines of force never intersect each other.

Electric lines of force are always perpendicular to an equipotential surface.

Electric Flux (ϕ_E)

Electric flux over an area is equal to the total number of electric field lines crossing this area.

Electric flux through a small area element dS is given by

$$\phi_E = \mathbf{E} \cdot d\mathbf{S}$$

where $\mathbf{E}$ = electric field intensity

and $d\mathbf{S}$ = area vector.

Its SI unit is N-m^{2}C^{-1}.

Gauss' Theorem

The electric flux over any closed surface is $\frac{1}{\varepsilon_0}$ times the total charge enclosed by that surface, i.e.

$$\phi_E = \oint_s \mathbf{E} \cdot d\mathbf{S} = \frac{1}{\varepsilon_0} \Sigma q$$

If a charge q is placed at the centre of a cube, then

total electric flux linked with the whole cube $= \frac{q}{\varepsilon_0}$

electric flux linked with one face of the cube $= \frac{q}{6\varepsilon_0}$

(i) **Electric Field at Any Point on the Axis of a Uniformly Charged Ring** A ring-shaped conductor with radius a carries a total charge Q uniformly distributed around it. Let us calculate the electric field at a point P that lies on the axis of the ring at a distance x from its centre.

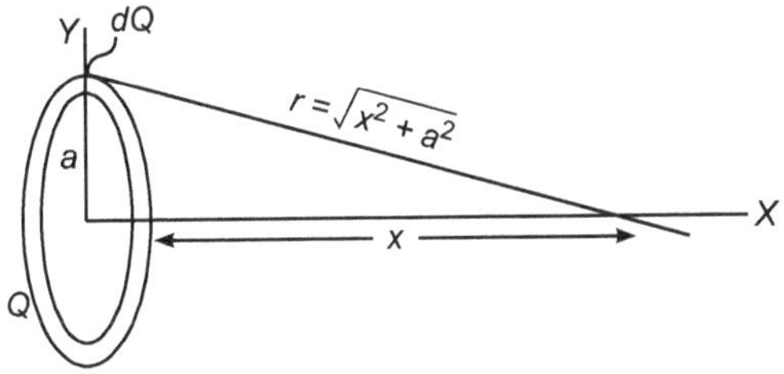

$$E_x = \frac{1}{4\pi\varepsilon_0} \cdot \frac{xQ}{(x^2 + a^2)^{3/2}}$$

The maximum value of electric field

$$E = \frac{1}{4\pi\varepsilon_0} \left(\frac{2Q}{3\sqrt{3}R^2} \right)$$

(ii) **Electric Field due to a Charged Spherical Shell**

(a) At an extreme point ($r > R$)

$$E = \frac{1}{4\pi\,\varepsilon_0} \frac{q}{r^2}$$

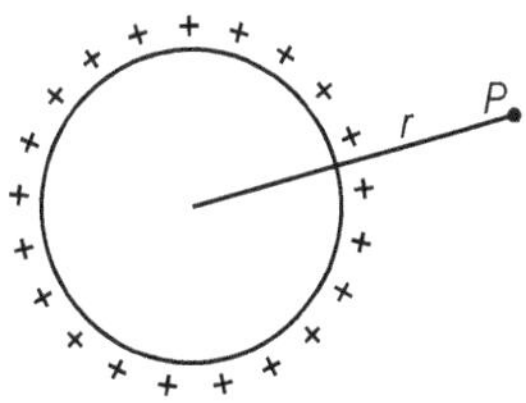

(b) At the surface of a shell ($r = R$)

$$E = \frac{1}{4\pi\varepsilon_0}\frac{q}{R^2}$$

(c) At an internal point ($r < R$)

$$E = 0$$

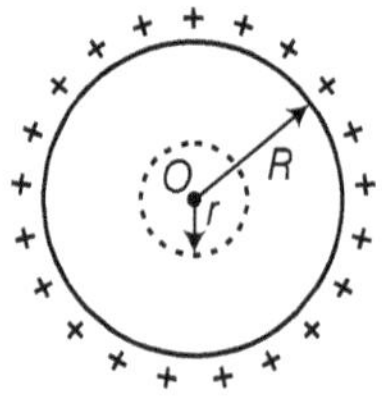

(iii) **Electric Field Intensity due to an Infinite Line Charge**

$$E = \frac{1}{2\pi\varepsilon_0}\frac{\lambda}{r}$$

where λ is linear charge density and r is distance from the line charge.

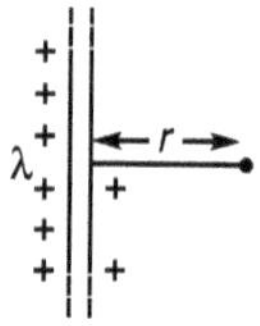

(iv) **Electric Field Near an Infinite Plane Sheet of Charge**

$$E = \frac{\sigma}{2\varepsilon_0}$$

where σ = surface charge density.

If infinite plane sheet has uniform thickness,

then $$E = \frac{\sigma}{\varepsilon_0}$$

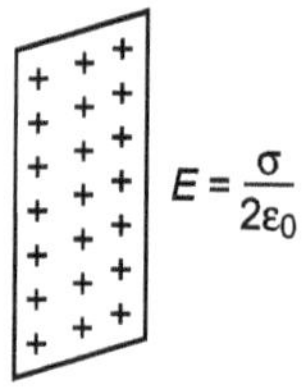

(v) **Electric Potential due to a Charged Conducting Spherical Shell**

(a) At an extreme point, ($r > R$)

$$V = \frac{1}{4\pi\varepsilon_0}\frac{q}{r}$$

(b) At the surface of a shell, ($r = R$)

$$V = \frac{1}{4\pi\varepsilon_0}\frac{q}{R}$$

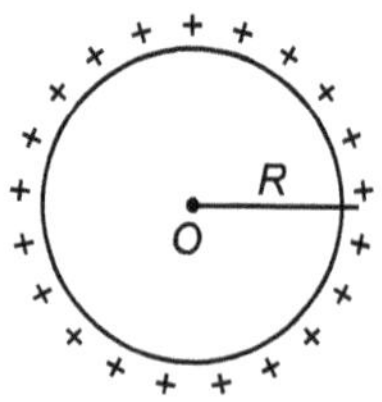

(c) At an internal point ($r < R$)

$$V = \frac{1}{4\pi\varepsilon_0}\frac{q}{R}$$

Therefore potential inside a charged conducting spherical shell is equal to the potential at its surface.

(vi) **Electric Field and Potential due to a Charged Non-Conducting Sphere**

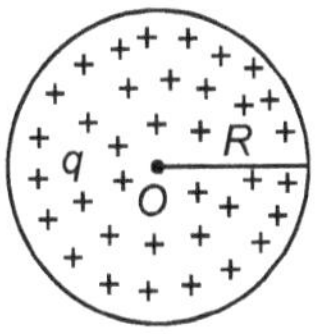

At an extreme point, ($r > R$)

(a) Electric field intensity $E = \frac{1}{4\pi\varepsilon_0}\frac{q}{r^2}$

(b) Electric potential $V = \frac{1}{4\pi\varepsilon_0}\frac{q}{r}$

On the surface, ($r = R$)

(a) Electric field intensity $E = \frac{1}{4\pi\varepsilon_0}\frac{q}{R^2}$

(b) Electric potential $V = \frac{1}{4\pi\varepsilon_0}\frac{q}{R}$

Inside the sphere, ($r < R$)

(a) Electric field intensity $E = \frac{1}{4\pi\varepsilon_0}\frac{qr}{R^3}$

(b) Electric potential $V = \frac{1}{4\pi\varepsilon_0}\frac{q(3R^2 - r^2)}{2R^3}$

At the centre of the sphere, ($r = 0$)

(a) Electric field intensity $E = 0$

(b) Electric potential $V = \frac{3}{2}\times\frac{1}{4\pi\varepsilon_0}\frac{q}{R}$

Electric Dipole

An electric dipole consists of two equal and opposite point charges separated by a very small distance. e.g. a molecule of HCl, a molecule of water etc.

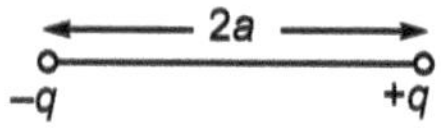

Electric Dipole Moment $p = q \times 2l$

Its SI unit is 'coulomb-metre' and its dimension is [LTA].

It is a vector quantity and its direction is from negative charge towards positive charge.

Electric Field Intensity and Potential due to an Electric Dipole

(i) **On Axial Line**

Electric field intensity $E = \frac{1}{4\pi\varepsilon_0} \frac{2pr}{(r^2 - a^2)^2}$

If $r >> 2a$, then $E = \frac{1}{4\pi\varepsilon_0} \frac{2p}{r^3}$

Electric potential $V = \frac{1}{4\pi\varepsilon_0} \frac{p}{(r^2 - a^2)}$

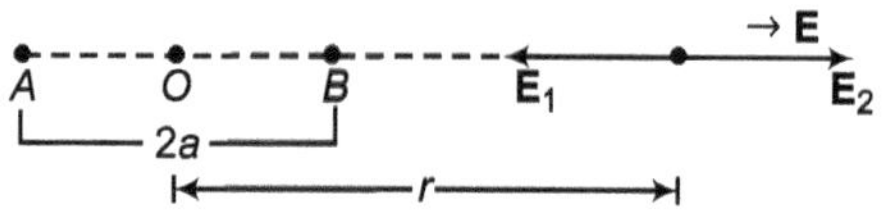

If $r >> 2a$, then $V = \frac{1}{4\pi\varepsilon_0} \frac{p}{r^2}$

(ii) **On Equatorial Line**

Electric field intensity

$$E = \frac{1}{4\pi\varepsilon_0} \frac{p}{(r^2 + a^2)^{3/2}}$$

If $r >> 2a$, then $E = \frac{1}{4\pi\varepsilon_0} \frac{p}{r^3}$

Electric potential $V = 0$

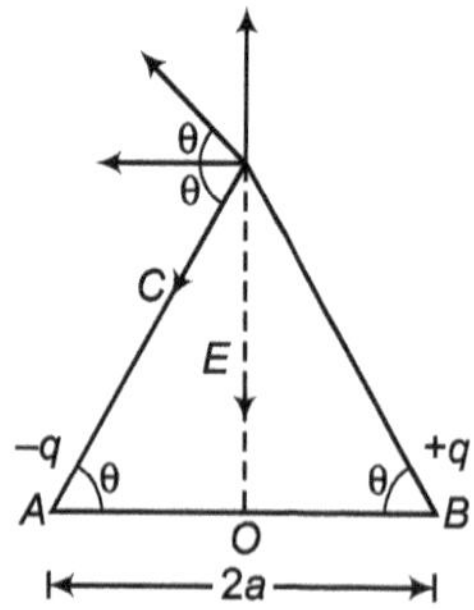

(iii) At any Point along a Line Making θ Angle with Dipole Axis

Electric field intensity $E = \frac{1}{4\pi \varepsilon_0} \frac{p\sqrt{(1 + 3\cos^2 \theta)}}{r^3}$

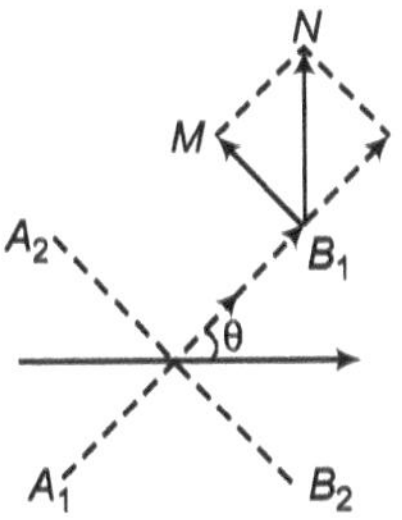

Electric potential $V = \frac{1}{4\pi \varepsilon_0} \frac{p \cos \theta}{(r^2 - a^2 \cos^2 \theta)}$

If $r >> 2a$, then $V = \frac{1}{4\pi \varepsilon_0} \frac{p \cos \theta}{r^2}$

Torque

Torque acting on an electric dipole placed in uniform electric field is given by

$$\tau = Ep \sin \theta \quad \text{or} \quad \tau = \mathbf{p} \times \mathbf{E}$$

When $\theta = 90°$, then $\tau_{max} = Ep$

When electric dipole is parallel to electric field, it is in stable equilibrium and when it is anti-parallel to electric field, it is in unstable equilibrium.

Work Done

Work done is rotating an electric dipole in a uniform electric field from angle θ_1 to θ_2 is given by

$$W = Ep(\cos \theta_1 - \cos \theta_2)$$

If initially it is in the direction of electric field, then work done in rotating through an angle θ

$$W = Ep(1 - \cos \theta).$$

Potential Energy

Potential energy of an electric dipole in a uniform electric field is given by $U = - pE \cos \theta$.

Dipole in Non-uniform Electric Field

When an electric dipole is placed in a non-uniform electric field, then a resultant force as well as a torque act on it.

Net force on electric dipole $= (qE_1 - qE_2)$, along the direction of greater electric field intensity.

Therefore electric dipole undergoes rotational as well as linear motion.

Potential Energy of Charge System

Two point charge system, contains charges q_1 and q_2 separated by a distance r is given by $U = \frac{1}{4\pi \varepsilon_0} \frac{q_1 q_2}{r}$

Three point charge system

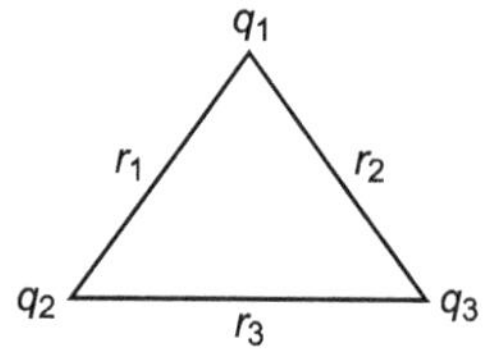

$$U = \frac{1}{4\pi \varepsilon_0} \cdot \left[\frac{q_1 q_2}{r_1} + \frac{q_2 q_3}{r_2} + \frac{q_3 q_1}{r_3} \right]$$

Important Points

- When charge is given to a soap bubble its size gets increased.
- In equilibrium for a charged soap bubble, pressure due to surface tension = electric pressure due to charging

$$\frac{4T}{r} = \frac{\sigma^2}{2\varepsilon_0}$$

or $$\frac{4T}{r} = \frac{1}{2\varepsilon_0} \left(\frac{q}{4\pi r^2} \right)^2$$

or $$q = 8\pi r \sqrt{2\varepsilon_0 rT}$$

where, r is radius of soap bubble and T is surface tension of soap bubble.

Behaviour of a Conductor in an Electrostatic Field

(i) Electric field at any point inside the conductor is zero.

(ii) Electric field at any point on the surface of charged conductor is directly proportional to the surface density of charge at that point, but electric potential does not depend upon the surface density of charge.

(iii) Electric potential at any point inside the conductor is constant and equal to potential.

Electrostatic Shielding

The process of protecting certain field from external electric field is called electrostatic shielding.

Electrostatic shielding is achieved by enclosing that region in a closed metallic chamber.

Dielectric

Dielectrics are of two types

Non-polar Dielectric

The non-polar dielectrics (like N_2, O_2, benzene, methane etc.) are made up of non-polar atoms/molecules, in which the centre of positive charge coincides with the centre of negative charge of the atom/molecule.

Polar Dielectric

The polar dielectric (like H_2O, CO_2, NH_3 etc) are made up of polar atoms/molecules, in which the centre of positive charge does not coincide with the centre of negative charge of the atom.

Dielectric Constant (k)

The ratio of the strength of the applied electric field to the strength of the reduced value of electric field on placing the dielectric between the plates of a capacitor is the dielectric constant. It is denoted by k (or ε_r). $k = \dfrac{E_0}{E}$

Polarisation (*P*)

The induced dipole moment developed per unit volume in a dielectric slab on placing it in an electric field is called polarisation. It is denoted by P. $P = \chi_e E$

where χ_e is known as electric susceptibility of the dielectric medium.

Capacitor

A **capacitor** is a device which is used to store huge charge over it, without changing its dimensions.

When an earthed conductor is placed near a charged conductor, then it decreases its potential and therefore more charge can be stored over it.

A capacitor is a pair of two conductors of any shape, close to each other and have equal and opposite charges.

Capacitance of a conductor $C = \frac{q}{V}$

Its SI unit is coulomb/volt or farad.

Its other units are $1\,\mu F = 10^{-6}\,F$

$$1\,\mu\mu F = 1\,pF = 10^{-12}\,F$$

Its dimensional formula is $[M^{-1}L^{-2}T^{4}A^{2}]$.

Capacitance of an Isolated Spherical Conductor

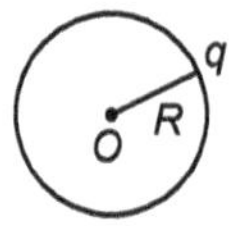

$$C = 4\pi\,\varepsilon_0 K\,R$$

For air $K = 1$

$$\therefore \quad C = 4\pi\,\varepsilon_0 R = \frac{R}{9 \times 10^9}$$

Parallel Plate Capacitor

The parallel plate capacitor consists of two metal plates parallel to each other and separated by a distance d.

Capacitance $C = \frac{KA\varepsilon_0}{d}$

For air capacitor $C_0 = \frac{A\varepsilon_0}{d}$

When a dielectric slab is inserted between the plates partially, then its capacitance.

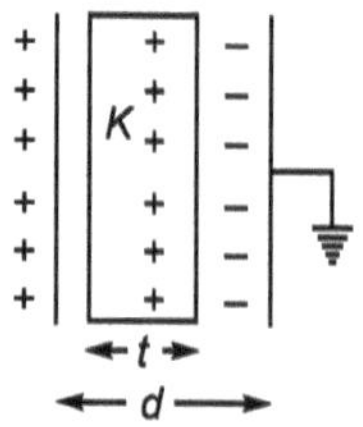

$$C = \frac{A\varepsilon_0}{\left(d - t + \frac{t}{K}\right)}$$

If a conducting (metal) slab is inserted between the plates, then

$$C = \frac{A\varepsilon_0}{(d - t)}$$

When more than one dielectric slabs are placed fully between the plates, then

$$C = \frac{A\varepsilon_0}{\left(\frac{t_1}{K_1} + \frac{t_2}{K_2} + \frac{t_3}{K_3} + \dots\right)}$$

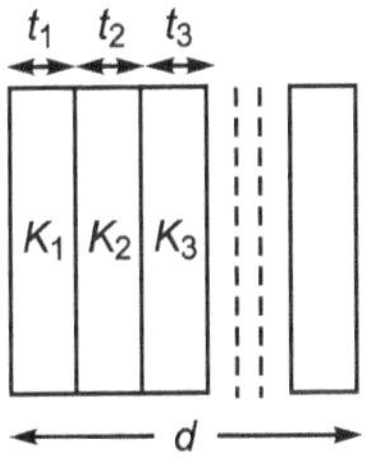

The plates of a parallel plate capacitor attract each other with a force,

$$F = \frac{Q^2}{2A\varepsilon_0}$$

When a dielectric slab is placed between the plates of a capacitor than charge induced on its side due to polarization of dielectric is

$$q' = q\left(1 - \frac{1}{K}\right)$$

Capacitors Combination

(i) **In Series**

Resultant capacitance $\frac{1}{C} = \frac{1}{C_1} + \frac{1}{C_2} + \frac{1}{C_3} + \dots$

In series charge is same on each capacitor, which is equal to the charge supplied by the source.

If $V_1, V_2, V_3, \dots$ are potential differences across the plates of the capacitors then total voltage applied by the source

$$V = V_1 + V_2 + V_3 + \dots$$

(ii) **In Parallel**

Resultant capacitance

$$C = C_1 + C_2 + C_3 + \dots$$

In parallel potential differences across the plates of each capacitor is same.

If $q_1, q_2, q_3, \dots$ are charges on the plate of capacitors connected in parallel, then total charge given by the source

$$q = q_1 + q_2 + q_3 + \dots$$

Electric potential energy of a charged conductor or a capacitor is given by,

$$U = \frac{1}{2} Vq = \frac{1}{2} CV^2 = \frac{1}{2} \frac{q^2}{C}$$

Redistribution of Charge

When two isolated charged conductors are connected to each other then charge is redistributed in the ratio of their capacitances.

$$\text{Common potential } V = \frac{q_1 + q_2}{C_1 + C_2} = \frac{C_1 V_1 + C_2 V_2}{C_1 + C_2}$$

$$\text{Energy loss } = \frac{1}{2} \frac{C_1 C_2 (V_1 - V_2)^2}{(C_1 + C_2)}$$

This energy is lossed in the form of heat in connecting wires.

When n small drops, each of capacitance C, charged to potential V with charge q, surface charge density σ and potential energy U coalesce to form a single drop.

Then for new drop,

Total charge $= nq$

Total capacitance $= n^{1/3} C$

Total potential $= n^{2/3} V$

Surface charge density $= n^{1/3} \sigma$

Total potential energy $= n^{2/3} U$

Van-de-Graaff Generator

It is a device used to build up very high potential difference of the order of few million volt.

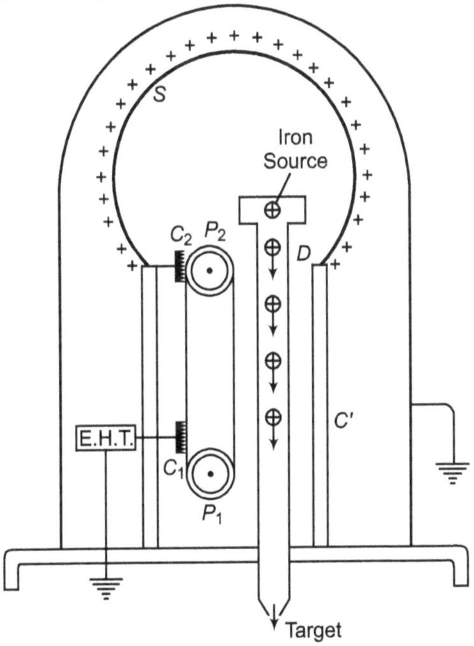

Its working is based on two points

(i) The action of sharp points (corona discharge)

(ii) Total charge given to a spherical shell resides on its outer surface.

Lightning Conductor

When a charged cloud passes by a tall building, the charge on the cloud passes to the earth through the building. This causes a big damage to the building. Thus to protect the tall building from lightning, the lightning conductors, (which are pointed metal roads) passes over the charge on the clouds to earth, thus protecting the building.

20

Current Electricity

Electric Current (I)

The rate of flow of charge through any cross-section of a wire is called electric current flowing through it.

Electric current $(I) = \frac{q}{t}$. Its SI unit is ampere (A).

The conventional direction of electric current is the direction of motion of positive charge.

The current is the same for all cross-sections of a conductor of non-uniform cross-section. Similar to the water flow, charge flows faster where the conductor is smaller in cross-section and slower where the conductor is larger in cross-section, so that charge rate remains unchanged.

If a charge q revolves in a circle with frequency f, the equivalent current,

$$i = qf$$

In a metallic conductor current flows due to motion of free electrons while in electrolytes and ionized gases current flows due to electrons and positive ions.

Types of Electric Current

According to its magnitude and direction electric current is of two types

(i) **Direct Current (DC)** Its magnitude and direction do not change with time. A cell, battery or DC dynamo are the sources of direct current.

(ii) **Alternating Current (AC)** An electric current whose magnitude changes continuously and changes its direction periodically is called alternating current. AC dynamo is the source of alternating current.

Thermal Velocity of Free Electrons

Free electrons in a metal move randomly with a very high speed of the order of 10^5 ms^{-1}. This speed is called thermal velocity of free electrons. Average thermal velocity of free electrons in any direction remains zero.

Drift Velocity of Free Electrons

When a potential difference is applied across the ends of a conductor, the free electrons in it move with an average velocity opposite to the direction of electric field, which is called drift velocity of free electrons.

Drift velocity $$v_d = \frac{eE\tau}{m} = \frac{eV\tau}{ml}$$

where, τ = relaxation time, e = charge on electron,
E = electric field intensity, l = length of the conductor,
V = potential difference across the ends of the conductor and
m = mass of electron.

Relation between electric current and drift velocity is given by

$$v_d = \frac{I}{An\,e}$$

Current Density

The electric current flowing per unit area of cross-section of a conductor is called current density.

Current density $(J) = \frac{I}{A} = nev_d$

Its SI unit is ampere metre^{-2} and dimensional formula is $[AT^{-2}]$. It is a vector quantity and its direction is in the direction of motion of positive charge or in the direction of flow of current.

Mobility

The drift velocity of electron per unit electric field applied is called mobility of electron.

Mobility of electron $(\mu) = \frac{v_d}{E}$

Its SI unit is $m^2s^{-1}V^{-1}$ and its dimensional forumla is $[M^{-1}T^2A]$.

Ohm's Law

If physical conditions of a conductor such as temperature remains unchanged, then the electric current (I) flowing through the conductor is directly proportional to the potential difference (V) applied across its ends.

$$I \propto V$$

or

$$V = IR$$

where, R is the electrical resistance of the conductor and $R = \frac{Ane^2\tau}{ml}$.

Ohmic Conductors

Those conductors which obey Ohm's law, are called ohmic conductors e.g. all metallic conductors are ohmic conductor.

For ohmic conductors V-I graph is a straight line.

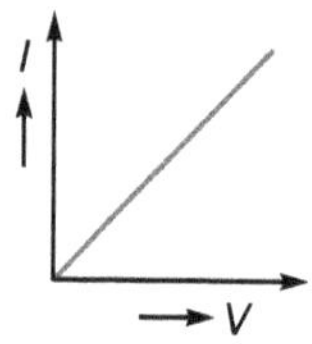

Non-ohmic Conductors

Those conductors which do not obey Ohm's law, are called non-ohmic conductors. e.g. diode valve, triode valve, transistor , vacuum tubes etc.

For non-ohmic conductors V-I graph is not a straight line.

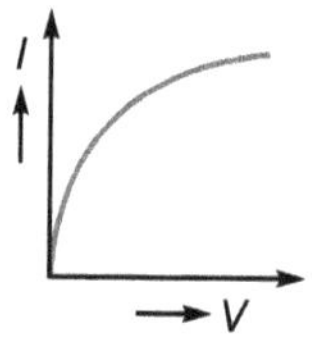

Electrical Resistance

The obstruction offered by any conductor in the path of flow of current is called its electrical resistance.

Electrical resistance, $R = \frac{V}{I}$

Its SI unit is ohm (Ω) and its dimensional formula is $[ML^2T^{-3}A^{-2}]$.

Electrical resistance of a conductor $R = \frac{\rho l}{A}$

where, l = length of the conductor, A = cross-section area and ρ = resistivity of the material of the conductor.

Resistivity

Resistivity of a material of a conductor is given by

$$\rho = \frac{m}{ne^2\tau}$$

where, n = number of free electrons per unit volume.

Resistivity of a material depend on temperature and nature of the material.

It is independent of dimensions of the conductor, *i.e.*, length, area of cross-section etc.

Resistivity of metals increases with increase in temperature as

$$\rho_t = \rho_0 (1 + \alpha t)$$

where ρ_0 and ρ_t are resistivity of metals at 0°C and t°C and α = temperature coefficient of resistivity of the material.

For metals α is positive, for some alloys like nichrome, manganin and constantan, α is positive but very low.

For semiconductors and insulators, α is negative.

Resistivity is low for metals, more for semiconductors and very high for alloys like nichrome, constantan etc.

> In magnetic field the resistivity of metals increases. But resistivity of ferromagnetic materials such as iron, nickel, cobalt etc decreases in magnetic field.

Electrical Conductivity

The reciprocal of resistivity is called electrical conductivity.

$$\text{Electrical conductivity } (\sigma) = \frac{1}{\rho} = \frac{l}{RA} = \frac{ne^2\tau}{m}$$

Its SI units is $\text{ohm}^{-1}\text{m}^{-1}$ or mho m^{-1} or siemen m^{-1}.

Relation between current density (J) and electrical conductivity (σ) is given by

$$J = \sigma E$$

where, E = electric field intensity.

Superconductors

When few metals are cooled, then below a certain critical temperature their electrical resistance suddenly becomes zero. In this state, these substances are called **superconductors** and this phenomena is called **superconductivity.**

Mercury become superconductor at 4.2 K, lead at 7.25 K and niobium at 9.2 K.

Colour Coding of Carbon Resistors

The resistance of a carbon resistor can be calculated by the code given on it in the form of coloured strips.

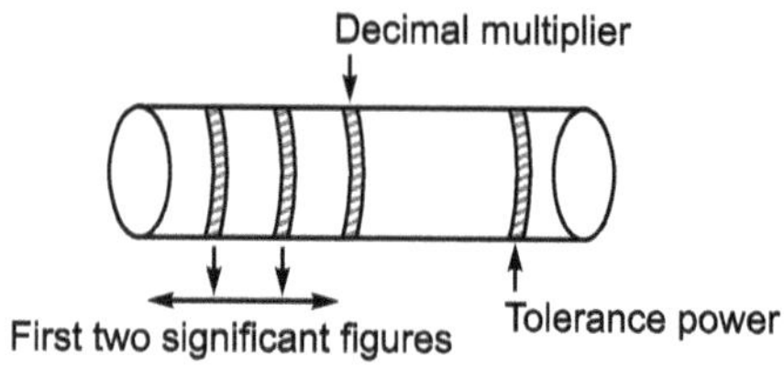

Colour coding

Colour	Figure
Black	0
Brown	1
Red	2
Orange	3
Yellow	4
Green	5
Blue	6
Violet	7
Grey	8
White	9

Tolerance power

Colour	Tolerance
Gold	5%
Silver	10%
No colour	20%

This colour coding can be easily learned in the sequence "B B ROY Great Bratain Very Good Wife".

Combination of Resistors

1. In Series

(i) Equivalent resistance, $R = R_1 + R_2 + R_3$

(ii) Current through each resistor is same.

(iii) Sum of potential differences across individual resistors is equal to the potential difference, applied by the source.

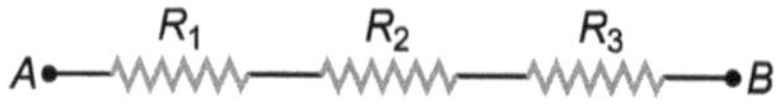

2. In Parallel

Equivalent resistance,

$$\frac{1}{R} = \frac{1}{R_1} + \frac{1}{R_2} + \frac{1}{R_3}$$

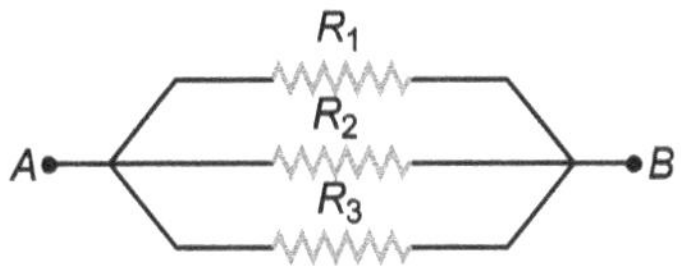

Potential difference across each resistor is same.

Sum of electric currents flowing through individual resistors is equal to the electric current drawn from the source.

Electric Cell

An electric cell is a device which converts chemical energy into electrical energy.

Electric cells are of two types

(i) **Primary Cells** Primary cells cannot be charged again. Voltic, Daniel and Laclanche cells are primary cells.

(ii) **Secondary Cells** Secondary cells can be charged again and again. Acid and alkali accumulators are secondary cells.

Electro-Motive-Force (emf) of a Cell

The energy given by a cell in flowing unit positive charge throughout the circuit completely one time, is equal to the emf of a cell.

Emf of a cell $(E) = \frac{W}{q}$.

Its SI unit is volt.

Terminal Potential Difference of a Cell

The energy given by a cell in flowing unit positive charge through the outer circuit one time from one terminal of the cell to the other terminal of the cell.

Terminal potential difference $(V) = \frac{W}{q}$

Its SI unit is also volt.

Internal Resistance of a Cell

The obstruction offered by the electrolyte of a cell in the path of electric current is called internal resistance (r) of the cell. Internal resistance of a cell

(i) Increases with increase in concentration of the electrolyte.

(ii) Increases with increase in distance between the electrodes.

(iii) Decreases with increase in area of electrodes dipped in electrolyte.

Relation between E, V and r

$$E = V + Ir$$

$$r = \left(\frac{E}{V} - 1\right) R$$

If cell is in charging state, then

$$E = V - Ir$$

Grouping of Cells

(i) **In Series** If n cells, each of emf E and internal resistance r are connected in series to a resistance R, then equivalent emf

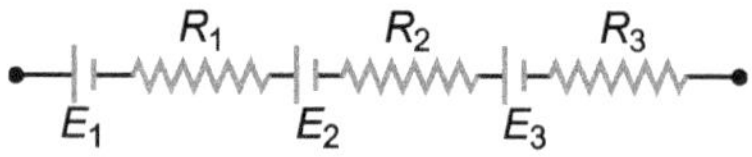

$$E_{eq} = E_1 + E_2 + \ldots + E_n = nE$$

Equivalent internal resistance $r_{eq} = r_1 + r_2 + \ldots + r_n = nr$

Current in the circuit $I = \dfrac{E_{eq}}{(R + r_{eq})} = \dfrac{nE}{(R + nr)}$

(ii) **In Parallel** If n cells, each of emf E and internal resistance r are connected to in parallel, then equivalent emf, $E_{eq} = E$

Equivalent internal resistance

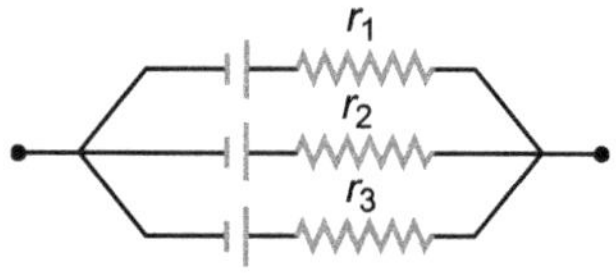

$$\frac{1}{r_{eq}} = \frac{1}{r_1} + \frac{1}{r_1} + \ldots + \frac{1}{r_n} = \frac{n}{r} \text{ or } r_{eq} = \frac{r}{n}$$

Current in the circuit, $I = \dfrac{E}{(R + r/n)}$

(iii) **Mixed Grouping of Cells** If n cells, each of emf E and internal resistance r are connected in series and such m rows are connected in parallel, then

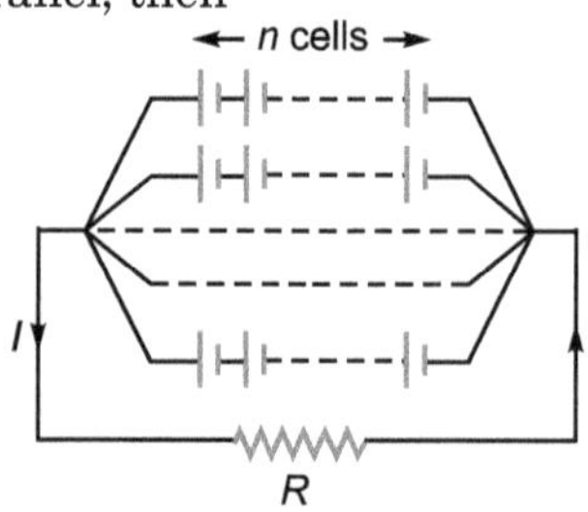

Equivalent emf, $E_{eq} = nE$

Equivalent internal resistance $r_{eq} = \frac{nr}{m}$

Current in the circuit, $I = \frac{nE}{\left(R + \frac{nr}{m}\right)}$

or $$I = \frac{mnE}{mR + nr}$$

Note *Current in this circuit will be maximum when external resistance is equal to the equivalent internal resistance, i.e.,*

$$R = \frac{nr}{m} \Rightarrow mR = nr$$

Kirchhoff's Laws

There are two Kirchhoff's laws for solving complicated electrical circuits

(i) **Junction Rule** The algebric sum of all currents meeting at a junction in a closed circuit is zero, i.e. $\Sigma I = 0$.

This law follows law of conservation of charge.

(ii) **Loop Rule** The algebric sum of all the potential differences in any closed circuit is zero, i.e.

$$\Sigma V = 0 \Rightarrow \Sigma E = \Sigma IR$$

This law follows law of conservation of energy.

Balanced Wheatstone Bridge

Wheatstone bridge is also known as a **meter bridge** or **slide wire bridge.**

This is an arrangement of four resistance in which one resistance is unknown and rest known. The Wheatston bridge as shown in figure. The bridge is said to be balanced when deflection in galvanometer is zero, *i.e.*, $i_g = 0$.

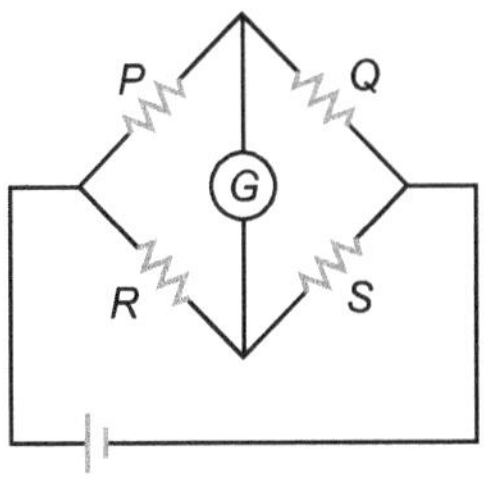

Principle of Wheatstone Bridge

$$\frac{P}{Q} = \frac{R}{S}$$

The value of unknown resistance S can found, as we know the value of P, Q and R. It may be remembered that the bridge is most sensitive, when all the four resistances are of the same order.

Meter Bridge

This is the simplest form of Wheatstone bridge and is specially useful for comparing resistance more accurately.

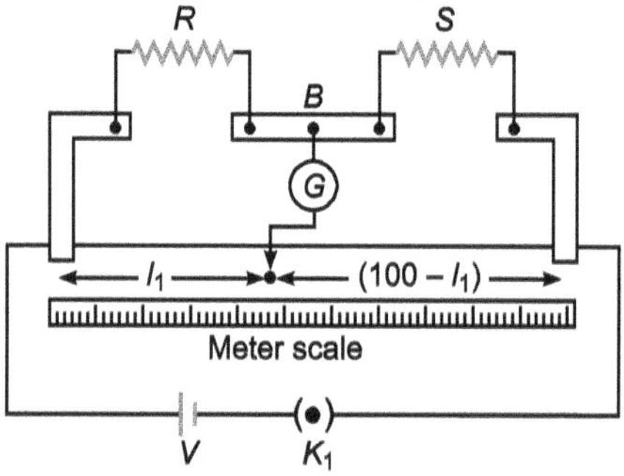

$$\frac{R}{S} = \frac{l_1}{(100 - l_1)}$$

where l_1 is the length of wire from one end where null point is obtained.

Potentiometer

Potentiometer is an ideal device to measure the potential difference between two points. It consists of a long resistance wire AB of uniform cross-section in which a steady direct current is set up by means of a battery.

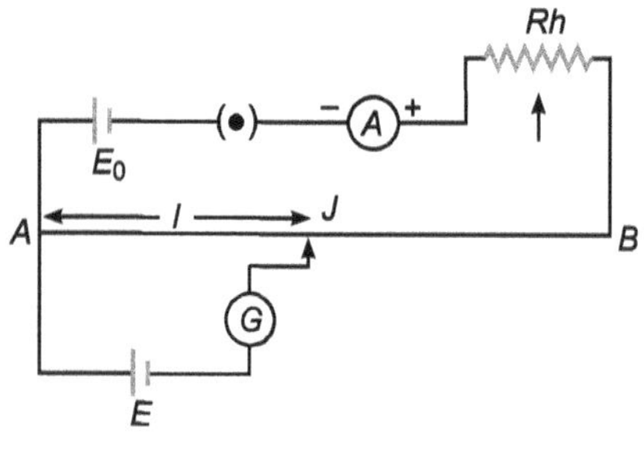

If R be the total resistance of potentiometer wire L its total length, then potential gradient, *i.e.*, fall in potential per unit length along the potentiometer will be

$$K = \frac{V}{L} = \frac{IR}{L}$$

$$= \frac{E_0 R}{(R_0 + R)L}$$

where, E_0 = emf of battery and R_0 = resistance inserted by means of rheostat Rh.

Determination of emf of a Cell using Potentiometer

If with a cell of emf E on sliding the contact point we obtain zero deflection in galvanometer G when contact point is at J at a length l from the end where positive terminal of cell have been joined, then fall in potential along length l is just balancing the emf of cell. Thus, we have

$$E = Kl$$

or
$$\frac{E_1}{E_2} = \frac{l_1}{l_2}$$

Determination of Internal Resistance of a Cell using Potentiometer

The arrangement is shown in figure. If the cell E is in open circuit and balancing length l_1, then

$$E = Kl_1$$

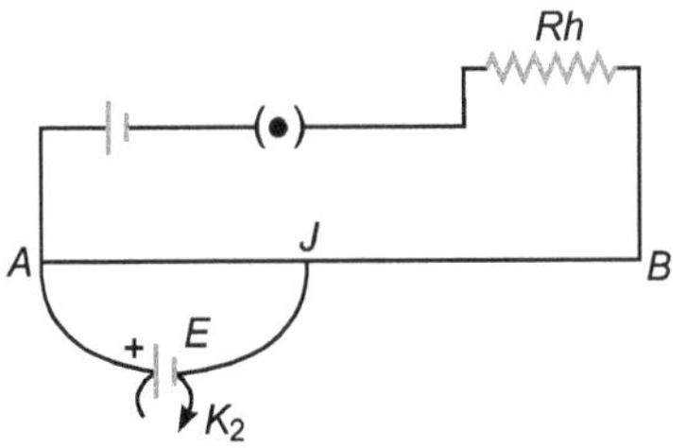

But if by inserting key K_2 circuit of cell is closed, then potential difference V is balanced by a length l_2 of potential, where

$$V = Kl_2$$

Internal resistance of cell

$$r = \frac{E - V}{V} R = \frac{l_1 - l_2}{l_2} R$$

Important Points

- Potentiometer is an ideal voltmeter.
- Sensitivity of potentiometer is increased by increasing length of potentiometer wire.
- If n identical resistances are first connected in series and then in parallel, the ratio of the equivalent resistance.

$$\frac{R_s}{R_p} = \frac{n^2}{1}$$

- If a skeleton cube is made with 12 equal resistance, each having a resistance R, then the net resistance across

  (a) the diagonal of cube $= \frac{5}{6}R$

  (b) the diagonal of a face $= \frac{3}{4}R$

  (c) along a side $= \frac{7}{12}R$

- If a resistance wire is stretched to a greater length, keeping volume constant, then

$$R \propto l^2 \Rightarrow \frac{R_1}{R_2} = \left(\frac{l_1}{l_2}\right)^2$$

and

$$R \propto \frac{1}{r^4} \Rightarrow \frac{R_1}{R_2} = \left(\frac{r_2}{r_1}\right)^4$$

where l is the length of wire and r is the radius of cross-section area of wire.

21

Heating and Chemical Effects of Current

Heating Effects of Current (Joule's Law)

When current I flows through a conductor of resistance R for a time t, then heat generated in it is given by

$$H = I^2Rt \text{ joule} = \frac{I^2Rt}{4.18} \text{ cal}$$

or

$$H = \frac{V^2}{R} t \text{ joule} = \frac{V^2t}{4.18\,R} \text{ cal}$$

where, V = potential difference applied across the ends of the conductor.

Electric Power

The electrical energy produced or consumed per unit time is called electric power.

$$\text{Electric power, } P = VI = I^2R = \frac{V^2}{R}$$

where, V is the potential difference across the conductor, I is current flowing through the conductor and R is the resistance.

Its SI unit is watt (W).

The other units of electric power are kilowatt and horse power.

$$1 \text{ kW} = 1000 \text{ W}$$

$$1 \text{ HP} = 746 \text{ W}$$

Electric Energy

The energy supplied by any source in maintaining the current in the electric circuit is called electric energy consumed by the electric circuit.

$$\text{Electric energy } (W) = Vq = VIt = I^2Rt = \frac{V^2}{R}t$$

Its SI unit is joule (J) but another unit is watt-hour. The bigger unit of electric energy is kilowatt hour (kWh). It is known as board of trade (BOT) unit.

$$\begin{aligned} 1 \text{ kilowatt hour} &= 1000 \text{ watt} \times 1 \text{ hour} \\ &= 1000 \text{ J/s} \times 3600 \text{ s} \\ &= 3.6 \times 10^6 \text{ J} \end{aligned}$$

1 Horse power = 746 watt

The electric energy consumed in kWh is given by

$$W \text{ (in kWh)} = \frac{V \text{ (in volt)} \times I \text{ (in ampere)} \times t \text{ (in hour)}}{1000}$$

> The electrical resistance of the wires supplying current is very small, therefore these wire do not heat up when current passes through them.
>
> The electrical resistance of filament of a lamp is very high, therefore it shows more heating effect, when electric current passes through it.
>
> A heater wire must be of high resistivity and of high melting point.

Electric Fuse

An electric fuse is a safety device used for protecting electric circuits from demaging it due to excess flow of current. It is made of tin-lead alloy (63% tin + 37% lead).

It should have high resistance and low melting point and should be connected in series with the live wire.

Maximum safe current which can be passed through a fuse wire is independent of its length. However, it depends on the radius r of wire as

$$I \propto r^{3/2}$$

Short Circuiting

When accidentily the live wire comes in contact with nutral wire, then resistance of the circuit decreases and a high current flows through the circuit. This phenomena is called short circuiting.

Overloading

When a high current flows through the wire which is beyond the rating of wire, then heating of wire takes place. This phenomena is called overloading.

Rating of Electrical Appliances

The values of power and voltage taken together for an electrical appliance is called rating of the appliance.

- When a 40 W and a 100 W bulbs are connected in series, then 40 W bulb will glow brighter than 100 W bulb.
- When a 40 W and a 100 W bulbs are connected in parallel, then 100 W bulb will glow brighter than 40 W bulb.
- In series, if any bulb gets fused, then others will not glow.
- In parallel, if any one bulb get fused, then others will continue to glow.
- All switches should be connected in series with a live wire.

Fusing of Bulb When it is Switched On

Usually filament bulbs get fused when they are switched on. This is because with rise in temperature the resistance of the bulb increases and becomes constant in steady state. So, the power consumed by the bulb $\left(\frac{V^2}{R}\right)$ initially is more than that in steady state and hence the bulb glows more brightly in beginning and may get fused.

Chemical Effect of Electric Current (Electrolysis)

When a direct current flows through a acidic or basic solution it dissociate into positive and negative ions. This phenomena is called electrolysis and these solution are called electrolytes.

Some Terms of Electrolysis

(i) **Anode** The electrode connected to the positive terminal of the battery is called an anode.

(ii) **Cathode** The electrode connected to the negative terminal of the battery is called a cathode.

(iii) **Anions** The ions carrying negative charge and move towards the anode in electrolysis are called anions.

(iv) **Cations** The ions carrying positive charge and move towards the cathode in electrolysis are called cations.

(v) **Voltameter** The vassel in which the electrolysis is carried out is called a voltameter.

(vi) **Ionisation** The phenomenon of separation of a molecule into oppositely charged ions is known as ionisation.

Faraday's Laws of Electrolysis

Faraday's laws of electrolysis are of two types :

First Law

The mass of the substance liberated during electrolysis at each electrode is directly proportional to the total charge passed through the electrolyte.

$$m \propto q \quad \text{or} \quad m = Zq = ZIt$$

where, Z = electrochemical equivalent of the substance deposited on electrode,

q = total charge passed through the electrolyte,

I = electric current and

t = time.

Second Law

The mass of each substance liberated at the electrodes in electrolysis by a given amount of change is proportional to the chemical equivalents of the substances.

$$m \propto E \quad \text{or} \quad \frac{m_1}{m_2} = \frac{E_1}{E_2}$$

where m_1 and m_2 are masses of the substance liberated on electrodes by passing same amount of current for the same time and E_1, E_2 are chemical equivalents of these substances.

$$\text{Chemical equivalent} = \frac{\text{atomic weight}}{\text{valency}}$$

$$\text{Faraday's constant } (F) = \frac{E}{Z}$$

Faraday's constant is equal to the amount of charge required to liberate one chemical equivalent (in gram) of mass of a substance at a electrode during electrolysis.

Faraday's constant, $(F) = Ne$

where, N = Avogadro's number and e = electronic charge.

or $$F = (6.0229 \times 10^{23}) \times (1.602 \times 10^{-19})$$

$$= 96487\,\text{C} \approx 96500\,\text{C}$$

Electroplating

The process of coating an object, that conduct electricity, with another metal is called electroplating. The article of cheep metal are coated with precious metal like silver and gold to make them look attractive.

Anodising

The process of coating aluminium with its oxide electrochemically to protect it against corrosion is called anodising.

Uses of Electrolysis

Electrolysis is used for local anaesthesia. For it current is passed through a nerve, due to which it becomes insensitive to pain.

Electrolysis is used for nerve stimulation of polio patients.

Electrochemical Cell

An electric cell is a device which maintains a continuous flow of charge in a circuit by a chemical reaction.

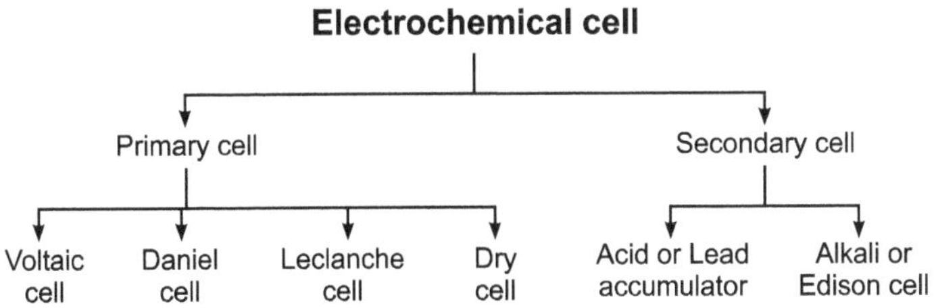

1. Primary Cell

Primary cell is an electrochemical cell, which once discharge cannot be put to use again by passing electric current from an external source.

(i) **Voltaic Cell** This cell was first discovered in 1791 by scientist Galvani and then it was modified in 1800 by scientist Volta. To set up a simple cell, take a glass vessel containing dilute sulphuric cell, acid as the electrolyte. Two rods, one of copper and the other of zinc are kept immersed in the electrolyte. These rods are kept separated from each other. They are called the **electrodes**. Due to chemical reaction a potential difference of 1.08 V is developed between these rod.

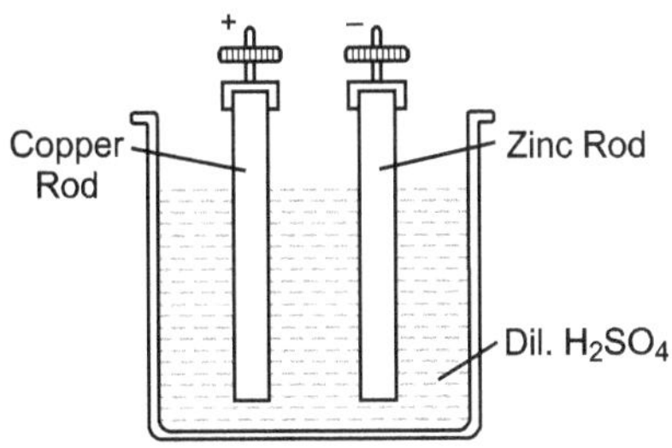

(ii) **Dainel Cell** It consists of a copper vessel in which a porous pot containing dilute sulphuric acid is placed. An amalgamated zinc rod (which acts as the negative electrode) is put in the porous pot. The copper vessel contains a concentrated solution of copper sulphate which acts as an oxidising agent. To maintain the strength of copper sulphate solution, crystals of the copper sulphate are placed on the perforated shelf, partly immersed in the solution.

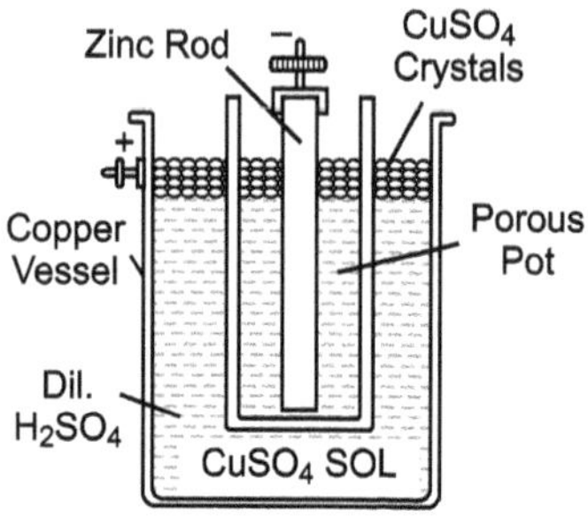

(iii) **Leclanche Cell** In this cell, anode is a carbon rod placed in a porous pot containing a mixture of powdered carbon and manganese dioxide. The carbon powder is used to make the manganese dioxide conducting or to increase the surface area of the carbon electrode. An amalgamated zinc rod is used as a negative electrode.

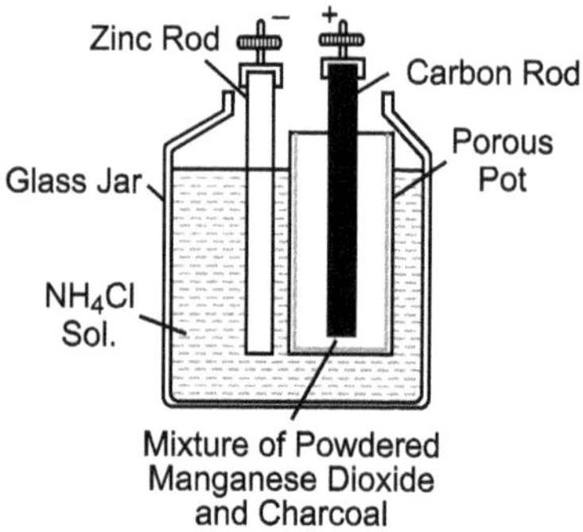

(iv) **Dry Cell** It is modified form of a Leclanche cell. The electrolyte in a dry cell is in the form of a jelly composed of starch, flour and ammonium chloride. The positive electrode is a carbon rod surrounded by a mixture of manganese dioxide and carbon contained in a cloth bag. This bag is placed in a zinc can and the space in between is filled with the ammonium chloride jelly. The zinc can itself acts as the negative electrode.

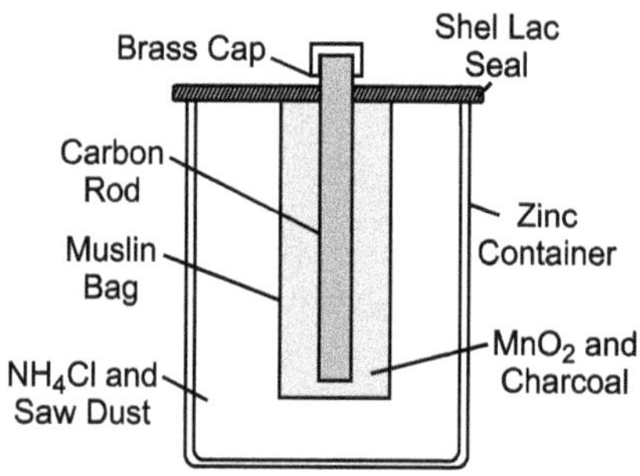

2. Secondary Cell

Secondary cell is an electrochemical cell, which has to be charged initially by passing electric current from an external source, so as to convert the stored chemical energy into electrical energy.

Lead accumulator, Ni-Fe or alkali accumulator are the secondary cells.

Comparison of lead accumulator and Ni-Fe cell

S.No	Lead accumulator	Ni-Fe cell
(i)	It can be used to draw strong current.	It cannot be used to draw a strong current.
(ii)	When it is completely charged its emf is 2.2 V.	When fully charged, its emf is 1.35 V.
(iii)	Its efficiency is nearly 70%.	Its efficiency is nearly 50%.

Thermoelectric Effect

Thermocouple

If two wires of different metals are joined at their ends so as to form two junctions, then the resulting arrangement is known as thermocouple.

Uses of Thermocouple

(a) Thermometer to measure temperature.

(b) Thermoelectric current-meter to measure current.

(c) Thermoelectric generator

(d) Thermoelectric refrigerator.

Seebeck Effect

When the two junctions of a thermocouple are kept at different temperatures, then an emf is produced between the junctions, which is called thermo emf and this phenomena is called Seebeck effect.

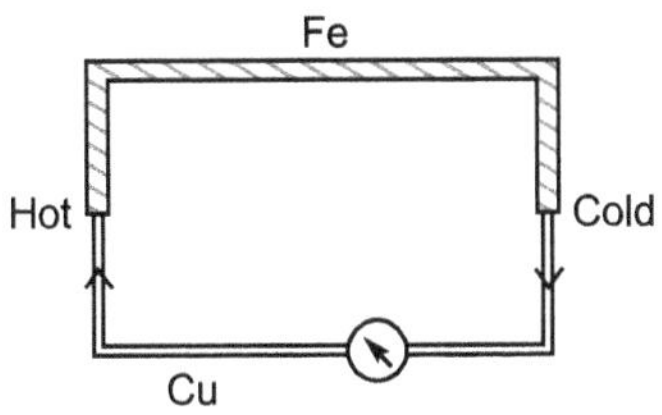

Thermoelectric Series or Seebeck Series

Seebeck form a series of metals for which thermoelectric current flows in a thermocouple through the hot junction from a metal occuring earlier, to a metal occuring later, in series.

The series is given below

Bi, Ni, Co, Pd, Pt, Cu, Mn, Hg, Pb, Sn, Au, Zn, Cd, Fe, Sb, Te, Fe, Mo, Cr

Neutral Temperature (T_n)

The temperature of the hot junction at which thermo emf in a thermocouple is maximum, is called neutral temperature. The value of neutral temperature is constant for thermocouple. Its value depends upon the nature of material forming the thermocouple and independent of temperature of cold junction.

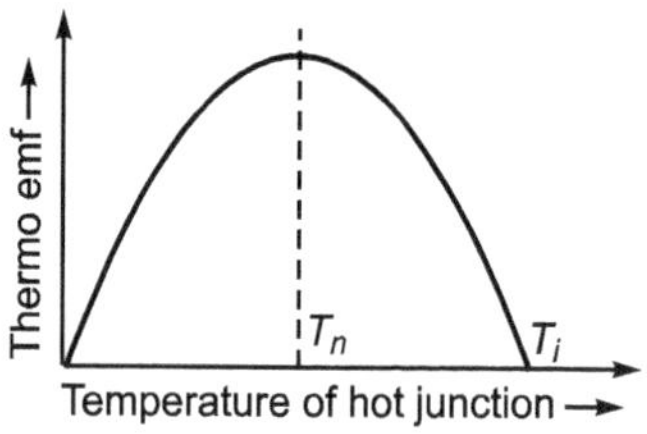

Temperature of Inversion (T_i)

The temperature of the hot junction at which the thermo emf in a thermocouple becomes zero and beyond it, reverses its direction is called temperature of inversion. Its value depends upon the temperature of the cold junction as well as the nature of the materials forming the thermocouple.

Relation between neutral temperature (T_n) and temperature of inversion (T_i) is given by

$$T_n = \frac{T_i + T_0}{2}$$

where T_0 is the temperature of **cold junction.**

When cold junction is at 0°C, then thermo emf changes with temperature of hot junction (T) as follows :

$$E = \alpha T + \frac{1}{2}\beta T^2$$

where α and β are constants depending upon nature of metals forming thermocouple.

Thermoelectric Power

The rate of change of thermo emf with temperature, is called thermoelectric power.

Thermoelectric power, $S = \dfrac{dE}{dt} = \alpha + \beta T$

Thermoelectric Thermometer

It is a device used to measure both low and high temperatures. Thermoelectric thermometer have much wider range of measurement of temperature (from –200°C to 1600°C). They are quite sentitive and can measure temperature accurately upto 0.05°C. Disadvantage of thermometer is that it does not give direct reading and hence it cannot be used in experiments on calorimetry.

Peltier Effect

When current is passed through a thermocouple, then heat is generated at its one junction and absorbed at another junction. This phenomena is called Peltier effect.

Peltier effect is the reverse effect of Seebeck effect.

Peltier Coefficient

The ratio of heat energy absorbed or evolved at a junction of a thermocouple to the charge flowing through it called Peltier coefficient.

Peltier coefficient, $$\pi = \frac{\text{Peltier heat}}{\text{Charge flowing}}$$

Its SI unit is 'JC^{-1}'.

Thomson's Effect

If two parts of a single conductor are maintained at different temperatures, then an emf is produced between them, which is called **Thomson's emf** and this phenomena is called Thomson's effect.

Thomson's Coefficient

The ratio of Thomson's emf and temperature difference between two points is called Thomson's coefficient.

Thomson's coefficient, $$(\sigma) = \frac{dV}{dT}$$

where, dV = potential difference between two points and

dT = temperature difference between the two same points.

Thermopile

It is a combination of a large number of thermocouples in series. It is use to detect the heat radiation and to note small variation or difference in temperature.

22

Magnetic Effect of Current

Oersted's Experiment

A magnetic field is produced in the surrounding of any current carrying conductor.

The direction of this magnetic field can be obtained by Ampere's swimming rule.

SI unit of magnetic field is Wm^{-2} or T (tesla).

The strength of magnetic field is called one tesla, if a charge of one coulomb, when moving with a velocity of 1 ms^{-1} along a direction perpendicular to the direction of the magnetic field experiences a force of one newton.

$$1 \text{ tesla (T)} = 1 \text{ weber metre}^{-2} \text{ (Wbm}^{-2}\text{)}$$

$$= 1 \text{ newton ampere}^{-1} \text{ metre}^{-1} \text{ (NA}^{-1} \text{ m}^{-1}\text{)}$$

CGS units of magnetic field are called gauss or oersted.

$$1 \text{ gauss} = 10^{-4} \text{ tesla.}$$

Maxwell's Cork Screw Rule

If a right handed cork screw is imagined to be rotated in such a direction that tip of the screw points in the direction of the current, then direction of rotation of corkscrew gives the direction of magnetic line of force.

The conventional sign for a magnetic field coming out of the plane and normal to it is denoted by $\odot$.

The magnetic field perpendicular to the plane in the downward direction is denoted by $\otimes$.

Ampere's Swimming Rule

If a man is swimming along the wire in the direction of current with his face turned towards the needle, so that the current enters through his feet, then north pole of the magnetic needle will be deflected towards his left hand.

Magnetic Field

The space in the surrounding of a magnet or any current carrying conductor in which its magnetic influence can be experienced.

Biot Savart's Law

The magnetic field produced by a current carrying element of length dl, carrying current I at a point separated by a distance r is given by

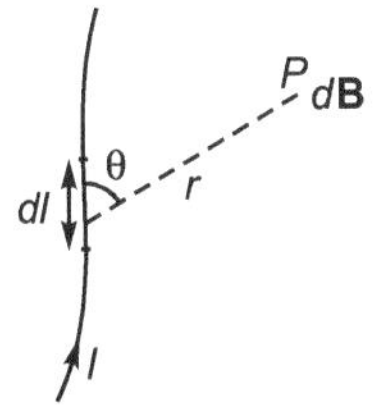

$$d\mathbf{B} = \frac{\mu_0}{4\pi}\frac{I(d\mathbf{l} \times \mathbf{r})}{r^3}$$

or

$$dB = \frac{\mu_0}{4\pi}\frac{Idl\sin\theta}{r^2}$$

where, θ is the angle between the direction of the current and μ_0 is absolute permeability of the free space.

SI unit of magnetic field is Wm^{-2} or (tesla) and CGS unit of magnetic field is gauss or oersted 1 gauss $= 10^{-4}$ tesla.

The direction of magnetic field $d\mathbf{B}$ is that of $I\, d\mathbf{l} \times \mathbf{r}$.

Magnetic Field Due to a Straight Current Carrying Conductor

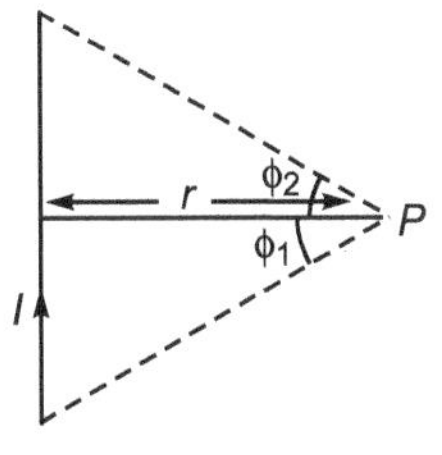

$$B = \frac{\mu_0}{4\pi}\frac{I}{r}(\sin\phi_1 + \sin\phi_2)$$

where ϕ_1 and ϕ_2 are angles, which the lines joining the two ends of the conductor to the observation point make with the perpendicular from the observation point to the conductor.

For infinite length conductor and observation point is near the centre of the conductor,

$$B = \frac{\mu_0}{4\pi} \frac{2I}{r}$$

For infinite length conductor and observation point is near one end of the conductor,

$$B = \frac{\mu_0}{4\pi} \frac{I}{r}$$

Magnetic Field Lines

- They are used to respresent magnetic field in a region.
- They are closed continuous curves.
- Tangent drawn at any point gives the direction of magnetic field.
- They cannot interact.
- Outside a magnet, they are directed from north to south pole and inside a magnet they are directed from south to north.

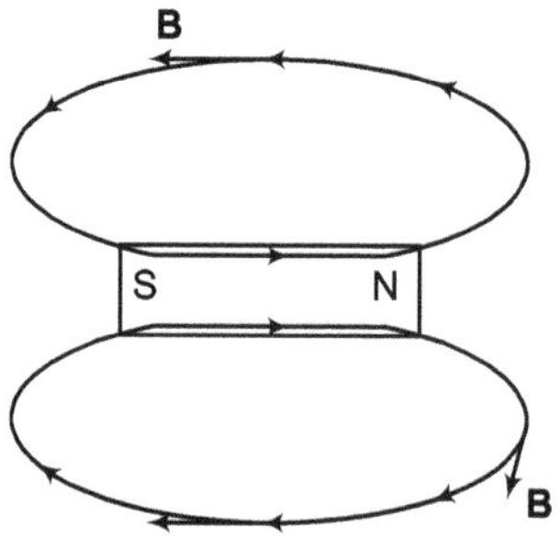

The magnetic field lines due to a straight current carrying conductor are concentric circles having centre at conductor and in a plane perpendicular to the conductor.

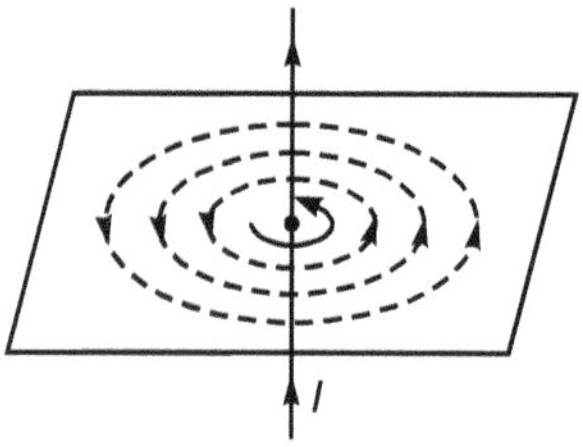

The direction of magnetic field lines can be obtained by Right Hand Thumb Rule.

Right Hand Thumb Rule

If we hold a current carrying conductor in the grip of the right hand in such a way that thumb points in the direction of current, then curling of fingers represents the direction of magnetic field lines.

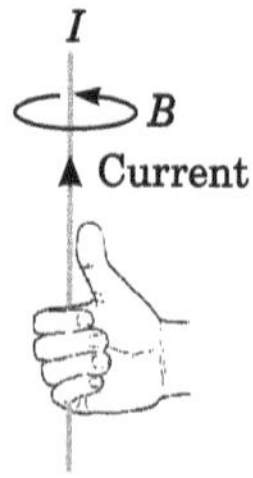

Magnetic Field on the Axis of a Current Carrying Circular Coil

Magnetic field at axis at a distance x from centre O.

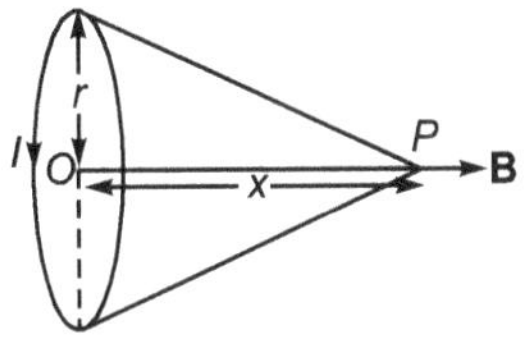

$$B = \frac{\mu_0 n I r^2}{2(r^2 + x^2)^{3/2}}$$

where, r = radius of the coil, n = number of turns in the coil and I = current.

At centre of the coil, $B = \frac{\mu_0 nI}{2r}$

If we look at one face of the coil and the direction of current flowing through the coil is **clockwise**, then that face has south polarity and if direction of current is **anti-clockwise**, then that face has north polarity.

Magnetic Dipole

Every current carrying loop is a magnetic dipole. It has two poles south (S) and north (N).

This is similar to a bar magnet.

Each magnetic dipole has some magnetic moment (M). The magnitude of M is,

$$|M| = NiA$$

where, N = number of turns in the loop,
i = current in the loop and
A = area of cross-section of the loop.

The current carrying loop behaves as a small magnetic dipole placed along the axis one face of the loop behaves as north pole while the other face of loop behaves as South pole.

Ampere's Circuital Law

The line integral of magnetic field induction **B** around any closed path in vacuum is equal to μ_0 times the total current threading the closed path, i.e.

$$\oint \mathbf{B} \cdot d\mathbf{l} = \mu_0 I$$

where **B** is the magnetic field, $d\mathbf{l}$ is small element, μ_0 is the absolute permeability of free space and I is the current.

Ampere's circuital law holds good for a closed path of any size and shape around a current carrying conductor because the relation is independent of distance form conductor.

Solenoid

A solenoid is a closely wound helix of insulated copper wire.

Magnetic field at a point well inside a long solenoid is given by

$$B = \mu_0 nI$$

where, n = number of turns per unit length and

I = current flowing through the solenoid.

Magnetic field at a point on one end of a long solenoid is given by

$$B = \frac{\mu_0 nI}{2}$$

Toroid

A toroidal solenoid is an anchor ring around which is large number of turns of a copper wire are wrapped.

A toroid is an endless solenoid in the form of a ring.

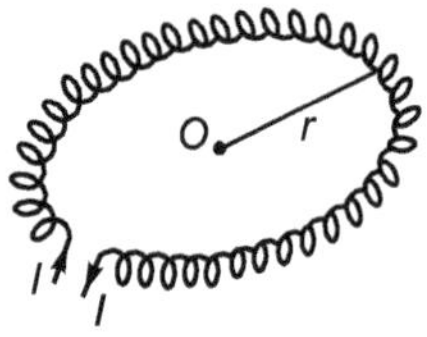

Magnetic field inside the turns of toroid is given by

$$B = \mu_0 nI$$

Magnetic field inside a toroid is constant and is always tangential to the circular closed path.

Magnetic field at any point inside the empty space surrounded by the toroid and outside the toroid, is zero, because net current enclosed by these space is zero.

Magnetic Field Due to a Current Carrying Long Circular Cylinder

Outside the cylinder ($r > R$)

$$B = \frac{\mu_0}{2\pi} \frac{I}{r}$$

Inside the cylinder when it is made of a thin metal sheet,

$$B = 0$$

Inside the cylinder when current is uniformely distributed throughout the cross-section of the cylinder ($r < R$)

$$B = \frac{\mu_0 \mu_r}{2\pi} \frac{Ir}{R^2}$$

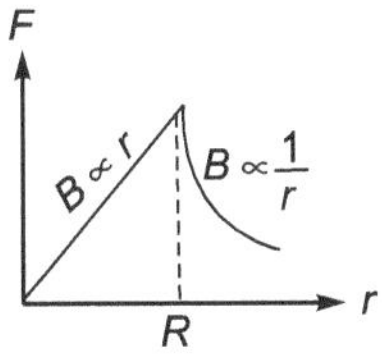

where, μ_0 and μ_r are permeabilities of free space and material of the cylinder, I is current flowing through the cylinder and r is radius of the cylinder.

Force Acting on a Charge Particle Moving in a Uniform Magnetic Field

$$\mathbf{F} = q\,(\mathbf{v} \times \mathbf{B})$$

or $$F = |\mathbf{F}| = Bqv \sin\theta$$

where, B = magnetic field intensity

q = charge on particle,

v = speed of the particle and

θ = angle between magnetic field and direction of motion.

This force is perpendicular to **B** as well as **v**.

Its direction can be obtained from **Fleming's left hand rule.**

Magnetic force acting on a current carrying conductor in a uniform magnetic field is given by

$$\mathbf{F} = I\,(\mathbf{I} \times \mathbf{B})$$

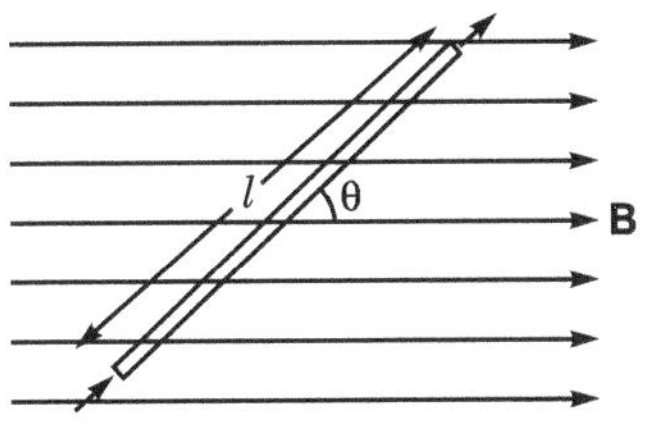

or $$F = BIl \sin\theta$$

Fleming's Left Hand Rule

If we stretch the thumb, the forefinger and the central finger of left hand in such a way that all three are perpendicular to each other, then if forefinger represents the direction of magnetic field, central finger represents the direction of current flowing through the conductor, then thumb will represent the direction of magnetic force.

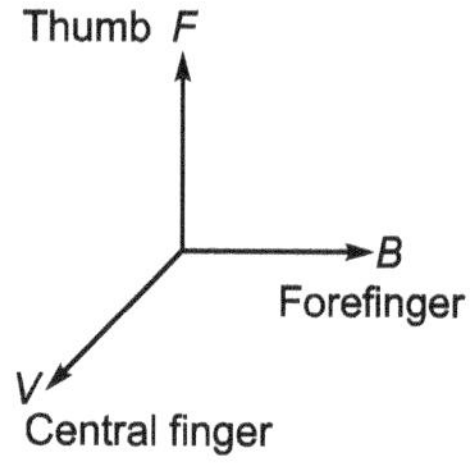

Lorentz Force

The total force experienced by a charge moving inside the electric and magnetic fields is called Lorentz force. It is given by $\mathbf{F} = q(\mathbf{E} + \mathbf{v} \times \mathbf{B})$

Motion of a Charged Particle in a Uniform Magnetic Field

When charged particle enter normal to the magnetic field it follows a circular path.

The radius of the path, $r = \dfrac{mv}{Bq}$

$\therefore \qquad r \propto mv$

and $\qquad r \propto \dfrac{1}{(q/m)}$

Time period, $\qquad T = \dfrac{2\pi m}{Bq}$

When charged particle enter magnetic field at any angle except 90°, then it follows helical path.

The radius of the path, $r = \dfrac{mv \sin\theta}{Bq}$

The distance travelled by the charged particle in one time period due to component of velocity $v \cos\theta$, is called pitch of the path.

$$\text{Pitch} = T \times v \cos\theta = \frac{2\pi mv \cos\theta}{Bq}$$

Cyclotron

Cyclotron is a device used to accelerate positively charged particles such as proton, deutron etc.

Principle of Cyclotron

A positively charged particle can be accelerated through a moderate electric field by crossing it again and again by use of strong magnetic field.

Radius of circular path, $r = \dfrac{mv}{Bq}$

Cyclotron frequency $\quad \nu = \dfrac{Bq}{2\pi m}$

where m and q are mass and charge of the positive ion and B is strength of the magnetic field.

Maximum kinetic energy gained by the particle.

$$E_{max} = \frac{B^2 q^2 r_0^2}{2m}$$

where, r_0 = maximum radius of circular path.

When a positive ion is accelerated by the cyclotron, it moves with greater and greater speed. As the speed of ion becomes comparable with that of light, the mass of the ion increases according to the relation.

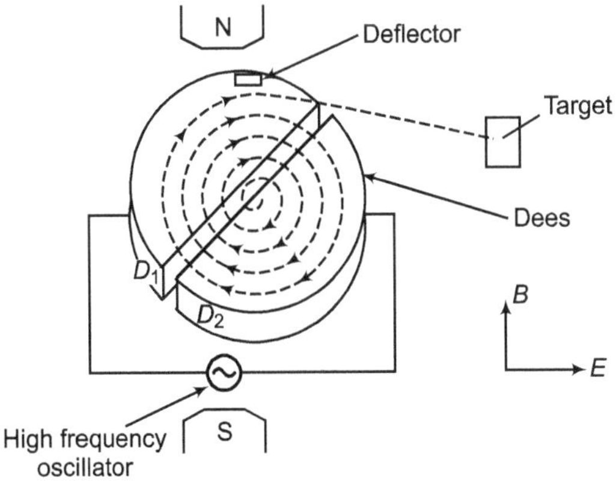

$$m = \frac{m_0}{\sqrt{1 - \dfrac{v^2}{c^2}}}$$

where, m = mass of the ion,
m_0 = maximum mass of the ion,
v = speed of ion and
c = speed of light.

Limitations of the Cyclotron

(i) Cyclotron cannot accelerated uncharged particle like neutron.

(ii) The positively charged particles having large mass i.e. ions cannot move at limitless speed in a cyclotron.

Force between Two Infinitely Long Parallel Current Carrying Conductors

Force between two long parallel current carrying conductors is given by

$$F = \frac{\mu_0}{2\pi} \frac{I_1 I_2}{r} l$$

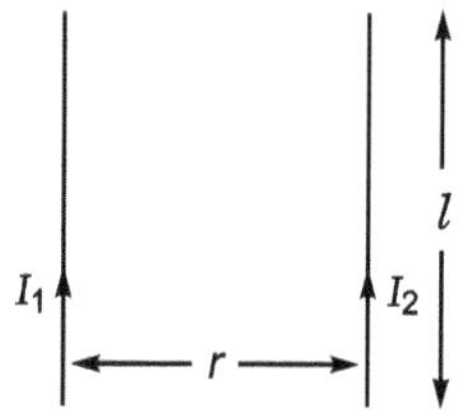

The force is attractive if current in both conductors is in same direction and repulsive if current in both conductors is in opposite direction.

If the currents is both parallel wires are equal and in same direction, then magnetic field at a point exactly half way between the wire is zero.

Torque acting on a Current Carrying Coil Placed Inside a Uniform Magnetic Field

Torque acting on a current carrying coil placed inside a uniform magnetic field is given by

$$\tau = NBIA \sin\theta$$

where, N = number of turns in the coil
B = magnetic field intensity
I = current in the coil and
A = area of cross-section of the coil
θ = angle between magnetic field and normal to the plane of the coil.

Moving Coil Galvanometer

It is a device used for the detection and measurement of the small currents.

In equilibrium, deflecting torque = restoring torque

$$NBIA = k\theta \quad \text{or} \quad I = \frac{k}{NBA}\theta$$

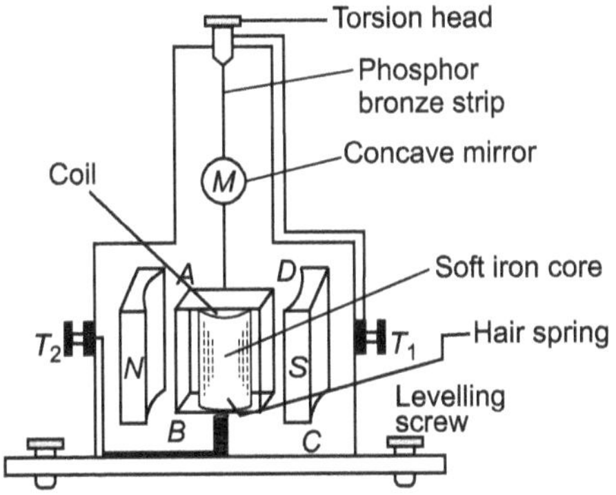

where, k = restoring torque per unit twist,
N = number of turns in the coil,
B = magnetic field intensity,
A = area of cross-section of the coil and
θ = angle of twist.

Current Sensitivity

The deflection produced per unit current in galvanometer is called its current sensitivity.

Current sensitivity

$$I_S = \frac{\theta}{I} = \frac{NBA}{K}$$

Voltage Sensitivity

The deflection produced per unit voltage applied across the ends of the galvanometer is called its voltage sensitivity.

Voltage sensitivity

$$V_s = \frac{\theta}{V} = \frac{NBA}{KR}$$

where R is the resistance of the galvanometer.

Therefore for a sensitive galvanometer

(i) N should be large (ii) B should be large
(iii) A should be large (iv) K should be small.

Ammeter

An ammeter is a low resistance galvanometer used for measuring the current in a circuit.

It is always connected in series.

Conversion of a Galvanometer into an Ammeter

A galvanometer can be converted into an ammeter by connecting a low resistance into its parallel.

If G is the resistance of a galvanometer and it give full scale deflection for current, I_g then required low resistance S, connected in its parallel for converting it into an ammeter of range I is given by

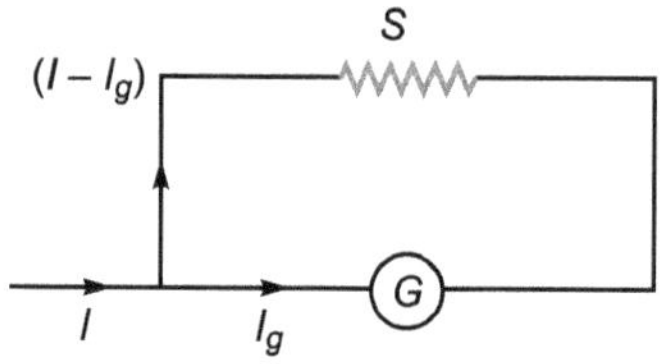

$$I_g \times G = (I - I_g) \times S$$

$$\Rightarrow \qquad S = \left(\frac{I_g}{I - I_g}\right) G$$

The resistance of an ideal ammeter is zero.

Voltmeter

A voltmeter is a high resistance galvanometer used for measuring the potential difference between two points.

It is always connected in parallel.

The resistance of an ideal voltmeter is infinity.

Conversion of a Galvanometer into a Voltmeter

A galvanometer can be converted into a voltmeter by connecting a high resistance into its series.

If a galvanometer of resistance G show full scale deflection for current I_g then required high resistance R, connected in series for converting it into a voltmeter of range V is given by

$$V = I_g (G + R)$$

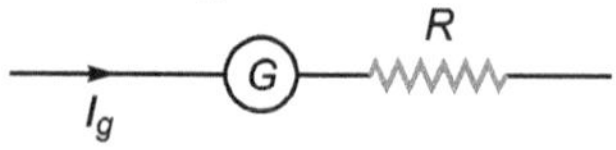

$$\Rightarrow \qquad R = \frac{V}{I_g} - G$$

23

Magnetism

The property of any object by virtue of which it can attract a piece of iron or steel is called **magnetism.**

Natural Magnet

A natural magnet is an ore of iron (Fe_3O_4), which attracts small pieces of iron, cobalt and nickel towards it.

Magnetite or lode stone is a natural magnet.

Artificial Magnet

A magnet which is prepared artificially is called an artificial magnet, e.g. a bar magnet, an electromagnet, a magnetic needle, a horse-shoe magnet etc.

According to molecular theory, every molecular of magnetic substance (whether magnetised or not) is a complete magnet itself.

The poles of a magnet are the two points near but within the ends of the magnet, at which the entire magnetism can be assumed to be concentrated.

The poles always occur in pairs and they are of equal strength. Like poles repel and unlike poles attract.

Properties of Magnet

(i) A freely suspended magnet always aligns itself into north-south direction.

(ii) Like magnetic poles repel and unlike magnetic poles attract each other.

(iii) Magnetic poles exist in pair.

Coulomb's Law

The force of interaction acting between two magnetic poles is directly proportional to the product of their pole strengths and inversely proportional to the square of the distance between them.

$$F = \frac{\mu_0}{4\pi} \cdot \frac{m_1 m_2}{r^2}$$

where $m_1, m_2 =$ pole strengths, $r =$ distance between poles and $\mu_0 =$ permeability of free space.

Magnetic Dipole

Magnetic dipole is an arrangement of two unlike magnetic poles of equal pole strength separated by a very small distance, e.g. a small bar magnet, a magnetic needle, a current carrying loop etc.

Magnetic Dipole Moment

The product of the distance $(2l)$ between the two poles and the pole strength of either pole is called magnetic dipole moment.

Magnetic dipole moment

$$\mathbf{M} = \mathbf{m}(2l)$$

Its SI unit is 'joule/tesla' or 'ampere-metre^2.

Its direction is from south pole towards north pole.

Magnetic Field Due to a Magnetic Dipole

(i) **On Axial Line**

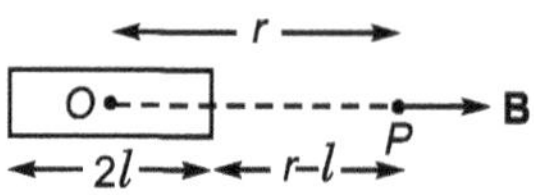

$$B = \frac{\mu_0}{4\pi} \frac{2Mr}{(r^2 - l^2)^2}$$

If $r >> l$, then

$$B = \frac{\mu_0}{4\pi} \frac{2M}{r^3}$$

(ii) **On Equatorial Line**

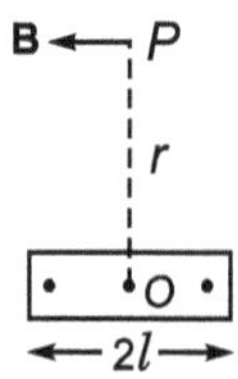

$$B = \frac{\mu_0}{4\pi} \frac{M}{(r^2 + l^2)^{3/2}}$$

If $r >> l$, then

$$B = \frac{\mu_0}{4\pi} \frac{M}{r^3}$$

Torque Acting on a Magnetic Dipole

When a magnetic dipole (**M**) is placed in a uniform magnetic field (**B**), then a torque acts on it, which is given by

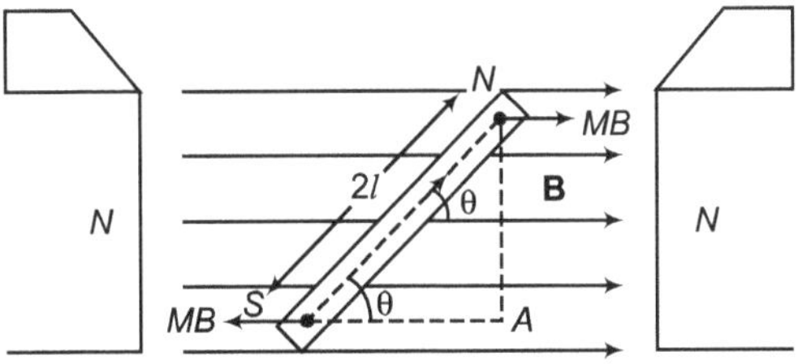

$$\tau = \mathbf{M} \times \mathbf{B}$$

or $$\tau = MB \sin\theta$$

where θ is the angle between the dipole axis and magnetic field.

Potential Energy of a Magnetic Dipole in a Uniform Magnetic Field

The work done in rotating the dipole against the action of the torque is stored as potential energy of the dipole.

Potential energy, $U = W = -MB\cos\theta = -\mathbf{M} \cdot \mathbf{B}$

Current Carrying Loop

A current carrying loop behaves as a magnetic dipole. If we look the upper face of the loop and current is flowing anti-clockwise, then it has a north polarity and if current is flowing clockwise, then it has a south polarity.

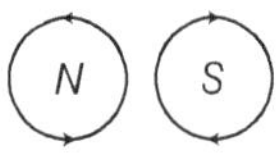

Magnetic dipole moment of a current carrying loop is given by

$$M = IA$$

For N such turns $M = NIA$

where I = current and A = area of cross-section of the coil.

Gauss' Law in Magnetism

Surface integral of magnetic field over any closed or open surface is always zero.

$$\oint_S \mathbf{B} \cdot d\mathbf{S} = 0$$

This law tells that the net magnetic flux through any surface is always zero.

> When in an atom any electron revolve in an orbit it is equivalent to a current loop. Therefore, atom behaves as a magnetic dipole.

Magnetic Moment of an Atom

Magnetic moment of an atom $M = \frac{1}{2} e\omega r^2$

where e = charge on an electron, ω = angular velocity of electron and r = radius of orbit.

or
$$M = n \frac{eh}{4\pi m}$$

where h = Planck's constant and m = mass of an electron

and $\frac{eh}{4\pi m} = \mu_B$, called Bohr magneton and its value is 9.27×10^{-24} A-m^2.

Earth's Magnetism

Earth is a huge magnet. There are three components of earth's magnetism

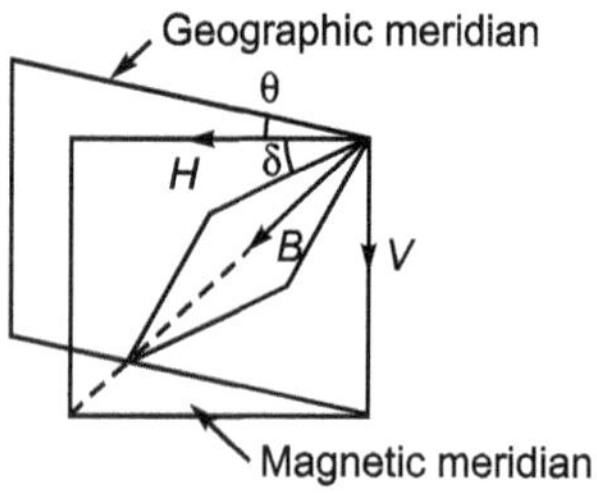

(i) **Magnetic Declination** (θ) The smaller angle subtended between the magnetic meridian and geographic meridian is called magnetic declination.

(ii) **Magnetic Inclination or Magnetic Dip** (δ) The smaller angle subtended between the magnetic axis and horizontal is called magnetic inclination on magnetic dip.

(iii) **Horizontal Component of Earth's Magnetic Field** (H) If B is the intensity of earth's magnetic field then horizontal component of earth's magnetic field $H = B \cos \delta$

It acts from south to north direction.

Vertical component of earth's magnetic field

$$V = B \sin \delta$$

$$\therefore \quad B = \sqrt{H^2 + V^2}$$

and

$$\tan \delta = \frac{V}{H}$$

Angle of dip is zero at magnetic equator and 90° at poles.

Magnetic Meridian

A vertical plane passing through the magnetic axis is called magnetic meridian.

Geographic Meridian

A vertical plane passing through the geographic axis is called geographic meridian.

Magnetic Map

Magnetic map is obtained by drawing lines on the surface of earth, which passes through different places having same magnetic elements.

The main lines drawn on earth's surface are given below

(i) **Isogonic Line** A line joining places of equal declination is called an isogonic line.

(ii) **Agonic Line** A line joining places of zero declination is called an agonic line.

(iii) **Isoclinic Line** A line joining places of equal inclination or dip is called an isoclinic line.

(iv) **Aclinic Line** A line joining places of zero inclination or dip is called an aclinic line.

(v) **Isodynamic Line** A line joining places of equal horizontal component of earth's magnetic field (H) is called an isodynamic line.

Magnetic Latitude

(i) If at any place, the angle of dip is δ and magnetic latitude is λ, then $\tan\delta = 2\tan\lambda$.

(ii) The total intensity of earth's magnetic field

$$I = I_0\sqrt{1 + 3\sin^2\lambda}$$

where $$I_0 = \frac{M}{R^3}$$

It is assumed that a bar magnet of earth has magnetic moment M and radius of earth is R.

> **Magnetic maps** are maps obtained by drawing lines passing through different places on the surface of earth, having the same value of a magnetic element.

Neutral Points

Neutral point of a bar magnet is a point at which the resultant magnetic field of a bar magnet and horizontal component of earth's magnetic field are zero.

When North pole of a bar magnet is placed towards South pole of the earth, then neutral point is obtained on axial line.

$$B = \frac{\mu_0}{4\pi}\frac{2Mr}{(r^2 - l^2)^2} = H$$

If $r >> l$, then $$B = \frac{\mu_0}{4\pi}\frac{2M}{r^3} = H$$

When north pole of a bar magnet is placed towards north pole of the earth, then neutral point is obtained on equatorial line

$$B = \frac{\mu_0}{4\pi}\frac{M}{(r^2 + l^2)^{3/2}} = H$$

If $r >> l$, then $$B = \frac{\mu_0}{4\pi}\frac{M}{r^3} = H$$

Tangent Law

When a bar magnet is freely suspended under the combined effect of two uniform magnetic fields of intensities B and H acting at 90° to each other, then it bar magnet comes to rest making an angle θ with the direction of H, then

$$B = H\tan\theta$$

Deflection Magnetometer

It is a device used to determine M and H. Its working is based on tangent law.

Deflection magnetometer can be used into two settings

(i) **Tangent A setting** In this setting the arms of the magnetometer are along east-west and magnet is parallel to the arms.

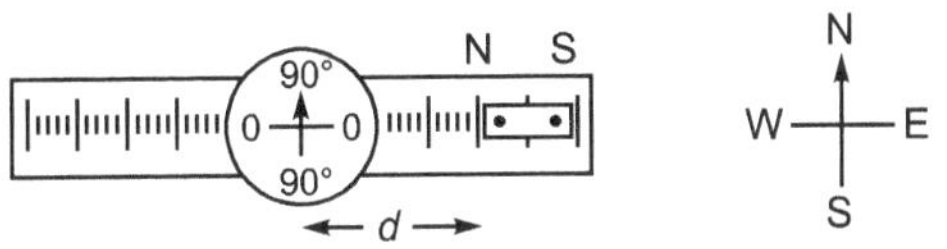

In equilibrium

$$B = H \tan\theta$$

$$\frac{\mu_0}{4\pi}\frac{2M}{d^3} = H \tan\theta$$

(ii) **Tangent B setting** In this setting the arms of the magnetometer are along north-south and magnet is perpendicular to these arms.

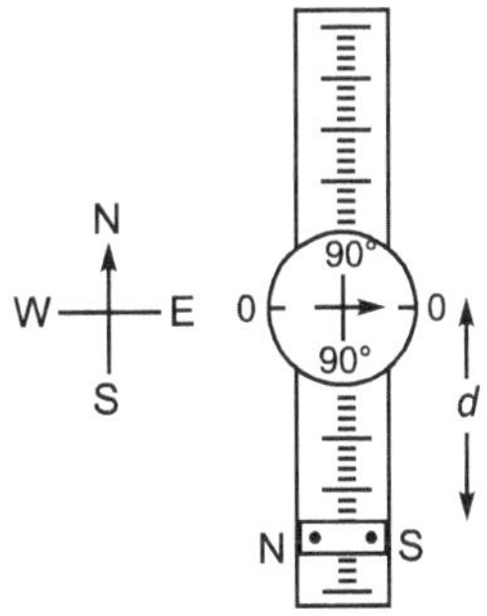

In equilibrium

$$B = H \tan\theta$$

$$\frac{\mu_0}{4\pi}\frac{M}{d^3} = H \tan\theta$$

In above setting the experiment can be performed in two ways

(a) **Deflection method** In this method one magnet is used at a time and deflection in galvanometer is observed.

Ratio of magnetic dipole moments of the magnets

$$\frac{M_1}{M_2} = \frac{\tan\theta_1}{\tan\theta_2}$$

where θ_1 and θ_2 are mean values of deflection for two magnets.

(b) **Null method** In this method both magnets are used at a time and no deflectioin condition is obtained. If magnets are at distances d_1 and d_2, then

$$\frac{M_1}{M_2} = \left(\frac{d_1}{d_2}\right)^3$$

Tangent Galvanometer

It is a device used for detection and measurement of low electric currents.

Its working is based on tangent law.

If θ is the deflection produced in galvanometer when I current flows through it, then

$$I = \frac{2R}{N\mu_0} H \tan\theta$$

$$= \frac{H}{G} \tan\theta = K \tan\theta$$

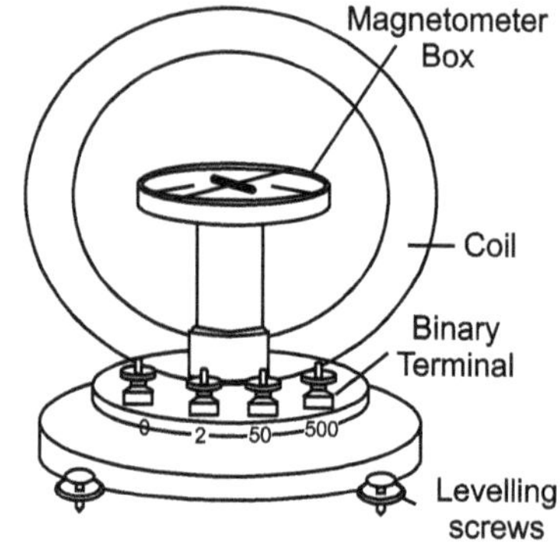

where $G = \frac{N\mu_0}{2R}$ is called **galvanometer constant** and $K = \frac{H}{G}$, is called **reduction factor** of tangent galvanometer.

Here N is number of turns in the coil and R is radius of the coil.

Tangent galvanometer is also called moving magnet type galvanometer.

Vibration Magnetometer

It is based on simple harmonic oscillations of a magnet suspended in uniform magnetic field.

Time period of vibrations is given by

$$T = 2\pi \sqrt{\frac{I}{MH}}$$

where, I = moment of inertia of the magnet,

M = magnetic dipole moment of the magnet and

H = horizontal component of earth's magnetic field.

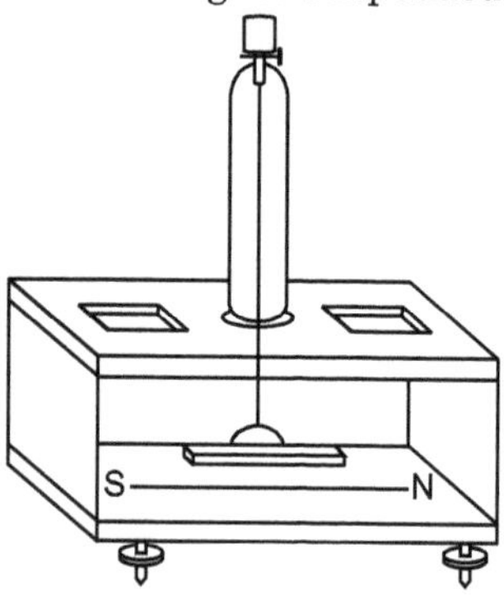

When two magnets of unequal size are placed one above the other and north poles of both magnets are towards geographic north then time period of oscillations is given by :

$$T_1 = 2\pi \sqrt{\frac{I_1 + I_2}{(M_1 + M_2) H}}$$

If north pole of first magnet and south pole of second magnet is towards geographic north, then time period of oscillations is given by :

$$T_2 = 2\pi \sqrt{\frac{(I_1 + I_2)}{(M_1 - M_2) H}}$$

Then,
$$\frac{M_1}{M_2} = \frac{T_2^2 + T_1^2}{T_2^2 - T_1^2}$$

Magnetic Flux

The number of magnetic lines of force passing through any surface is called magnetic flux linked with that surface.

Magnetic flux $(\phi) = \mathbf{B} \cdot \mathbf{A} = BA \cos \theta$

where B is magnetic field intensity or magnetic induction, A is area of the surface.

Its unit is 'weber'.

Magnetic Induction

The magnetic flux passing through per unit normal area, is called magnetic induction.

$$\text{Magnetic induction } (B) = \frac{\phi}{A}$$

Its unit is 'weber/metre^2' or 'tesla'.

Magnetic of Material

To describe the magnetic properties of materials, following terms are required

(i) **Magnetic Permeability** It is the ability of a material to permit the passage of magnetic lines of force through it.

$$\text{Magnetic permeability } (\mu) = \frac{B}{H}$$

where B is magnetic induction and H is magnetising force or magnetic intensity.

(ii) **Magnetising Force or Magnetic Intensity** The degree up to which a magnetic field can magnetise a material is defined in terms of magnetic intensity.

$$\text{Magnetic intensity } (H) = \frac{B}{\mu}$$

(iii) **Intensity of Magnetisation** The magnetic dipole moment developed per unit volume of the material is called intensity of magnetisation.

Intensity of magnetisation $(\mathbf{I}) = \dfrac{\mathbf{M}}{V} = \dfrac{m}{A}$

where V = volume and A = area of cross-section of the specimen.

Magnetic induction $B = \mu_0 (H + I)$

(iv) **Magnetic Susceptibility** (χ_m) The ratio of the intensity of magnetisation (I) induced in the material to the magnetising force (H) applied, is called magnetic susceptibility.

$$\text{Magnetic susceptibility } (\chi_m) = \frac{I}{H}$$

Relation between Magnetic Parmeability and Susceptibility is given by :

$$\mu = \mu_0 (1 + \chi_m)$$

Classification of Magnetic Materials

On the basis of their magnetic properties magnetic materials are divided into three categories

(i) Diamagnetic substances

(ii) Paramagnetic substances

(iii) Ferromagnetic substances

S. No.	Diamagnetic substances	Paramagnetic substances	Ferromagnetic substances
1.	These substances when placed in a magnetic field, acquire feeble magnetism opposite to the direction of the magnetic field. 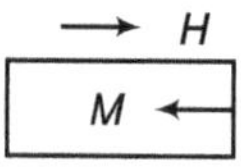	These substances when placed in a magnetic field, acquire feeble magnetism in the direction of the magnetic field. 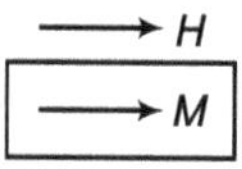	These substances when placed in a magnetic field are strongly magnetised in the direction of the field. 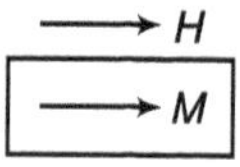
2.	These substances are repelled by a magnet.	These substances are feebly attracted by a magnet.	These substances are strongly attracted by a magnet.

Contd.

S. No.	Diamagnetic substances	Paramagnetic substances	Ferromagnetic substances
3.	When a diamagnetic solution is poured into a U- tube and one arm is placed between the poles of strong magnet, the level of solution in that arm is lowered. 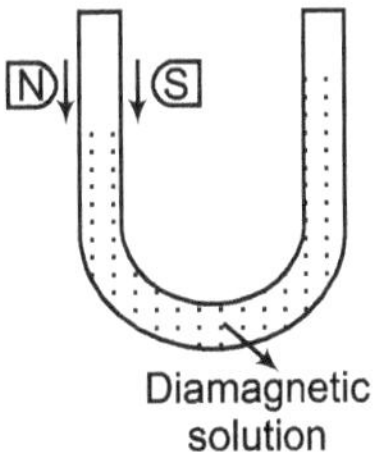	The level of the paramagnetic solution in that arm rises. 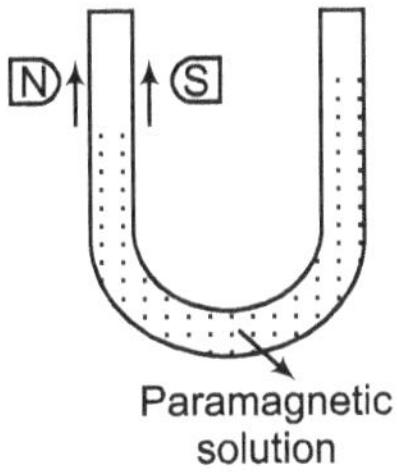	No liquid is ferromagnetic.
4.	If a rod of diamagnetic material is suspended freely between two magnetic poles, its axis becomes perpendicular to the magnetic field. 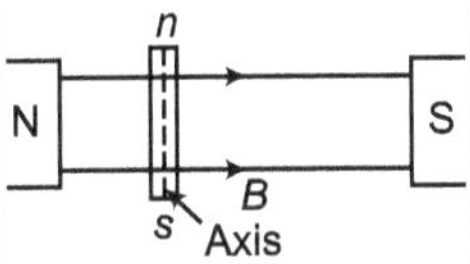	Paramagnetic rod becomes parallel to the magnetic field. 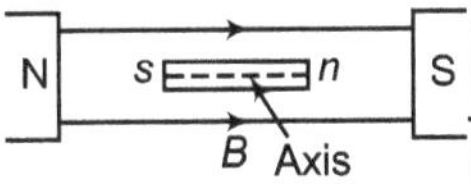	Ferromagnetic rod also becomes parallel to the magnetic field. 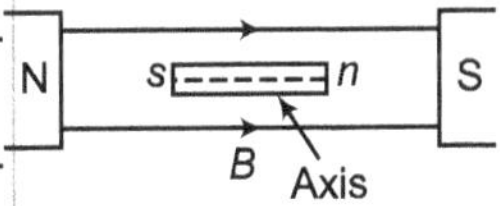
5.	In non-uniform magnetic field, the diamagnetic substances are attracted towards the weaker fields, i.e. they move from stronger to weaker magnetic field.	In non-uniform magnetic field they move from weaker to stronger part of the magnetic field slowly.	In non-uniform magnetic field they move from weaker to stronger magnetic field rapidly.
6.	Their permeability is less than one ($\mu < 1$).	Their permeability is slightly greater than one ($\mu > 1$).	Their permeability is much greater than one ($\mu >> 1$).

Contd.

S. No.	Diamagnetic substances	Paramagnetic substances	Ferromagnetic substances
7.	Their susceptibility is small and negative. Their susceptibility is independent of temperature.	Their susceptibility is small and positive. Their susceptibility is inversely proportional to absolute temperature which is Curie's law, i.e. $\chi \propto 1/T$	Their susceptibility is large and postitve. They also follow Curie law. i.e. $(\chi \propto 1/T)$. At Curie temperature ferromagnetic substances change into paramagnetic substances.
8.	Shape of diamagnetic liquid in a glass crucible and kept over two magnetic poles. 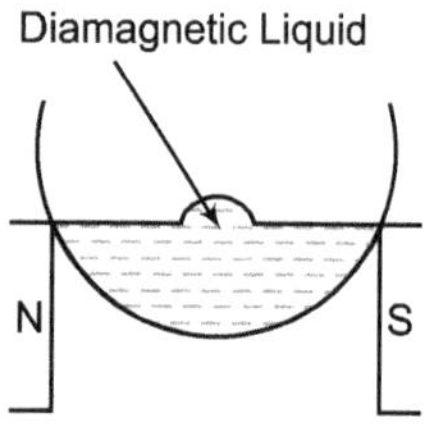	Shape of paramagnetic liquid in a glass crucible and kept over two magnetic poles. 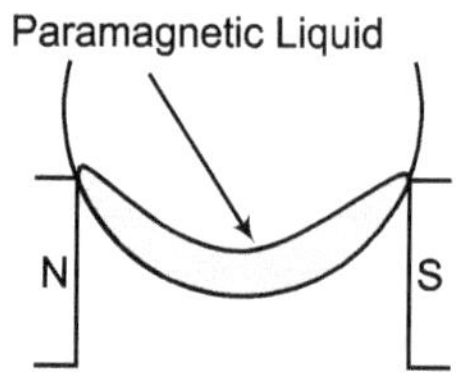	No liquid is ferromagnetic.
9.	In these substances the magnetic lines of force are farther than in air.	In these substances the magnetic lines of force are closer than in air.	In these substances magnetic lines of force are much closer than in air.
10.	The resultant magnetic moment of these substances is zero.	These substances have a permanent magnetic moment.	These substances also have a permanent magnetic moment.

The atoms of a paramagnetic substance contain even number of electrons.

The atoms or molecules of a paramagnetic substance do not possess any net magnetic moment.

The atoms of a paramagnetic material contain odd number of electrons.

Every atom or molecule of a paramagnetic substance has its own magnet moment, i.e. its each atom or molecule is a tiny magnet.

In a ferromagnetic substance, there are several tiny regions called **domains**. Each domain contain approximately 10^{10} atoms.

Each domain is a strong magnet as all atoms or molecules in a domain have same direction of magnetic moment.

Curie Law in Magnetism

The magnetic susceptibility of a paramagnetic substance is inversely proportional to its absolute temperature.

$$\chi_m \propto \frac{1}{T}$$

$\Rightarrow \quad \chi_m T = \text{constant}$

where χ_m = magnetic susceptibility of a para magnetic substance and T = absolute temperature.

Hysteresis

The lagging of intensity of magnetisation (I) or magnetic induction (B) behind magnetising field (H), when a specimen of a magnetic substance is taken through a complete cycle of magnetisation is called hysteresis.

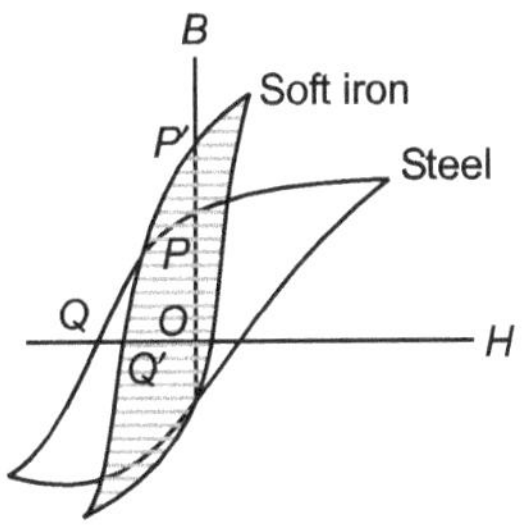

Retentivity or Residual Magnetism

The value of the intensity of magnetisation of a material, when the magnetising field is reduced to zero is called retentivity or residual magnetism of the material.

Coercivity

The value of the reverse magnetising field that should be applied to a given sample in order to reduce its intensity of magnetisation or magnetic induction to zero is called coercivity.

Permanent Magnets

Commonly steel is used to make a permanent magnet because steel has high residual magnetism and high coercivity.

Electromagnets

Electromagnets are made of soft iron because area of hysteresis loop for soft iron is small. Therefore, energy loss is small for a cycle of magnetisation and demagnetisation.

Permanent magnets are made by the materials such as steel, for which residual magnetism as well as coercivity should be high. Electromagnets are made by the materials such as soft iron for which residual magnetism is high, coercivity is low and hysteresis loss is low.

Important Points

- Magnetic length $= \frac{5}{6} \times$ geometric length of magnet.
- About 90% of magnetic moment is due to spin motion of electrons and remaining 10% of magnetic moment is due to the orbital motion of electrons.
- When a magnet having magnetic moment M is cut into two equal parts.

(i) Parallel to its length

$$M' = \frac{m}{2} \times l = \frac{M}{2}$$

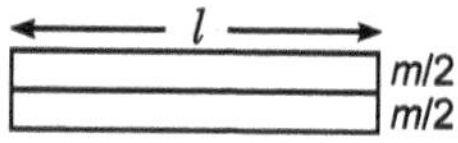

(ii) Perpendicular to its length

$$M' = m \times \frac{l}{2} = \frac{M}{2}$$

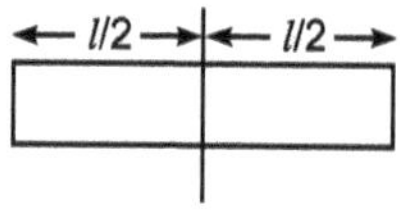

- When a magnet of length l, pole strength m and of magnetic moment M is turned into a semicircular arc. Then its new magnetic moment

$$M' = m \times 2R = m \times 2 \times \frac{l}{\pi} \quad (\pi R = l)$$

$$= \frac{2M}{\pi} \qquad (\because M = m \times l)$$

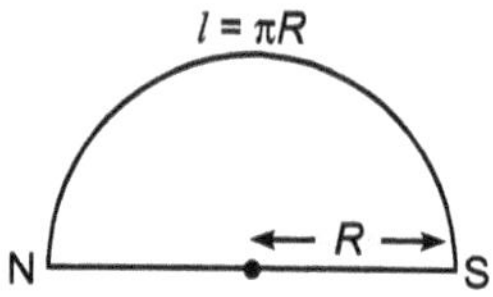

- A thin magnet of moment M is turned into an arc of 90°. Then new magnetic moment

$$M' = \frac{2\sqrt{2}M}{\pi}$$

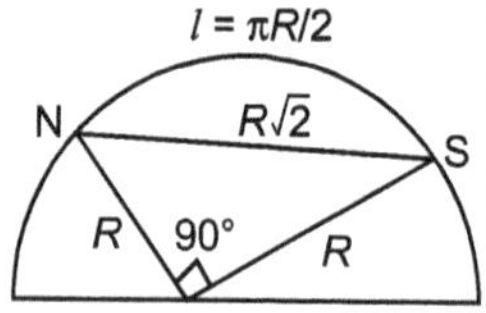

- A thin magnet of moment M is turned at mid point 90°. Then new magnet moment

$$M' = \frac{M}{\sqrt{2}}$$

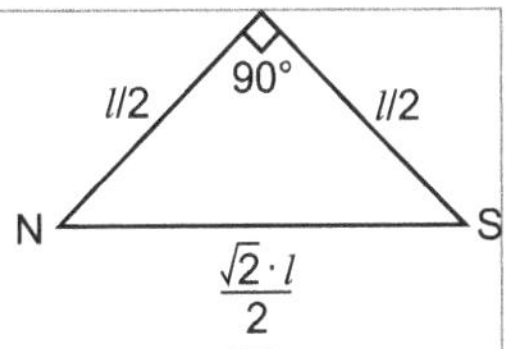

- A thin magnet of moment M is turned into an arc of 60°. Then new magnetic moment

$$M' = \frac{3M}{\pi}$$

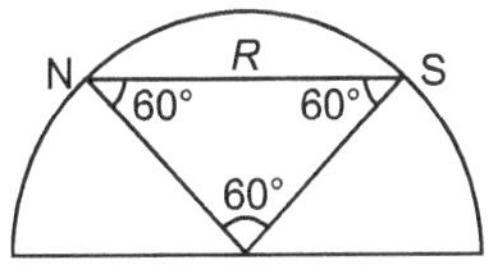

- A thin magnet of moment M is bent at mid point at angle 60°. Then new magnetic moment.

$$M' = \frac{M}{2}$$

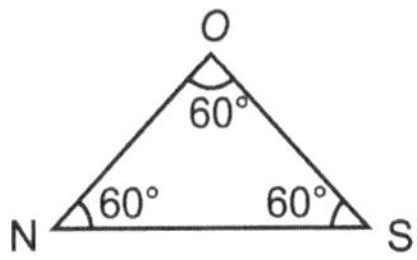

- Original magnet MOS is bent at O, the mid point at 60°. All sides are equal.
- The mutual interaction force between two small magnets of moments M_1 and M_2 is given by

$$F = K \frac{6M_1M_2}{d^4} \text{ in end-on position.}$$

Here, d denotes the separation between magnets.

- Magnetic length $= \frac{5}{6} \times$ geometric length of magnet.
- Cause of diamagnetism is orbital motion and cause of paramagnetism is spin motion of electrons. Cause of ferromagnetism lies in formation of domains.
- The perpendicular bisector of magnetic axis is known as neutral axis of magnet. Magnetism at neutral axis is zero and at poles is maximum.
- For steel coercivity is large. However, rententivity is comparatively smaller in case of steel. So, steel is used to make permanent magnets.
- For soft iron, coercivity is very small and area of hysteresis loop is small. So, soft iron is an ideal material for making electromagnets.

24

Electromagnetic Induction

Whenever the magnetic flux linked with an electric circuit changes, an emf is induced in the circuit. This phenomenon is called **electromagnetic induction.**

Faraday's Laws of Electromagnetic Induction

(i) Whenever the magnetic flux linked with a circuit changes, an induced emf is produced in it.

(ii) The induced emf lasts so long as the change in magnetic flux continues.

(iii) The magnitude of induced emf is directly proportional to the rate of change in magnetic flux, i.e.

$$E \propto \frac{d\phi}{dt} \quad \Rightarrow \quad E = -\frac{d\phi}{dt}$$

where constant of proportionality is one and negative sign indicates Lenz's law.

Here, magnetic flux $(\phi) = NBA \cos\theta$, SI unit of $\phi =$ weber, tesla-m^2

CGS unit of $\phi =$ maxwell, 1 weber $= 10^8$ maxwell,

Dimensional formula of magnetic flux

$$[\phi] = [ML^2 T^{-2} A^{-1}].$$

Lenz's Law

The direction of induced emf or induced current is always in such a way that it opposes the cause due to which it is produced.

Lenz's law is in accordance with the conservation of energy.

Note *To apply Lenz's law, you can remember RIN or ⊗ In (when the loop lies on the plane of paper).*

(i) **RIN** In RIN, R stands for right, I stands for increasing and N for north pole (anticlockwise). It means, if a loop is placed on the right side of a straight current-carrying conductor and the current in the conductor is increasing, then induced current in the loop is anticlockwise ().

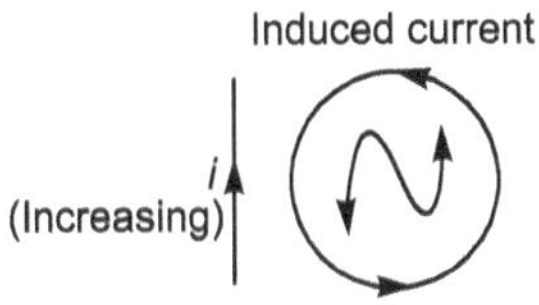

(ii) ⊗ **IN** In ⊗ IN suppose the magnetic field in the loop is perpendicular to paper inwards ⊗ and this field is increasing, then induced current in the loop is anticlockwise ().

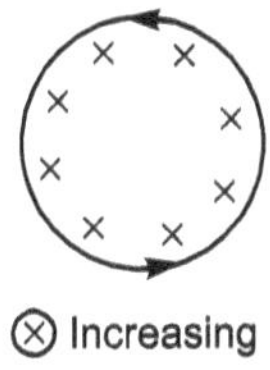

Motional Emf

If a rod of length l moves perpendicular to a magnetic field B, with a velocity v, then induced emf produced in it given by

$$E = \mathbf{B} \cdot \mathbf{v} \times \mathbf{l} = Bvl$$

If a metallic rod of length l rotates about one of its ends in a plane perpendicular to the magnetic field, then the induced emf produced across its ends is given by

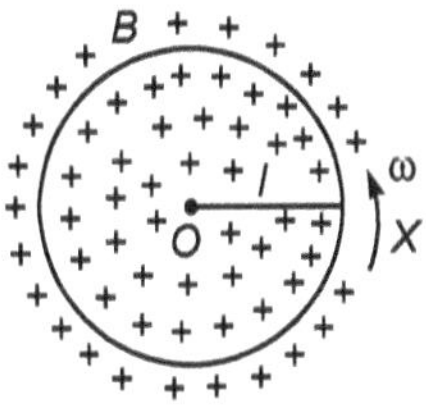

$$E = \frac{1}{2} B\omega l^2 = BAf$$

where, ω = angular frequency of rotation, $A = \pi l^2$ = area of circle and f = frequency of rotation.

If a metallic disc of radius r rotates about its own centre in a plane perpendicular to the magnetic field B, then the induced emf produced between the centre and the edge is given by

$$E = \frac{1}{2} B\omega r^2 = BAf$$

where, ω = angular velocity of rotation, f = frequency of rotation and $A = \pi r^2$ = area of disc.

The direction of induced current in any conductor can be obtained from Fleming's right hand rule.

A rectangular coil moves linearly in a field when coil moves with constant velocity in a uniform magnetic field, flux and induced emf will be zero.

A rod moves at an angle θ with the direction of magnetic field, with velocity $E = -Blv \sin \theta$.

An emf is induced

(i) When a magnet is moved with respect to a coil.

(ii) When a conductor falls freely in East-West direction.

(iii) When an aeroplane flies horizontally.

(iv) When strength of current flowing in a coil is increased or decreased, induced current is developed in the coil in same or opposite direction.

(v) When a train moves horizontally in any direction.

Energy Consideration

If a rod of length l moves perpendicular to a magnetic field B, with a velocity v, then power required to do this, is

$$P = Fv \text{ or } P = \frac{B^2 l^2 v^2}{r}$$

The agent that does this work is mechanical. This mechanical energy is dissipated as joule heat and is given by

$$P_J = \frac{B^2 l^2 v^2}{r}$$

Thus, mechanical energy, which was required to move the rod, is converted into electrical energy and then to thermal energy.

Fleming's Right Hand Rule

If we stretch the thumb, the forefinger and the central finger of right hand in such a way that all three are perpendicular to each other, then if thumb represent the direction of motion, the forefinger represent the direction of magnetic field, then central finger will represent the direction of induced current.

If R is the electrical resistance of the circuit, then induced current in the circuit is given by $I = \frac{E}{R}$.

If induced current is produced in a coil rotated in uniform magnetic field, then

$$I = \frac{NBA\,\omega \sin \omega t}{R} = I_0 \sin \omega t$$

where, $I_0 = NBA\omega$ = peak value of induced current,

N = number of turns in the coil ,

B = magnetic induction,

ω = angular velocity of rotation and

A = area of cross-section of the coil.

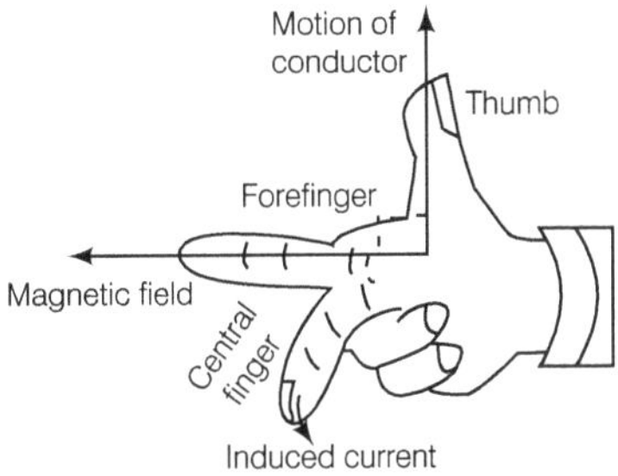

Eddy Currents

If a piece of metal is placed in a varying magnetic field or rotated with high speed in a uniform magnetic field, then induced current set up in the piece are like whirepool of air, called eddy currents.

The magnitude of eddy currents is given by $i = -\frac{e}{R} = \frac{d\phi / dt}{R}$, where R is the resistance.

Eddy currents are also known as Facault's current.

Self-Induction

The phenomena of production of induced emf in a circuit due to change in current flowing in its own, is called self-induction.

Coefficient of Self-Induction

The magnetic flux linked with a coil.

$$\phi = LI$$

where, L = coefficient of self-induction.

The induced emf in the coil,

$$E = -L\frac{dI}{dt}$$

It unit of self-induction is henry (H) and its dimensional formula is $[ML^2T^{-2}A^{-2}]$.

Self-inductance of a long solenoid is given by normal text

$$L = \frac{\mu_0 N^2 A}{l} = \mu_0 n^2 A l$$

where, N = total number of turns in the solenoid,

l = length of the coil, n = number of turns in the coil and

A = area of cross-section of the coil.

If core of the solenoid is of any other magnetic material, then

$$L = \frac{\mu_0 \mu_r N^2 A}{l}$$

Self-inductance of a toroid, $L = \dfrac{\mu_0 N^2 A}{2\pi r}$

where, r = radius of the toroid.

Energy stored in an inductor, $E = \dfrac{1}{2} L I^2$

Mutual Induction

The phenomena of production of induced emf in a circuit due to the change in magnetic flux in its neighbouring circuit, is called mutual induction.

Coefficient of Mutual Induction

If two coils are coupled with each, other then magnetic flux linked with a coil (secondary coil)

$$\phi = MI$$

where, M is coefficient of mutual induction and I is current flowing through primary coil.

The induced emf in the secondary coil,

$$E = -M \frac{dI}{dt}$$

where, $\dfrac{dI}{dt}$ is the rate of change of current through primary coil.

The unit of coefficient of mutual induction is henry (H) and its dimension is $[ML^2T^{-2}A^{-2}]$.

The coefficient of mutual induction depends on geometry of two coils, distance between them and orientation of the two coils.

Coefficient of Coupling

Two coils are said to be coupled if full a part of the fuse produced by one links with the other.

$K = \sqrt{\frac{M}{L_1 L_2}}$, where L_1 and L_2 are coefficients of self-induction of the two coils and M is coefficient of mutual induction of the two coils.

Coefficient of coupling is maximum ($K = 1$) in case (a), when coils are coaxial and minimum in case (b), when coils are placed a right angles.

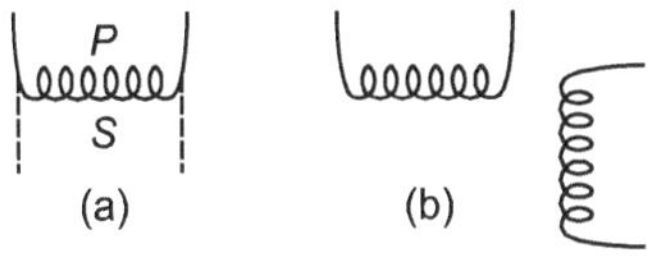

Mutual inductance of two long coaxial solenoids is given by

$$M = \frac{\mu_0 N_1 N_2 A}{l}$$

$$= \mu_0 n_1 n_2 A l$$

where, N_1 and N_2 are total number of turns in both coils, $n_1 n_2$ are number of turns per unit length in coils, A is area of cross-section of coils and l is length of the coils.

Grouping of Coils

(a) When three coils of inductances L_1, L_2 and L_3 are connected in series and the coefficient of coupling $K = 0$, as in series, then $L = L_1 + L_2 + L_3$

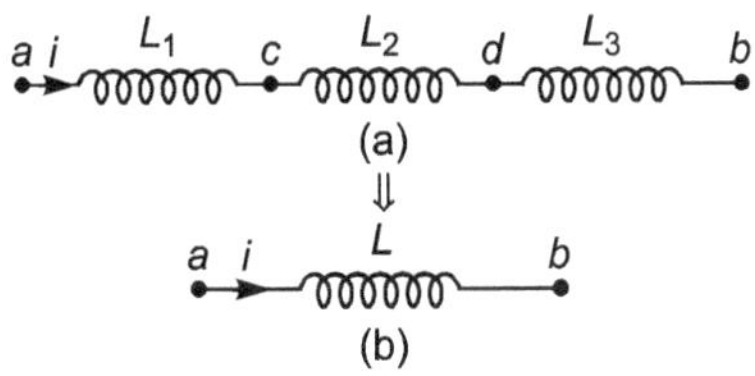

(b) When three coils of inductances L_1, L_2 and L_3 are connected in parallel and the coefficient of coupling $K = 0$ as in parallel, then

$$\frac{1}{L} = \frac{1}{L_1} + \frac{1}{L_2} + \frac{1}{L_3}$$

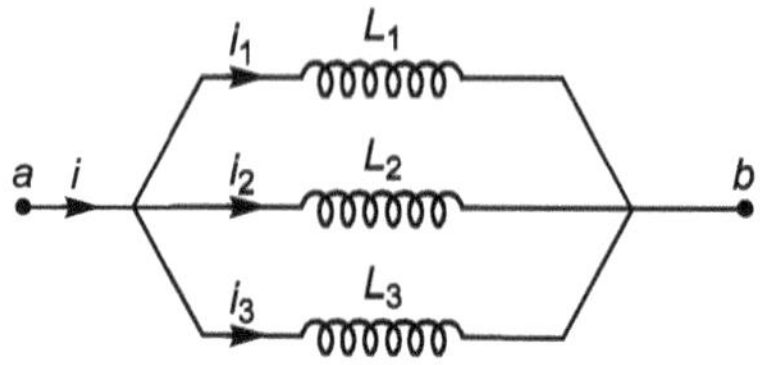

(a)

⇓

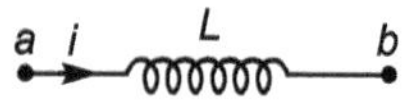

(b)

If coefficient of coupling $K = 1$, then

(i) **In series**

(a) If current in two coils are in the same direction, then

$$L = L_1 + L_2 + 2M$$

(b) If current in two coils are in opposite directions, then

$$L = L_1 + L_2 - 2M$$

(ii) **In parallel**

(a) If current in two coils are in same direction, then

$$L = \frac{L_1 L_2 - M^2}{L_1 + L_2 + 2M}$$

(b) If current in two coils are in opposite directions, then

$$L = \frac{L_1 L_2 - M^2}{L_1 + L_2 - 2M}$$

25 Alternating Currents

Transient Current

An electric current which vary for a small finite time, while growing from zero to maximum or decaying from maximum to zero is called a transient current.

Growth of Current in an Inductor

Growth of current in an inductor at any instant of time t is given by

$$I = I_0 (1 - e^{-Rt/L})$$

where, I_0 = maximum current, L = self-inductance of the inductor and R = resistance of the circuit.

Here $\frac{R}{L} = \tau$, is called time constant of a L-R circuit.

Time constant of a L-R circuit is the time in which current in the circuit grows to 63.2% of the maximum value of current.

Decay of current in an inductor at any time t is given by

$$I = I_0 \, e^{-Rt/L}$$

Time constant of a L-R circuit is the time in which current decays to 36.8% of the maximum value of current.

Charging and Discharging of a Capacitor

The instantaneous charge on a capacitor on charging at any instant of time t is given by

$$q = q_0 \, [1 - e^{-t/RC}]$$

where $RC = \tau$, is called time constant of a R-C circuit.

The instantaneous charge on a capacitor in discharging at any instant of time t is given by $q = q_0 e^{-t/RC}$.

Time constant of a R-C circuit is the time in which charge in the capacitor grows to 63.8% or decay to 36.8% of the maximum charge on capacitor.

Alternating Current

An electric current whose magnitude changes continuously with time and changes its direction periodically, is called an alternating current.

The instantaneous value of alternating current at any instant of time t is given by

$$I = I_0 \sin \omega t$$

where, I_0 = peak value of alternating current.

The variation of alternating current with time is shown in graph given below

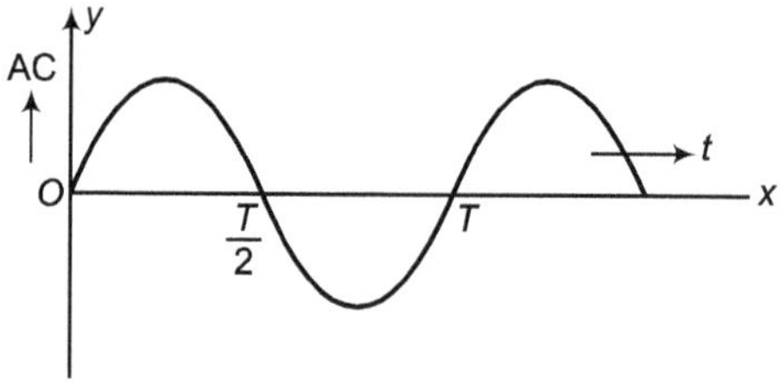

Mean or average value of alternating current for first half cycle

$$I_m = \frac{2I_0}{\pi} = 0.637\, I_0$$

Mean or average value of alternating current for next half cycle

$$I'_m = -\frac{2I_0}{\pi} = -\,0.637\, I_0$$

Mean or average value of alternating current for one complete cycle = 0.

Root mean square value of alternating current

$$I_V = I_{\text{rms}} = \frac{I_0}{\sqrt{2}} = 0.707\, I_0$$

where, I_0 = peak value of alternating current.

Root mean square value of alternating voltage

$$V_{\text{rms}} = \frac{V_0}{\sqrt{2}} = 0.707\, V_0$$

Reactance

The opposition offered by an inductor or by a capacitor in the path of flow of alternating current is called reactance.

Reactance is of two types

(i) **Inductive Reactance** (X_L) Inductive reactance is the resistance offered by an inductor.

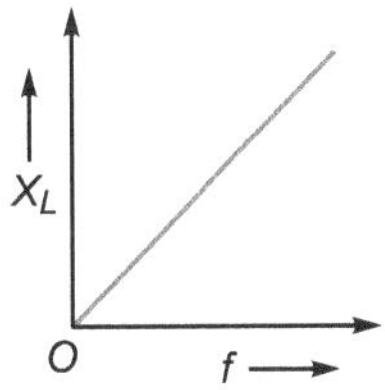

Inductive reactance $(X_L) = L\omega = L2\pi f = \frac{L2\pi}{T}$

Its unit is ohm. $X_L \propto f$

For direct current, $X_L = 0$ $(\because f = 0)$

(ii) **Capacitive Reactance** (X_C) Capacitive reactance is the resistance offered by an inductor.

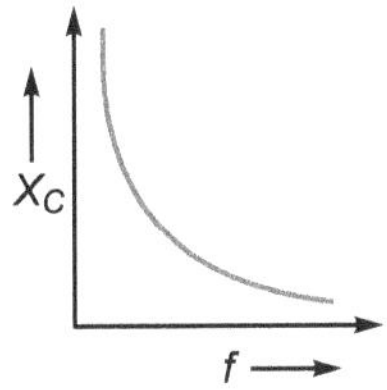

Capacitive reactance,

$$X_C = \frac{1}{C\omega} = \frac{1}{C2\pi f} = \frac{T}{C\,2\pi}$$

Its unit is ohm $X_C \propto \frac{1}{f}$

For direct current, $X_C = \infty$ $(\because f = 0)$

Impedance

The opposition offered by an AC circuit containing more than one out of three components L, C and R, is called impedance (Z) of the circuit.

Impedance of an AC circuit, $Z = \sqrt{R^2 + (X_L - X_C)^2}$

Its SI unit is ohm.

Power in an AC Circuit

The power is defined as the rate at which work is being in the circuit.

The average power in an AC circuit,

$$P_{av} = V_{rms}\, i_{rms} \cos\theta$$
$$= \frac{V}{\sqrt{2}} \frac{i}{\sqrt{2}} \cos\theta = \frac{Vi}{2} \cos\theta$$

where, $\cos\theta = \frac{\text{Resistance}(R)}{\text{Impedance}(Z)}$ is called the power factor of AC circuit.

Current and Potential Relations

Here, we will discuss current and potential relations for different AC circuits.

(i) **Pure Resistive Circuit (R Circuit)**

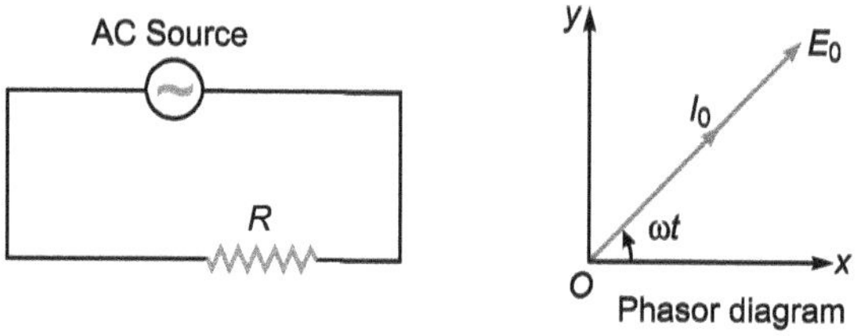

(a) Alternating emf, $E = E_0 \sin \omega t$

(b) Alternating current, $I = I_0 \sin \omega t$

(c) Alternating emf and alternating current both are in the same phase.

(d) Average power decay, $(\overline{P}) = E_V \cdot I_V$

(e) Power factor, $\cos \theta = 1$

(ii) **Pure Inductive Circuit (L Circuit)**

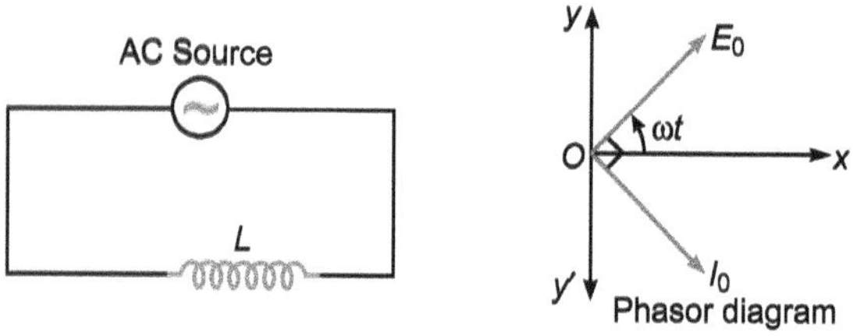

(a) Alternating emf, $E = E_0 \sin \omega t$

(b) Alternating current, $I = I_0 \sin (\omega t - \pi/2)$

(c) Alternating current lags behind alternating emf by $\frac{\pi}{2}$.

(d) Inductive reactance, $X_L = L\omega = L 2\pi f$

(e) Average power decay, $(\overline{P}) = 0$

(f) Power factor, $\cos \theta = \cos 90° = 0$

(iii) **Pure Capacitive Circuit**

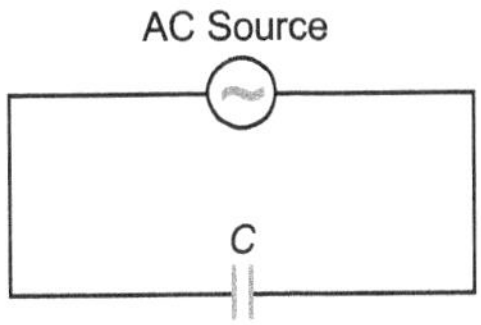

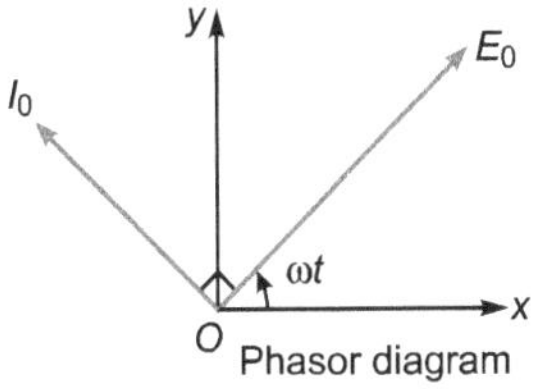

Phasor diagram

(a) Alternating emf, $E = E_0 \sin \omega t$

(b) Alternating current, $I = I_0 \sin(\omega t + \pi/2)$

(c) Alternating current leads the alternating emf by $\frac{\pi}{2}$.

(d) Capacitive reactance, $X_C = C\omega = C2\pi f$

(e) Avearge power decay, $(\overline{P}) = 0$

(f) Power factor, $\cos\theta = \cos 90° = 0$

(iv) ***R-C* Circuit**

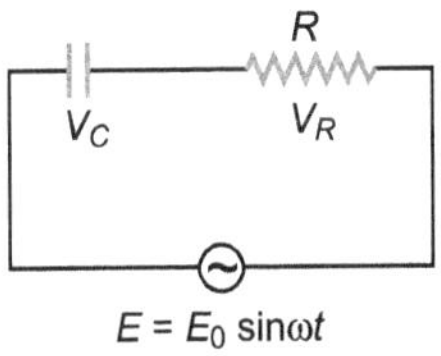

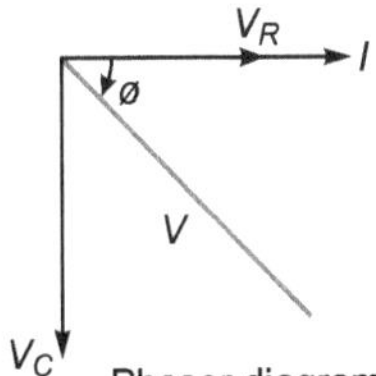

Phasor diagram

$$E = E_0 \sin \omega t$$

$$I = \frac{E_0}{2} \sin(\omega t - \phi)$$

$$Z = \sqrt{R^2 + \left(\frac{1}{\omega C}\right)^2}$$

$$\tan\phi = \frac{-\frac{1}{\omega C}}{R}$$

Current leading the voltage by ϕ

$$V^2 = V_R^2 + V_C^2$$

(v) ***L-C* Circuit**

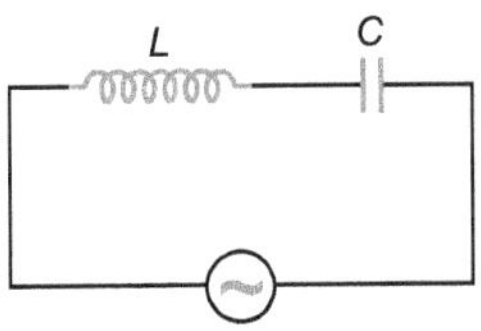

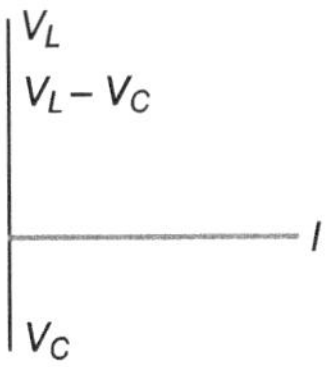

Phasor diagram

$$E = E_0 \sin \omega t, \; I = \frac{E}{2} \sin(\omega t - \phi)$$

$$Z = X_L - X_C \text{ and } \tan \phi = \frac{X_L - X_C}{0}$$

- For $X_L > X_C, \phi = \frac{\pi}{2}$ and for $X_L < X_C, \phi = -\frac{\pi}{2}$
- If $X_L = X_C$ at $\omega = \frac{1}{\sqrt{LC}}, Z = 0$

(vi) ***L-C-R* Circuit**

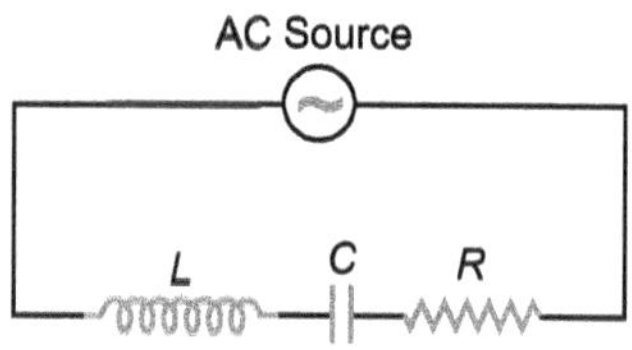

(a) Alternating emf, $E = E_0 \sin \omega t$

(b) Alternating current, $I = I_0 \sin(\omega t \pm \theta)$

(c) Alternating current lags leads behind alternating emf by θ.

(d) Resultant voltage, $V = \sqrt{V_R^2 + (V_L - V_C)^2}$

(e) Impedance, $Z = \sqrt{R^2 + (X_L - X_C)^2}$

(f) Power factor, $\cos \theta = \frac{R}{Z} = \frac{R}{\sqrt{R^2 + (X_L - X_C)^2}}$

(g) Average power decay, $(\overline{P}) = E_V I_V \cos \theta$

Resonance in AC Circuit

The condition in which current is maximum or impedance is minimum in an AC circuit, is called resonance.

(i) **Series Resonance Circuit**

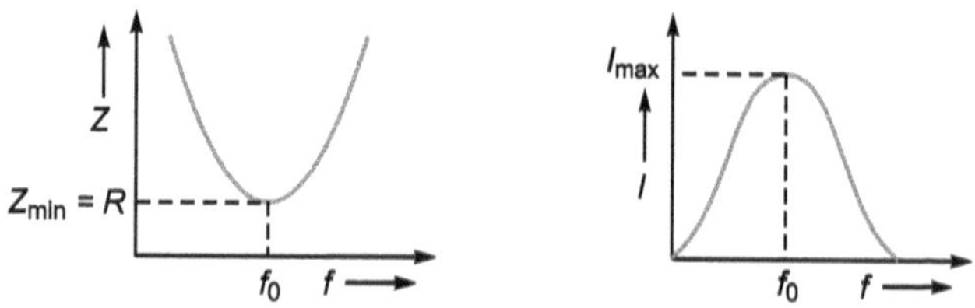

In this circuit components L, C and R are connected in series.

At resonance, $X_L = X_C$

Resonance frequency, $f = \dfrac{1}{2\pi\sqrt{LC}}$

At resonance impedance is minimum and equal to the resistance, i.e. $Z = R$.

If resonance maximum current flows through the circuit.

Q-factor or sharpness at resonance

$$Q = \frac{1}{R}\sqrt{\frac{L}{C}}$$

A series resonance circuit is also known as acception circuit.

(ii) **Parallel Resonance Circuit**

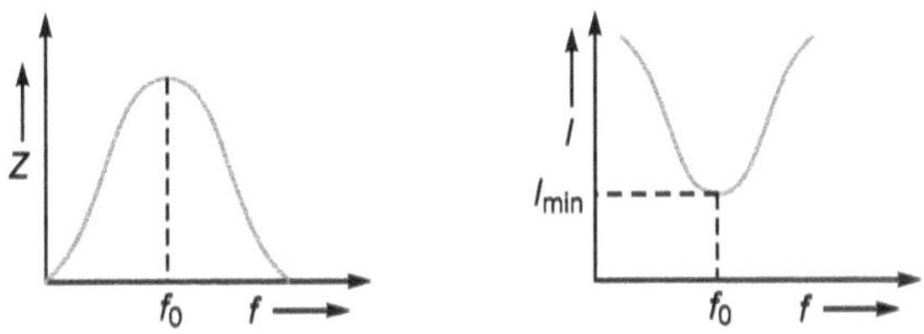

In this circuit L and C are connected in parallel with each other.

At resonance, $X_L = X_C$

Impedance (Z) of the circuit is maximum.

Current in the circuit is minimum.

Wattless Current

Average power is given by

$$P_{av} = E_{rms}\, I_{rms} \cos\phi$$

Here, the $I_{rms} \cos\phi$ contributes for power dissipation. Therefore, it is called wattless current.

AC Generator or Dynamo

It is a device which converts mechanical energy into alternating current energy.

Its working is based on electromagnetic induction.

The induced emf produced by the AC generator is given by

$$e = NBA\omega \sin \omega t = e_0 \sin \omega t$$

There are four main parts of an AC generator

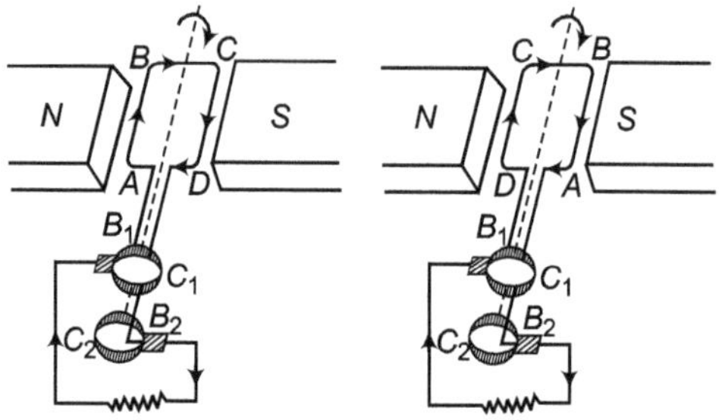

Working of AC dynamo

(i) **Armature** It is rectangular coil of insulated copper wire having a large number of turns.

(ii) **Field Magnets** These are two pole pieces of a strong electromagnet.

(iii) **Slip Rings** These are two hollow metallic rings.

(iv) **Brushes** These are two flexible metals or carbon rods, which remains slightly in contact with slip rings.

Note *An DC generator or dynamo contains split rings or commutator inspite of slip rings.*

DC Motor

It is a device which converts electrical energy into mechanical energy. Its working is based on the fact that when a current carrying coil is placed in uniform magnetic field a torque acts on it.

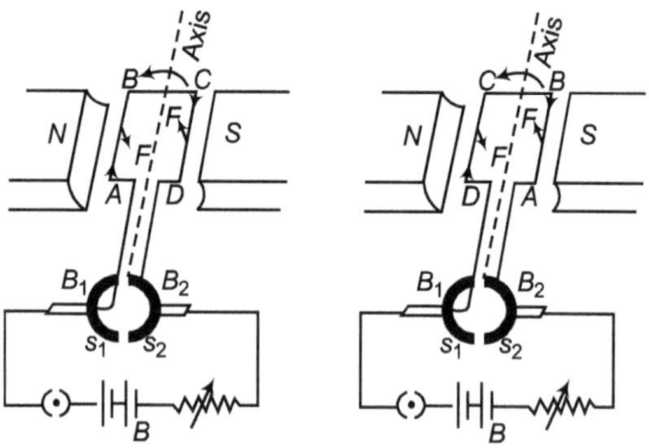

Torque acting on a current carrying coil placed in uniform magnetic field

$$\tau = NBIA \sin\theta$$

When armature coil rotates a back emf is produced in the coil.

Efficiency of a motor,

$$\eta = \frac{\text{Back emf}}{\text{Applied emf}} = \frac{E}{V}$$

Transformer

It is a device which can change a low voltage of high current into a high voltage of low current and *vice-versa.*

Its working is based on mutual induction.

There are two types of transformers.

(i) **Step-up Transformers** It converts a low voltage of high current into a high voltage of low current.

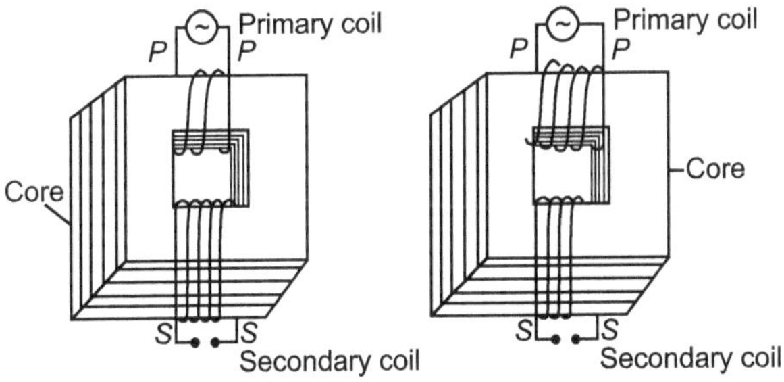

(b) Step-up transformer (b) Step-down transformer

In this transformer,

$$N_S > N_P, E_S > E_P$$

and $I_P > I_S$

(ii) **Step-down Transformer** It converts a high voltage of low current into a low voltage of high current.

In this transformer,

$$N_P > N_S, E_P > E_S \text{ and } I_P < I_S$$

Transformation Ratio

Transformation ratio,

$$K = \frac{N_S}{N_P} = \frac{E_S}{E_P} = \frac{I_P}{I_S}$$

For step-up transformer, $K > 1$

For step-down transformer, $K < 1$

Energy Losses in a Transformer

The main energy losses in a transformer are given below

(i) Iron loss

(ii) Copper loss

(iii) Flux loss

(iv) Hysteresis loss

(v) Humming loss

Important Points

- Transformer does not operate on direct current. It operates only on alternating voltages at input as well as at output.
- Transformer does not amplify power as vacuum tube.
- Transformer, a device based on mutual induction converts magnetic energy into electrical energy.
- Efficiency, $\eta = \dfrac{\text{Output power}}{\text{Input power}}$

  Generally efficiency ranges from 70% to 90%.
- A choke coil is a pure inductor. Average power consumed per cycle is zero in a choke coil.
- A DC motor connects DC energy from a battery into mechanical energy of rotation.
- An AC dynamo/generator produces are energy from mechanical energy of rotation of a coil.
- An induction coil generates high voltages of the order of 10^5 V from a battery. It is based on the phenomenon of mutual induction.

26

Ray Optics

Light

Light is a form of energy which produces the sensation of sight on our eyes.

Sources of light are of three types-thermal sources, gas discharge sources and luminescent sources.

Photometry is a branch of ray optics which deals with the measurement of light energy.

Characteristics of Light

Light waves are electromagnetic waves, whose nature is transverse.

The speed of light in vacuum is 3×10^8 m/s but it is different in different media.

The speed and wavelength of light change when it travels from one medium to another but its frequency remains unchanged.

Important Terms

(i) **Luminous Objects** The objects which emits its own light, are called luminous objects, e.g. sun, other stars, an oil lamp etc.

(ii) **Non-Luminous Objects** The objects which do not emit its own light but become visible due to the reflection of light falling on them, are called non-luminous objects, e.g. moon, table, chair, trees etc.

(iii) **Ray of Light** A straight line drawn in the direction of propagation of light is called a ray of light.

(iv) **Beam of Light** A bundle of the adjacent light rays is called a beam of light.

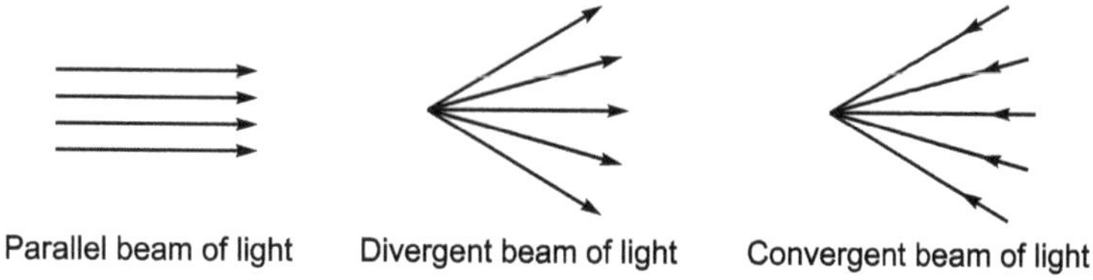

(v) **Image** If light ray coming from an object meets or appear to meet at a point after reflection or refraction, then this point is called image of the object.

(vi) **Real Image** The image obtained by the real meeting of light rays, is called a real image.

Real image can be obtained on a screen.

Real image is inverted.

(vii) **Virtual Image** The image obtained when light rays are not really meeting but appears to meet only, is called a virtual image.

Reflection of Light

The rebouncing back of light rays into the same medium on striking a highly polished surface such as a mirror, is called reflection of light.

Laws of Reflection

There are two laws of reflection.

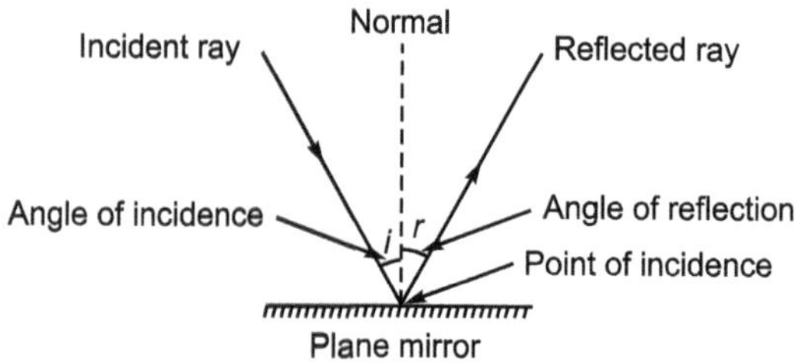

(i) The incident ray, the reflected ray and the normal at the point of incidence all three lie in the same plane.

(ii) The angle of incidence (i) is always equal to the angle of reflection (r).

Types of Reflection

(i) **Regular Reflection** When a parallel beam of reflected light rays is obtained for a parallel beam of incident light rays after reflection from a plane reflecting surface, then such type of reflection is called regular reflection.

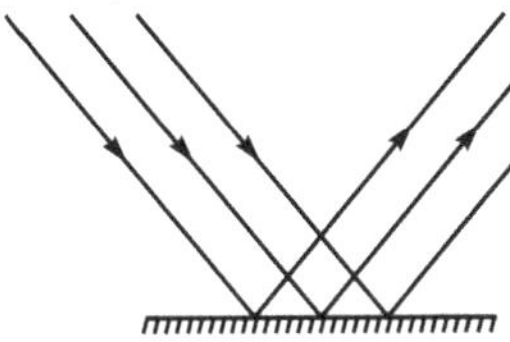

(ii) **Irregular or Diffused Reflection** When a non-parallel beam of reflected light rays is obtained for a parallel beam of incident light rays after reflection from a surface, then such type of reflection is called irregular or diffused reflection.

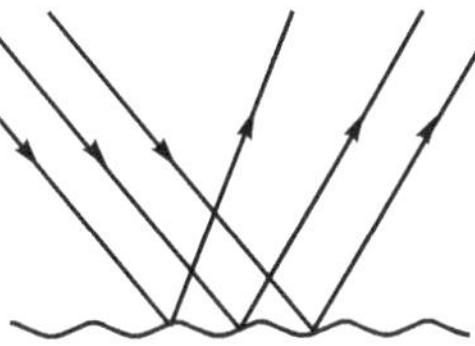

Mirror

A smooth and highly polished reflecting surface is called a mirror.

(i) **Plane Mirror** A highly polished plane surface is called a plane mirror.

Different properties of image formed by plane mirror

$$\text{Size of image} = \text{Size of object}$$

$$\text{Magnification} = \text{Unity}$$

$$\text{Distance of image} = \text{Distance of object}$$

A plane mirror may form a virtual as well as real image.

A man may see his full image in a mirror of half height of man.

When two plane mirror are held at an angle θ, the number of images of an object placed between them is given as below

(a) $n = \left(\frac{360°}{\theta} - 1\right)$, if $\frac{360°}{\theta}$ is even an integer.

(b) $n =$ integral part of $\frac{360°}{\theta}$, when $\frac{360°}{\theta}$ is odd an integer.

A plane mirror may form a real image, when the pencil of light incident on the mirror is convergent. Children, during their play form an image of sun as wall by a strip of plane mirror.

Kaleidoscope and periscope employ the principle of image formation by plane mirror.

If keeping an object fixed a plane mirror is rotated in its plane by an angle θ, then the reflected ray rotates in the same direction by an angle 2θ.

Focal length as well as radius of curvature of a plane mirror is infinity. Power of a plane mirror is zero.

An image formed by a plane mirror is virtual, erect, laterally inverted, of same size as that of object and at the same distance as the object from the mirror.

(ii) **Spherical Mirror** A highly polished curved surface whose reflecting surface is a cut part of a hollow at glass sphere is called a spherical mirror. Spherical mirrors are of two types

(a) **Concave Mirror** A spherical mirror whose bent in surface is reflecting surface, is called a concave mirror.

(b) **Convex Mirror** A spherical mirror whose buldging out surface is reflecting surface, is called a convex mirror.

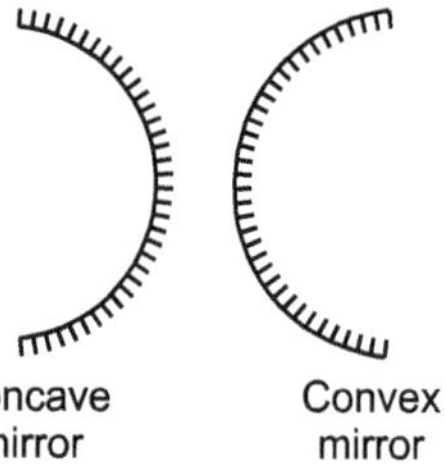

Some Terms Related to Spherical Mirrors are Given Below

(i) **Centre of Curvature** (C) It is the centre of the sphere of which the mirror or lens is a part.

(ii) **Radius of Curvature** (R) The radius of the hollow sphere of which the mirror is a part, is called radius of curvature.

(iii) **Pole** (P) The central point of the spherical mirror is called its pole (P).

(iv) **Focus** (F) When a parallel beam of light rays is incident on a spherical mirror, then after reflection it meets or appears to meet at a point on principal axis, which is called focus of the spherical mirror.

(v) **Focal Length** (f) The distance between the pole and focus is called focal length (f).

Relation between focal length and radius of curvature is given by

$$f = \frac{R}{2}$$

The power of a mirror is given as $P = \frac{1}{f}$ (metre)

(vi) **Mirror formula** $\frac{1}{f} = \frac{1}{v} + \frac{1}{u}$

where, f = focal length of the mirror, u = distance of the object and v = distance of the image.

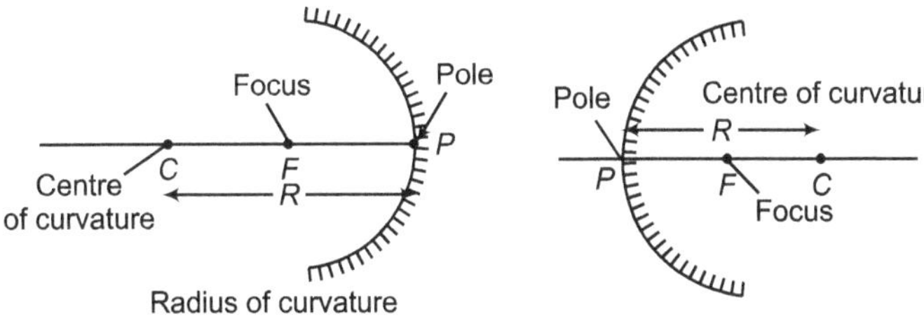

Newton's formula for a concave mirror

$$f = \sqrt{x_1 x_2}$$

$$\Rightarrow \quad f^2 = x_1 x_2$$

where, x_1 and x_2 are the distances of object and image from the focus.

Linear Magnification

The ratio of height of image (I) formed by a mirror to the height of the object (O) is called linear magnification (m).

Linear magnification $(m) = \frac{I}{O} = -\frac{v}{u}$

Areal and Axial Magnification

The ratio of area of image to the area of object is called areal magnification.

Areal magnification $= m^2 = \frac{\text{Area of image}}{\text{Area of object}} = \frac{v^2}{u^2}$

When a small sized object is placed linearly along the principle axis, then its longitudinal or axial magnification is given by

$$\text{Axial magnification} = -\frac{dv}{du} = \left(\frac{v}{u}\right)^2 = \left(\frac{f}{f-u}\right)^2 = \left(\frac{f-v}{f}\right)^2$$

Sign Convention for Spherical Mirrors

(i) All distances are measured from the pole of the mirror.

(ii) Distances measured in the direction of incident light rays are taken as positive.

(iii) Distances measured in opposite direction to the incident light rays are taken as negative.

(iv) Distances measured above the principal axis are positive.

(v) Distances measured below the principal axis are negative.

Lateral Inversion

In the image formed by a plane mirror the right side of the object appears as left side and *vise-versa*. This phenomena is called lateral inversion.

When object is placed between pole and focus of a concave mirror, then its virtual, erect and magnified image is formed.

A convex mirror forms a virtual, erect and diminished image for all conditions of object.

The focal length of concave mirror is taken negative and for a convex mirror taken as positive.

Refraction of Light

The deviation of light rays from its path when it travels from one transparent medium to another transparent medium, is called refraction of light.

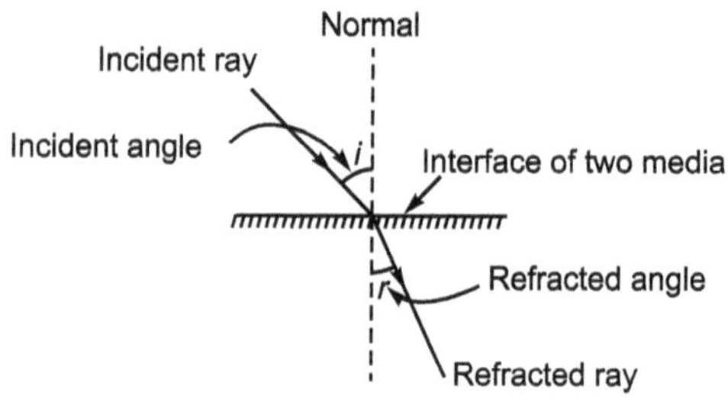

Cause of Refraction

The speed of light is different in different media.

Laws of Refraction

(i) The incident ray, the refracted ray and the normal at the point of incidence, all three lies in the same plane.

(ii) The ratio of sine of angle of incidence to the sine of angle of refraction is constant for a pair of two media,

i.e. $$\frac{\sin i}{\sin r} = \text{constant } ({}_1\mu_2)$$

where ${}_1\mu_2$ is called **refractive index** of second medium with respect to first medium.

This law is also called **Snell's law.**

Refractive Index

The ratio of speed of light in vacuum (c) to the speed of light in any medium (v) is called refractive index of the medium.

Refractive index of a medium,

$$\mu = \frac{c}{v}$$

Refractive index of water $= \frac{4}{3} = 1.33$

Refractive index of glass $= \frac{3}{2} = 1.50$

When light is reflected by a denser medium, phase difference of π radian or path difference of $\frac{\lambda}{2}$ or time difference $\frac{T}{2}$ is produced. This is known as Stoke's law. Distance x travelled by light in a medium of refractive index μ is equal to distance (μx) travelled in vacuum.

Time taken by light to transverse a thickness x of time $= \frac{\mu x}{c}$, where c = velocity of light in vacuum.

Relative Refractive Index

The refractive index of second medium with respect to first medium

$${}_1\mu_2 = \frac{v_1}{v_2} = \frac{\text{height of object}}{\text{height of image}}$$

Cauchy's Formula

Refractive index of a medium $\mu = A + \frac{B}{\lambda^2} + \frac{C}{\lambda^4} + ...$

where, λ = wavelength of light and A, B, C ..... are constant.

$\therefore$ Refractive index $\mu \propto \frac{1}{\lambda^2}$

Critical Angle

The angle of incidence in a denser medium for which the angle of refraction in rarer medium becomes 90°, is called critical angle (C).

Critical angle for diamond = 24°

Critical angle for glass = 42°

Critical angle for water = 48°

Refractive index of denser medium

$$\mu = \frac{1}{\sin C}$$

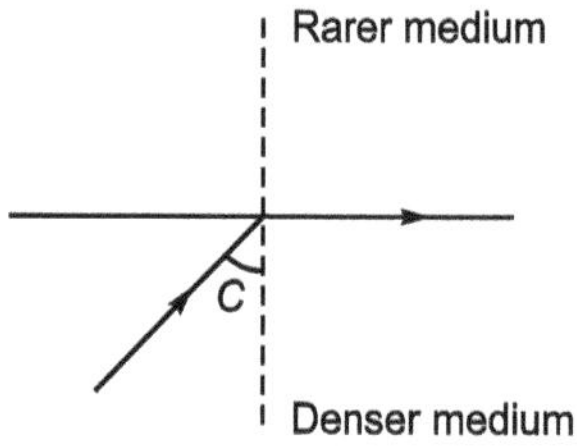

Total Internal Reflection (TIR)

When a light ray travelling from a denser medium towards a rarer medium is incident at the interface at an angle of incidence greater than critical angle, then light rays reflected back in to the denser medium. This phenomena is called **TIR**.

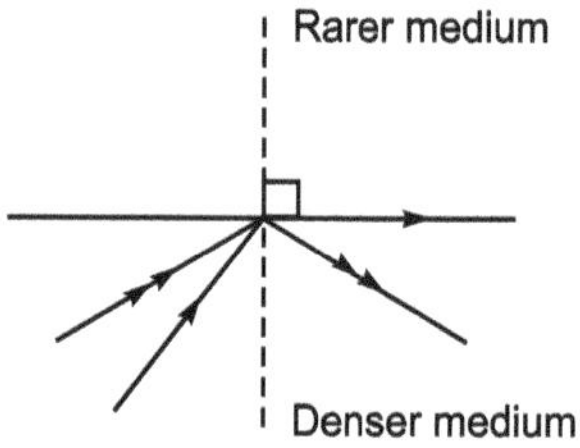

Critical angle increases with temperature.

The refractive index is maximum for violet colour of light and minimum for red colour of light, i.e. $\mu_V > \mu_R$, therefore critical angle is maximum for red colour of light and minimum for violet colour of light, i.e. $C_V < C_R$.

Total internal reflection occurs if angle of incidence in denser medium exceeds critical angle. Mirage is an optical illusion observed in deserts and roads on a hot day when the air near the ground is holler and hence rarer than the air above.

Optical Fibres

Optical fibres are also based on the phenomenon of total internal reflection. Optical fibres consist of several thousands of very long fine quality fibres of glass or quartz. The diameter of each fibre is of the order of 10^{-4} cm with refractive index of material being of the order of 1.5. Optical fibres are used in transmission and reception of electrical signals by converting them first into light signals.

For Refraction at a Convex or Concave Spherical Surface

$$\frac{\mu}{v} - \frac{1}{u} = \frac{(\mu - 1)}{R}$$

where, μ = refractive index, u = distance of object, v = distance of image and R = radius of curvature of the spherical surface.

Lens

A lens is a uniform transparent medium bounded between two spherical or one spherical and one plane surface.

Convex Lens

A lens which is thinner at edges and thicker at middle is called a convex or converging lens.

Concave Lens

A lens which is thicker at edges and thinner at middle, is called a concave or diverging lens.

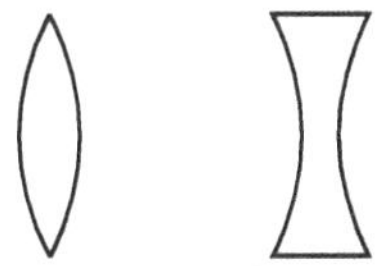

Lens Formula

$$\frac{1}{f} = \frac{1}{v} - \frac{1}{u}$$

where, f = focal length of the lens, u = distance of object, v = distance of image.

Lens Maker's formula

$$\frac{1}{f} = (\mu - 1)\left(\frac{1}{R_1} - \frac{1}{R_2}\right)$$

where, μ = refractive index of the material of the lens and R_1 and R_2 are radii of curvature of the lens.

Power of a Lens

The reciprocal of the focal length of a lens, when it is measured in metre, is called power of a lens.

Power of a lens, $(P) = \dfrac{1}{f\,(\text{metre})}$

Its unit is diopter (D).

The power of a convex (converging) lens is positive and for a concave (diverging) lens it is negative.

Focal Length of a Lens Combination

(i) When lenses are in contact $\dfrac{1}{F} = \dfrac{1}{f_1} + \dfrac{1}{f_2}$

Power of the combination $P = P_1 + P_2$

(ii) When lenses are separated by a distance d

$$\frac{1}{F} = \frac{1}{f_1} + \frac{1}{f_2} - \frac{d}{f_1 f_2}$$

Power of the combination

$$P = P_1 + P_2 - dP_1P_2$$

Linear Magnification

$$m = \frac{I}{O} = \frac{v}{u}$$

For a small sized object placed linearly along the principal axis, its axial (longitudinal) magnification is given by

$$\text{Axial magnification} = -\frac{dv}{du} = \left(\frac{v}{u}\right)^2$$

$$= \left(\frac{f}{f+u}\right)^2 = \left(\frac{f-v}{f}\right)^2$$

Focal Length of a Convex Lens by Displacement Method

Focal length of the convex lens

$$f = \frac{a^2 - d^2}{4a}$$

where, a = distance between the image pin and object pin and

d = distance between two positions of lens.

The distance between the two pins should be greater than four times the focal length of the convex lens, i.e. $a > 4f$.

Height of the object $\quad O = \sqrt{I_1 I_2}$

Cutting of a Lens

(i) If a symmetrical convex lens of focal length f is cut into two parts along its optic axis, then focal length of each part (a plano convex lens) is $2f$. However, if the two parts are joined as shown in figure, the focal length of combination is again f.

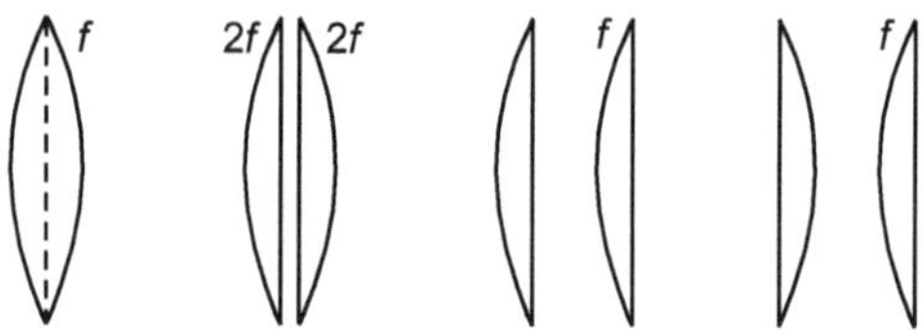

(ii) If a symmetrical convex lens of focal length f is cut into two parts along the principal axis, then focal length of each part remains unchanged as f. If these two parts are joined with curved ends an one side, focal length of the combination is $\frac{f}{2}$. But on joining two parts in opposite sense the net focal length becomes ∞.

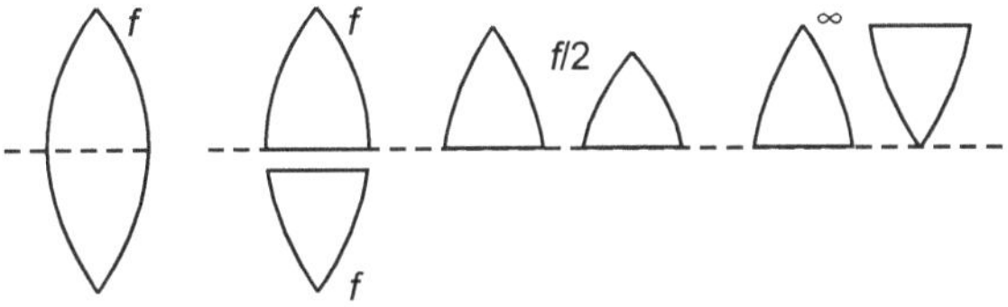

Prism

Prism is uniform transparent medium bounded between two refracting surfaces, inclined at an angle.

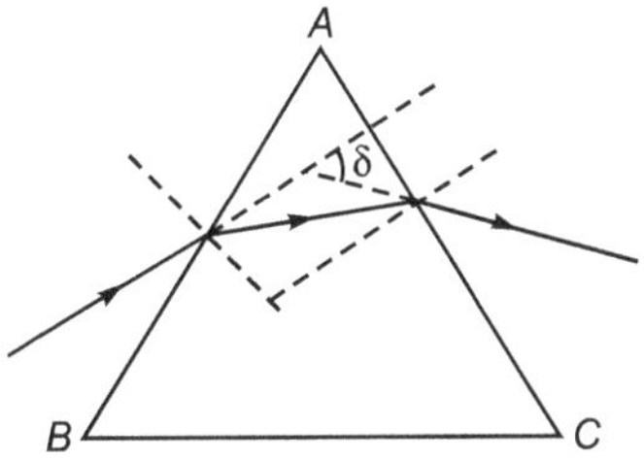

Angle of Deviation

The angle subtended between the direction of incident light ray and emergent light ray from a prism is called angle of deviation (δ).

Prism Formula

The refractive index of material of prism

$$\mu = \frac{\sin\left(\frac{A+\delta_m}{2}\right)}{\sin\left(\frac{A}{2}\right)}$$

where, A = prism angle and δ_m = minimum angle of deviation.

For very small angle prism

$$\delta_m = (\mu - 1)A$$

Dispersion of Light

The splitting of white light into its constituent colours in the sequence of VIBGYOR, on passing through a prism, is called dispersion of light.

The refractive index $\mu_V > \mu_R$, therefore violet colour deviates most and red colour deviates least, i.e. $\delta_V > \delta_R$.

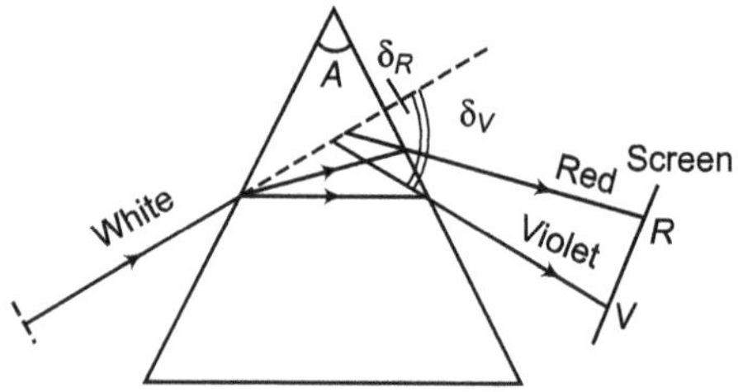

Angular Dispersion

The angle subtended between the direction of emergent violet and red rays of light from a prism is called angular dispersion.

Angular dispersion $(\theta) = \delta_V - \delta_R = (\mu_V - \mu_R) A$

where δ_V and δ_R are angle of deviation.

Dispersive Power

$$W = \frac{\theta}{\delta_Y} = \frac{(\mu_V - \mu_R)}{(\mu_Y - 1)}$$

where $\mu_Y = \dfrac{\mu_V + \mu_R}{2}$, is mean refractive index.

Human Eye

Human eye is an optical instrument which forms real image of the objects on retina.

Retina contains lakhs of cone and rod cells which are sensitive for colours of light and intensities of light respectively.

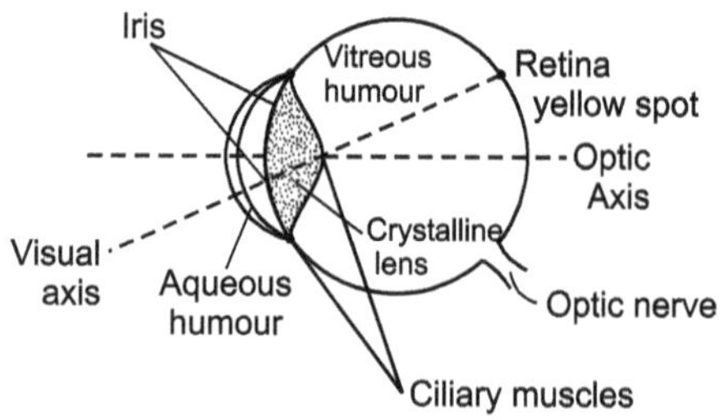

Ciliary muscles change the focal length of eye lens. This power of eye is called power of accomodation of eye.

Different defects of vision of human eye are described below

(i) **Myopia or Short-Sightedness** It is a defect of eye due to which a person can see near by objects clearly but cannot see far away objects clearly.

In this defect, the far point of eye shifts from infinity to a nearer distance.

This defect can be removed by using a concave lens of appropriate power.

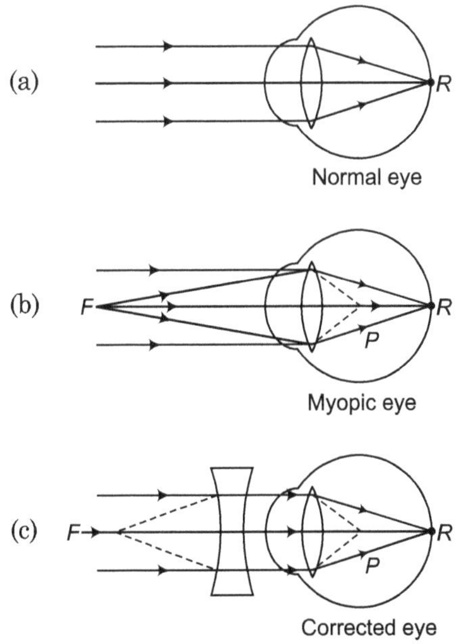

(ii) **Hypermetropia or Long-Sightedness** In this defect, a person can see far away objects clearly but cannot see near by objects clearly.

In this defect the near point of eye shifts away from the eye.

This defect can be removed by using a convex lens of appropriate power.

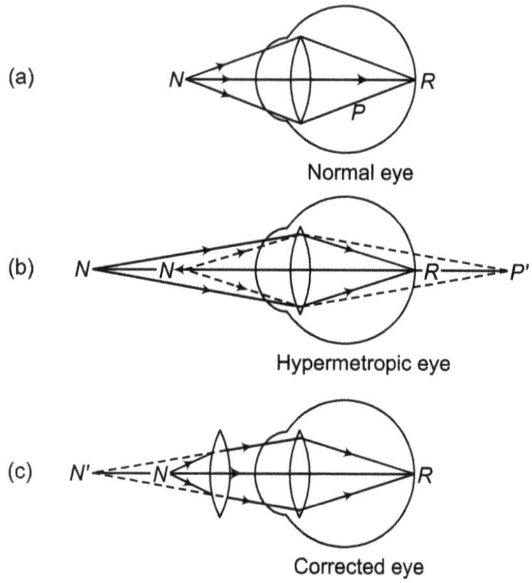

(iii) **Astigmatism** In this defect, a person cannot focus on horizontal and vertical lines at the same distance at the same time.

This defect can be removed by using suitable cylindrical lenses.

(iv) **Colour Blindness** In this defect, a person is unable to distinguish between few colours.

The reason of this defect is the absence of cone cells sensitive for few colours.

This defect cannot be removed.

(v) **Cataract** In this defect, an opaque white membrane is developed on cornea due to which person losts power of vision partially or completely.

This defect can be removed by removing this membrane through surgery.

Camera

A photograph camera consists of a light proof box, at one end of which a converging lens system is fitted. A light sensitive film is fixed at the other end of the box, opposite to the lens system. A real inverted image of the object is formed on the film by the lens system.

f-Number for a Camera The f-number represent the size of the aperture.

$$f\text{-number} = \frac{\text{Focal length of the lens } (F)}{\text{Diameter of the lens } (d)}$$

Generally 2, 2.8, 4, 5.6, 8, 11, 22, 32 are f-numbers.

The amount of light (L) entering the camera is directly proportional to the area (A) of the aperture, i.e.

$$L \propto A \propto d^2$$

$$\text{Brightness of image} \propto \frac{d^2}{F^2}$$

where, d = diameter of the lens and F = focal length of the lens.

Exposure time is the time for which light is incident of photographic film.

Simple Microscope

It is used for observing magnified images of objects. It is consists of a converging lens of small focal length.

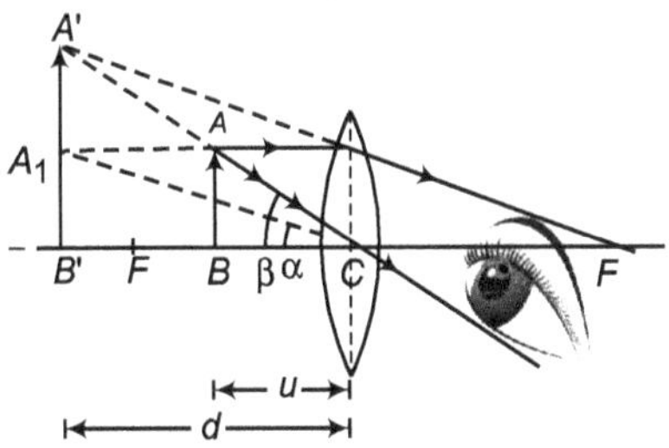

Magnifying Power

(i) When final image is formed at least distance of distinct vision (D), then $M = 1 + \frac{D}{f}$

where, f = focal length of the lens.

(ii) When final image is formed at infinity, then $M = \frac{D}{f}$.

Compound Microscope

It is a combination of two convex lenses called objective lens and eye piece separated by a distance. Both lenses are of small focal lengths but $f_o < f_e$, where f_o and f_e are focal lengths of objective lens and eye piece respectively.

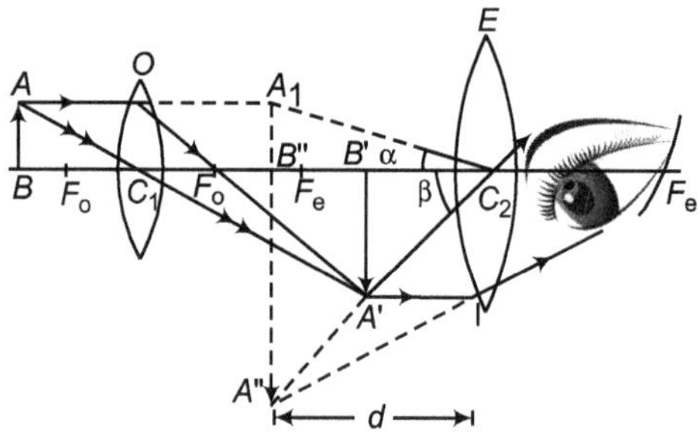

Magnifying Power

(i) When final image is formed at least distance of distinct vision (D), then

$$M = \frac{v_o}{u_o}\left(1 + \frac{D}{f_e}\right)$$

where, v_o = distance of image, formed by objective lens and
u_o = distance of object from the objective

(ii) When final image is formed at infinity, then

$$M = \frac{v_o}{u_o} \cdot \frac{D}{f_e}$$

Astronomical Telescope

It is also a combination of two lenses, called objective lens and eyepiece, separated by a distance. It is used for observing distinct images of heavenly bodies like stars, planets etc.

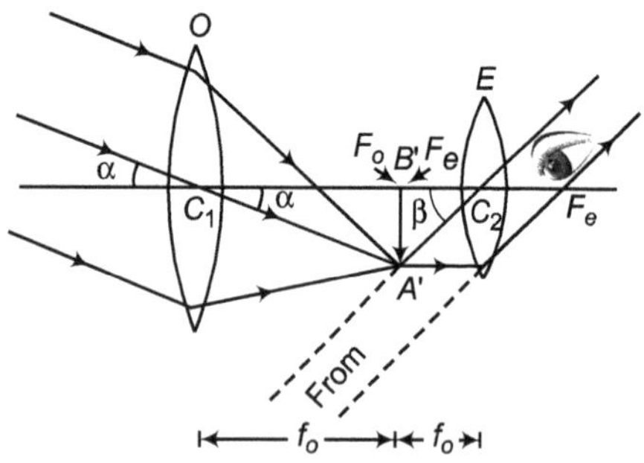

Objective lens is a convex lens of large aperture and large focal length while eyepiece is a convex lens of small aperture and small focal length.

Magnifying Power

(i) When final image is formed at least distance of distinct vision (D), then $M = \frac{f_o}{f_e}\left(1 + \frac{D}{f_e}\right)$, where f_o and f_e are focal lengths of objective and eyepiece respectively.

Length of the telescope $(L) = (f_o + u_e)$

where, u_e = distance of object from the eyepiece.

(ii) When final image is formed at infinity,

then $$M = \frac{f_o}{f_e}.$$

Length of the telescope $(L) = f_o + f_e$

For large magnifying power of a telescope f_o should be large and f_e should be small.

For large magnifying power of a microscope; $f_o < f_e$ but f_e should be small.

Resolving Power

The ability of an optical instrument to produce separate and clear images of two near by objects, is called its resolving power.

Limit of Resolution

The minimum distance between two near by objects which can be just resolved by the instrument, is called its limit of resolution (d).

Resolving power of a microscope $= \frac{1}{d} = \frac{2\mu \sin\theta}{\lambda}$.

where, d = limit of resolution,

λ = wavelength of light used,

μ = refractive index of the medium between the objects and objective lens

and θ = half of the cone angle.

Resolving power of a telescope $= \frac{1}{d\theta} = \frac{d}{1.22\,\lambda}$

where, $d\theta$ = limit of resolution

λ = wavelength of light used and

d = diameter of aperture of objective

Aberration of Lenses

The image formed by the lens suffer from following two main drawbacks

(i) **Spherical Aberration** Aberration of the lens due to which all the rays passes through the lens are not focussed at a single point and the image of a point object placed on the axis is blurred, is called spherical aberration.

It can be reduced by using

(a) lens of large focal lengths
(b) plano-convex lenses
(c) crossed lenses
(d) combining convex and concave lens

(ii) **Chromatic Aberration** Image of a white object formed by lens is usually coloured and blurred. This defect of the image produced by lens is called chromatic aberration.

Scattering of Light

When light passes through a medium in which particles are suspended whose size is of the order of wavelength of light, then light on striking these particles, deviated in different directions. These phenomena is called scattering of light.

According to the Lord Rayleigh, the intensity of scattered light

$$I \propto \frac{1}{\lambda^4}$$

Therefore, red colour of light is scattered least and violet colour of light is scattered most.

Daily Life Examples of Scattering of Light

(i) Blue colour of sky.
(ii) Red colour of signals of danger.
(iii) Black colour of sky in the absence of atmosphere
(iv) Red colour of the time of sun rise and sun set.
(v) The human eye is most sentitive to yellow colour.

27

Wave Optics

Wave optics describes the connection between waves and rays of light. According to wave theory of light, the light is a form of energy which travels through a medium in the form of transverse wave motion. The speed of light in a medium depends upon the nature of medium.

Newton's Corpuscular Theory

Light consists of very small invisible elastic particles which travel in vacuum with a speed of 3×10^8 m/s.

The theory could explain reflection and refraction.

The size of corpuscular of different colours of light are different.

It could not explain interference, diffraction, polarisation, photoelectric effect and Compton effect. The theory failed as it could not explain why light travels faster in a rarer medium than in a denser medium.

Wavefront

A wavefront is defined as the continuous locus of all the particles of a medium, which are vibrating in the same phase.

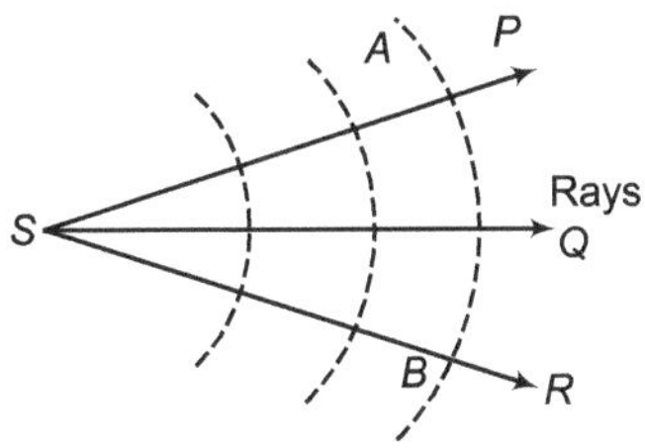

S = source of light, AB = wavefront and SP, SQ and SR are rays of light.

These are three types

(i) Spherical wavefront

(ii) Cylindrical wavefront

(iii) Plane wavefront

Huygen's Wave Theory

Light travel in a medium in the form of wavefront.

A wavefront is the locus of all the particles vibrating in same phase.

All particles on a wavefront behaves as a secondary source of light, which emits secondary wavelets.

The envelope of secondary wavelets represents the new position of a wavefront.

When source of light is a point source, the wavefront is spherical. Amplitude (A) is inversely proportional to distance (x) i.e. $A \propto \frac{1}{x}$.

$$\therefore \quad \text{Intensity } (I) \propto (\text{Amplitude})^2$$

When source of light is linear, the wavefront is cylindrical.

$$\text{Amplitude (A)} \propto \frac{1}{\sqrt{x}}$$

$$\therefore \quad \text{Intensity} \propto (\text{Amplitude})^2 \propto \frac{1}{x}$$

Huygen's Principle

(i) Every point on given wavefront (called primary wavefront) acts as a fresh source of new disturbance called secondary wavelets.

(ii) The secondary wavelets travels in all the directions with the speed of light in the medium.

(iii) A surface touching these secondary wavelets tangentially in the forward direction at any instant gives the new (secondary) wave front of that instant.

Reflection and Refraction of Wavefront

Reflection $\quad BC = AD$ and $\angle i = \angle r$

Refraction $\quad \frac{BC}{AD} = \frac{v_1}{v_2} = \frac{\sin i}{\sin r} = \frac{\mu_2}{\mu_1}$

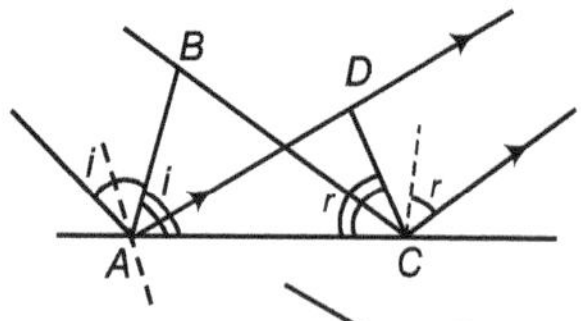

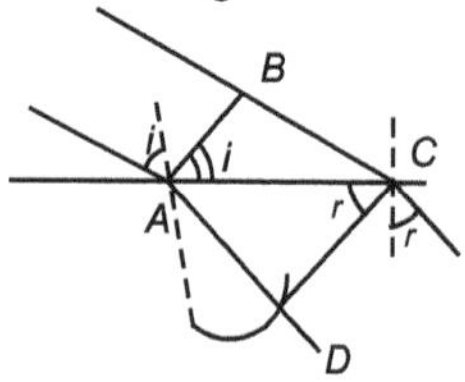

Doppler's Effect

The phenomenon of apparent change in frequency (or wavelength) of the light due to relative motion between the source of light and the observer is called **Doppler's effect.**

Doppler's shift

Apparent wavelength > actual wavelength

So, spectrum of the radiation from the source of light shifts towards the red end of spectrum. This is called **red shift**.

$$\text{Doppler's shift, } \Delta\lambda = \lambda \cdot \frac{v}{c}$$

where, v = speed of source w.r.t. stationary observer,

c = speed of light.

Maxwell's Electromagnetic Wave Theory

(i) Light waves are electromagnetic waves which do not require a material medium for their propagation.

(ii) Due to transverse nature, light wave undergoes polarisation.

(iii) The velocity of electromagnetic wave in vacuum is $c = \frac{1}{\sqrt{\mu_0 \varepsilon_0}}$.

(iv) The velocity of electromagnetic waves in medium is less than that of light, $v < c$

$$v = \frac{1}{\sqrt{\mu_0 \varepsilon_0 \varepsilon_r \mu_r}} = \frac{c}{\sqrt{\mu_0 \varepsilon_r}}$$

(v) The velocity of electromagnetic waves in a medium depend upon the electric and magnetic properties of the medium.

where, μ_0 = absolute magnetic permeability and

ε_0 = absolute electrical permittivity of free space.

(vi) It failed to explain the phenomenon of photoelectric effect, Compton effect and Raman effect.

Max Planck's Quantum Theory

(i) Light emits from a source in the form of packets of energy called quanta or photon.

(ii) The energy of a photon is $E = h\nu$, where h is Planck's constant and ν is the frequency of light.

(iii) Quantum theory could explain photoelectric effect, Compton effect and Raman effect.

(iii) Quantum theory failed to explain interference, diffraction and polarisation of light.

de-Broglie's Dual Theory

Light waves have dual nature, wave nature according to Maxwell's electromagnetic wave theory and particle nature according to Max-Planck's quantum theory.

Two natures of light are like the two faces of a coin. In any one phenomena only its one nature appears.

$$\text{Energy of photon} = h\nu = \frac{hc}{\lambda}$$

where, h = Planck's constant = 6.6×10^{-34} J/s

$$\text{de-Broglie wave equation is } \lambda = \frac{h}{p} = \frac{h}{mv}$$

where, h denotes Planck's constant.

Superposition of Waves

When two similar waves propagate in a medium simultaneously, then at any point the resultant displacement is equal to the vector sum of displacement produced by individual waves.

$$y = y_1 + y_2$$

Interference of Light

When two light waves of similar frequency having a zero or constant phase difference propagate in a medium simultaneously in the same direction, then due to their superposition maximum intensity is obtained at few points and minimum intensity at other few points. This phenomenon of redistribution of energy due to superposition of waves is called interference of light waves.

The interference taking place at points of maximum intensity is called **constructive interference.**

The interference taking place at points of minimum intensity is called **destructive interference.**

Fringe Width

The distance between the centres of two consecutive bright or dark fringes is called the fringe width.

The angular fringe width is given by $\theta = \frac{\lambda}{d}$.

where, λ is the wavelength of light d is the distance between two coherent sources.

Conditions for Constructive and Destructive Interference

For Constructive Interference

Phase difference, $\phi = 2n\pi$

Path difference, $\Delta x = n\lambda$

where, $n = 0, 1, 2, 3, \ldots$

For Destructive Interference

Phase difference, $\phi = (2n - 1)\pi$

Path difference, $\Delta x = \dfrac{(2n-1)\lambda}{2}$

where, $n = 1, 2, 3, \ldots$

If two waves of exactly same frequency and of amplitude a and b interfere, then **amplitude of resultant wave** is given by

$$R = \sqrt{a^2 + b^2 + 2ab\cos\phi}$$

where, ϕ is the phase difference between two waves.

$$R_{max} = (a + b)$$

$$R_{min} = (a - b)$$

Intensity of wave,

$$\therefore \quad I = a^2 + b^2 + 2ab\cos\phi$$

$$= I_1 + I_2 + 2\sqrt{I_1 I_2}\cos\phi$$

where, I_1 and I_2 are intensities of two waves.

$$\therefore \quad \frac{I_1}{I_2} = \frac{a^2}{b^2} = \frac{\omega_1}{\omega_2}$$

where, ω_1 and ω_2 are width of slits.

Energy remains conserved during interference.

Interference fringe width, $\beta = \dfrac{D\lambda}{d}$

where, D = distance of screen from slits, λ = wavelength of light and d = distance between two slits.

Distance of nth bright fringe from central fringe, $x_n = \dfrac{nD\lambda}{d}$.

Distance of nth dark fringe from central fringe, $x'_n = \dfrac{(2n-1)D\lambda}{2d}$.

Coherent Sources of Light

The sources of light emitting light of same wavelength, same frequency having a zero or constant phase difference are called coherent sources of light.

When a transparent sheet of refractive index μ and of thickness t is introduced in one of the path of interfering waves, then fringe pattern shifts in that direction by a distance Y

$$Y = \frac{D}{d}(\mu - 1)\,t$$

$$= \frac{\beta}{\lambda}(\mu - 1)\,t$$

where, β = fringe width.

Fresnel's Biprism

It is a combination of two prisms of very small refracting angles placed base to base. It is used to obtain two coherent sources from a single light source.

Lyod's Mirror

The shape of interference fringes are usually hyperbolic.

When screen is held at 90° to the line joining foci of the hyperbola, the fringes are circular.

When distance of screen (D) is very large compare to the distance between the slits (d), the fringes are straight.

Diffraction

The bending of light waves around the corners of an obstacle or aperture is called diffraction of light.

The phenomenon of diffraction is divided mainly in the following two classes

(a) Fresnel class

(b) Fraunhofer class

S.No.	Fresnel Class	Fraunhofer Class
1.	The source is at a finite distance.	The source is at infinite distance.
2.	No opticals are required.	Opticals are in the form of collimating lens and focusing lens are required.
3.	Fringes are not sharp and well defined.	Fringes are sharp and well defined.

Fraunhofer Diffraction at a Single Slit

Linear width of central maximum $\dfrac{2D\lambda}{a} = \dfrac{2f\lambda}{a}$

Angular width of central maximum $= \dfrac{2\lambda}{a}$

where, λ = wavelength of light, a = width of single slit, D = distance of screen from the slit and f = focal length of convex lens.

For Secondary Minima

(a) Path difference $= n\lambda$

(b) Linear distance $= \dfrac{nD\lambda}{a} = \dfrac{nf\lambda}{a}$

(c) Angular spread $= \dfrac{n\lambda}{a}$

where, $n = 1, 2, 3, \ldots$

For Secondary Maxima

(a) Path difference $= \dfrac{(2n+1)\lambda}{2}$

(b) Linear distance $= \dfrac{(2n+1)D\lambda}{2a} = \dfrac{(2n+1)f\lambda}{2a}$

(c) Angular spread $= \dfrac{(2n+1)\lambda}{2a}$

Important Points

- A soap bubble or oil film on water appears coloured in white light due to interference of light reflected from upper and lower surfaces of soap bubble or oil film.
- In interference fringe pattern, all bright and dark fringes are of same width.
- In diffraction fringe pattern central bright fringe is brightest and widest and remaining secondary maximas are of gradually decreasing intensities.
- The difference between interference and diffraction is that the interference is the superposition between the wavelets coming from two coherent sources while the diffraction is the superposition between the wavelets coming from the single wavefront.

Polarisation

The phenomenon of restructing of electric vectors of light into a single direction is called **polarisation.** Ordinary light has electric vectors in all possible directions in a plane perpendicular to the direction of propagation of light.

When ordinary light is passed through a tourmaline, calcite or quartz crystal the transmitted light have electric vectors in a particular direction parallel to the axis of crystal. This light is plane polarised light.

A plan e containing the vibrations of polarised light is called **plane of vibration.**
A plane perpendicular to the plane of vibration is called **plane of polarisation.**

Polarisation can take place only in transverse waves.

Nicol Prism

A nicol prism is an optical device which is used for producing plane polarised light and analysing light the same.

The nicol prism consists of two calcite crystal cut at 68° with its principal axis joined by a glue called Canada balsam.

Law of Malus

When a beam of completely plane polarised light is incident on an analyser, the intensity of transmitted light from analyser is directly proportional to the square of the cosine of the angle between plane of transmission of analyser and polariser, i.e.

$$I \propto \cos^2 \theta$$

When ordinary light is incident on a polariser the intensity of transmitted light is half of the intensity of incident light.

When a polariser and analyser are perpendicular to each other, then intensity of transmitted light from analyser becomes O.

Brewster's Law

When unpolarised light is incident at an angle of polarisation (i_p) on the interface separating air from a medium of refractive index μ, then reflected light becomes fully polarised, procided $\mu = \tan i_p$

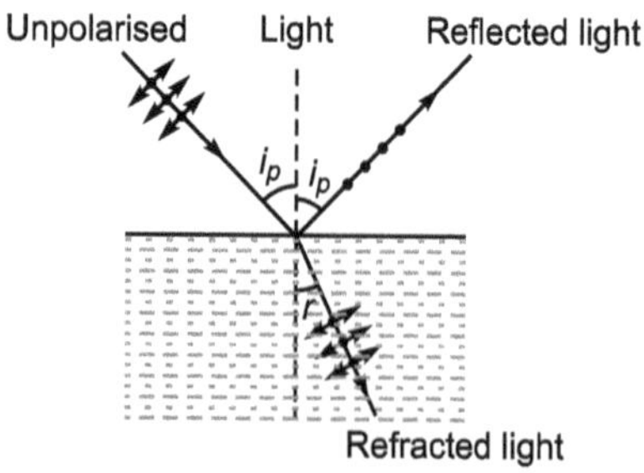

If angle of polarisation is i_p and angle of refraction is μ, then

$$i_p + r = 90°$$

$$\text{Refractive index, } \mu = \tan i_p = \frac{1}{\sin C}$$

where, C = critical angle.

Double Refraction

When unpolarised light is incident on a calcite or quartz crystal it splits up into two refracted rays, one of which follows laws of refraction, called **ordinary ray** (O-ray) and other do not follow laws of refraction, called **extraordinary ray** (E-ray). This phenomenon is called double refraction.

Dichroism

Few double refracting crystals have a property of absorbing one of the two refracted rays and allowing the other to emerge out. This property of crystal is called dichroism.

Polaroid

It is a polarising film mounted between two glass plates. It is used to produce polarised light.

A polaroid is used to avoid glare of light in spectacles.

Uses of Polaroid

(i) Polaroids are used in sun glasses. They protect the eyes from glare.

(ii) The polaroids are used in window panes of a train and especially of an aeroplane. They help to control the light entering through the window.

(iii) The pictures taken by a stercoscopic camera. When seen with the help of polaroid spectacles, create three dimensional effect.

(iv) The wind shield of an automobile is made of polaroid. Such a mind shield protects the eyes of the driver of the automobile from the **dazzling light** of the approaching vehicles.

28

Electrons, Photons and X-rays

Cathode Rays

Cathode rays are the stream of fast moving electrons. These rays are produced in a discharge tube at a pressure below 0.01 mm of mercury.

Properties of Cathode Rays

(i) Cathode rays are not electromagnetic rays.

(ii) Cathode rays are deflected by electric field and magnetic field.

(iii) Cathode rays produce heat in metals when they fall on them.

(iv) Cathode rays can pass through thin aluminium or gold foils without puncturing them.

(v) Cathode rays can produce physical and chemical change.

(vi) Cathode ray travel in straight line with high velocity momentum and energy and cast shadow of objects placed in their path.

(vii) On striking the target of high atomic weight and high melting point, they produce X-rays.

(viii) Cathode rays produce fluorescence and phosphorescence in certain substance and hence affect photographic plate.

(ix) When any charge particle move in a field where magnetic and electric fields are present, without any deviation, then

Magnetic force = Electrostatic force

$$Bev = Ee \text{ or } v = \frac{E}{B}$$

(x) Specific charge of cathode rays means the ratio of charge and mass.

(xi) Specific charge of electron was determined by JJ Thomson using perpendicular magnetic and electric field applied on a beam of electrons, at the same place.

(xii) Specific charge of electron $\frac{e}{m} = \frac{E^2}{2VB^2}$

where, E = electric field, B = magnetic field and V = potential difference applied across ends of tube.

(xii) The value of specific charge of an electron is 1.7589×10^{11} C/kg.

(xiv) Millikan measured the charge of an electron through his popular oil drop experiment.

(xv) The charge of the electron as determined by Millikan was found to be 1.602×10^{-19} C.

Positive Rays

Positive rays were discovered by Goldstein. Positive rays are moving positive ions of gas filled in the discharge tube. The mass of these particles is nearly equal to the mass of the atoms of gas.

(i) These consist of fast moving positively charged particles.
(ii) These rays are deflected in magnetic and electric fields.
(iii) These rays travel in straight line.
(iv) Speed of positive rays is less than that of cathode rays.
(v) These rays can produce fluorescence and phosphorescence.

Electron Emission

It is the phenomenon of emission of electron from the surface of a metal. The electron emission can be obtained from the following process

(i) Thermionic
(ii) Photoelectric emission
(iii) Field emission
(iv) Secondary emission

Photon

Photons are the packets of energy emitted by a source of radiation.

The energy of each photon is, $E = h\nu$

where, h is Planck's constant and ν is frequency of radiation.

The dimension of Planck's constant is $[ML^2T^{-1}]$ and its SI unit is Js.

The rest mass of a photon is zero.

The momentum of a photon, $p = \frac{h\nu}{c} = \frac{h}{\lambda}$

Dynamic or kinetic mass of photon, $m = \frac{h\nu}{c^2} = \frac{h}{c\lambda}$

where, c is speed of light in vacuum and λ is wavelength of radiation.

Photons are electrically neutral.

A body can radiate or absorb energy in whose number multiples of a quantum $h\nu$, $2h\nu$, $3h\nu \ldots nh\nu$, where n is positive integer.

Photoelectric Effect

The phenomenon of emission of electrons from a metal surface, when radiations of suitable frequency is incident on it, is called photoelectric effect.

Terms Related to Photoelectric Effect

(i) **Work Function** (ϕ) The minimum amount of energy required to eject one electron from a metal surface, is called its work function. Its dimension is $[ML^2T^{-2}]$ and unit is J; eV.

(ii) **Threshold Frequency** (ν_0) The minimum frequency of light which can eject photoelectron from a metal surface is called threshold frequency of that metal.

Its dimension is $[T^{-1}]$ and unit is Hz.

(iii) **Threshold Wavelength** (λ_{max}) The maximum wavelength of light which can eject photoelectron from a metal surface is called threshold wavelength of that metal.

Relation between work function, threshold frequency and threshold wavelength

$$\phi = h\nu_0 = \frac{hc}{\lambda_{max}}$$

Laws of Photoelectric Effect

1. For a given metal and frequency of incident light, the photoelectric current (the rate of emission of photoelectrons) is directly proportional to the intensity of incident light.

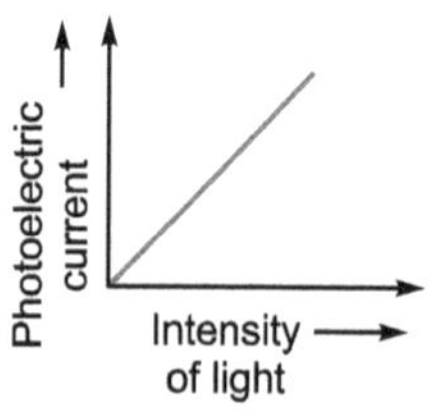

2. For a given metal, there is a certain minimum frequency, called **threshold frequency**, below which there is no emission of photo electrons takes place.
3. Above threshold frequency the maximum kinetic energy of photo electrons depends upon the frequency of incident light.
4. The photoelectric emission is an instantaneous process.

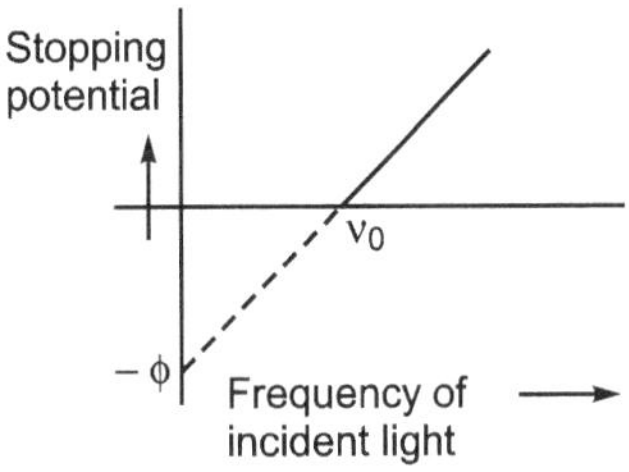

Einstein's Photoelectric Equation

The maximum kinetic energy of photoelectrons

$$(E_k)_{\max} = h\nu - \phi = h(\nu - \nu_0)$$

where, ν is frequency of incident light and ν_0 is threshold frequency.

Stopping Potential

The minimum negative potential given to anode plate at which photoelectric current becomes zero is called stopping potential (V_0).

Maximum kinetic energy of photo electrons

$$(E_k)_{\max} = \frac{1}{2} mv_{\max}^2 = eV_0$$

Dimension of stopping potential is $[ML^2T^{-3}A^{-1}]$ and unit is volt.

Compton Effect

When a monochromatic beam of X-falls on a target containing free electrons, it is scattered. As a result, the electrons recoil and scattered radiation has wavelength longer than incident one. This effect is called Compton effect.

(i) $\lambda' - \lambda = \lambda$ = Compton shift $\Delta\lambda = \dfrac{h}{m_0c}(1 - \cos\phi)$

where m_0 is rest mass of an electron and c is the speed of light $\dfrac{h}{m_0c}$.

Compton shift $\Delta\lambda$ is maximum, when $\phi = 180°$.

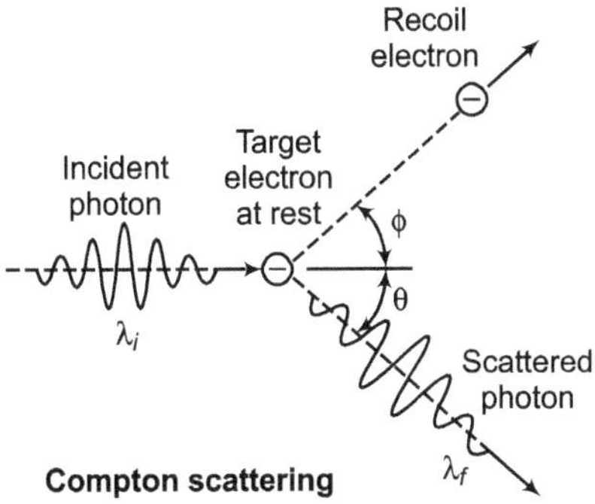

Compton scattering

(ii) Kinetic energy of recoil electron,

$$E_k = \frac{hc}{\lambda} - \frac{hc}{\lambda'}$$

(iii) Direction of recoil electron,

$$\tan\theta = \frac{\lambda \sin\phi}{\lambda' - \lambda \cos\phi}$$

(iv) Compton wavelength of electron,

$$= \frac{h}{m_0 c} = 0.024 \text{ Å}$$

(v) Maximum Compton shift

$$(\Delta\lambda)_{max} = \frac{2h}{m_0 c} = 0.048 \text{ Å}$$

Matter Waves on de-Broglie Waves

A wave is associated with every moving particle, called matter or de-Broglie wave.

de-Broglie Wavelength

If a particle of mass m is moving with velocity v, then wavelength of de-Broglie wave associated with it is given by

$$\lambda = \frac{h}{p} = \frac{h}{mv}$$

de-Broglie wavelength of an electron is given by

$$\lambda = \frac{h}{mv} = \frac{h}{\sqrt{2meV}} = \frac{12.27}{\sqrt{V}} \text{ Å}$$

where, m = mass of electron, e = electronic charge and V = potential difference with which electron is accelerated.

Davisson and Germer proves the existence of de-Broglie waves associated with an electron in motion.

Davisson-Germer Experiment

The wave nature of the material particles as predicted by de-Broglie was confirmed by Davisson and Germer (1927) in united states and by GP Thomson (1928) in scotland.

This experiment verified the wave nature of electron using Ni crystal.

Davisson and Germer found that the intensity of scattered beam of electrons was not the same but different at different angles of scattering. It is maximum for diffracting angle 50° at 54 V potential difference.

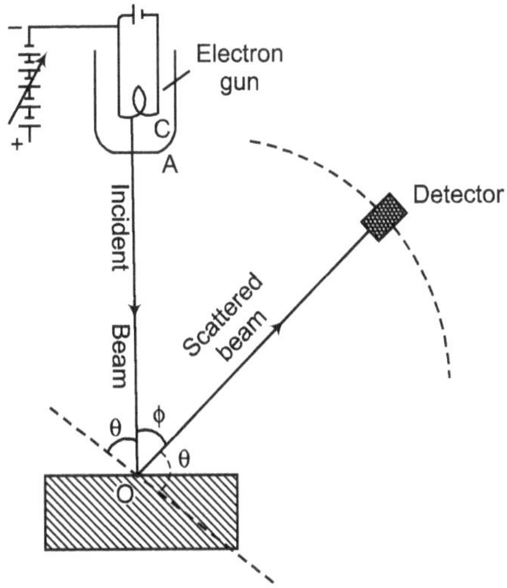

X-rays

When cathode rays strike on a heavy metal of high melting point, then a very small fraction of its energy converts in to a new type of waves, called X-rays.

Properties of X-rays

X-rays were discovered by **Roentgen.**

(i) X-rays are electromagnetic waves of wavelengths ranging from 0.1 Å to 100 Å and frequencies ranging from 10^{16} Hz to 10^{18} Hz.

(ii) Soft X-rays have greater wavelength and lower frequency.

(iii) Hard X-rays have lower wavelength and higher frequency.

(iv) X-rays are produced by coolidge tube.

(v) Molybdenum and tungsten provide suitable targets. These elements have large atomic number and high melting point for the purpose.

(vi) The intensity of X-rays depends on the heating voltage or filament current.

(vii) The kinetic energy of X-ray photons depends upon the voltage applied across the ends of coolidge tube.

(viii) Energy of X-ray photon is given by $E = h\nu = \frac{hc}{\lambda}$.

(ix) If total energy of fast moving electron transfer to X-ray photon, then its energy, $eV = h\nu = \frac{hc}{\lambda}$.

(x) Wavelength of emitted X-rays is given by $\lambda = \frac{hc}{eV}$.

where, h = Planck's constant, c = speed of light, e = electronic charge and V = potential difference applied across the ends of the tube.

(xi) Absorption of X-rays

$I = I_0 e^{-\mu x}$, where I_0 = initial intensity of X-rays, I = final intensity of emergent X-rays, x = thickness of material and μ = absorption coefficient.

Diffraction of X-rays

X-rays can be diffracted by crystals following **Bragg's law**. According to this

$$2d \sin\theta = n\lambda$$

where, $n = 1, 2, 3, \ldots$, and d = spacing of crystal planes, θ = angle of diffraction.

X-rays Spectrum

The energy spectrum of X-rays is a line spectrum, containing following series :

(i) ***K*-series** When electrons of any higher orbit ($n = 2, 3, 4, \ldots$) jump to first orbit ($n = 1$) then K-series of X-rays are produced.

(ii) ***L*-series** When electrons of higher orbit ($n = 3, 4, 5, \ldots$) jump to second orbit ($n = 2$), then L-series of X-rays are produced.

(iii) ***M*-seires** When electrons of higher orbit ($n = 4, 5, 6, \ldots$) jump to third orbit ($n = 3$), then M-series of X-rays are produced.

First lines of these series are called $K_\alpha, L_\alpha, M_\alpha$. Second lines of these series are called $K_\beta, L_\beta, M_\beta$.

Moseley's Law

The frequency of X-ray is given by

$$\nu = a(Z - b)^2$$

where, a and b are constants and Z is atomic number of element.
Frequency of X-rays

$$\nu \propto Z^2$$

29

Atomic Physics

Dalton's Atomic Theory

All elements are consists of very small invisible particles, called atoms. Atoms of same element are exactly same and atoms of different element are different.

Thomson's Atomic Model

Every atom is uniformly positive charged sphere of radius of the order of 10^{-10} m, in which entire mass is uniformly distributed and negative charged electrons are embedded randomly. The atom as a whole is neutral.

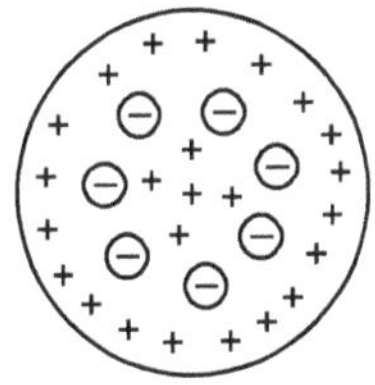

Limitations of Thomson's Atomic Model

1. It could not explain the origin of spectral series of hydrogen and other atoms.
2. It could not explain large angle scattering of α-particles.

Rutherford's Atomic Model

On the basis of this experiment, Rutherford made following observations

(i) The entire positive charge and almost entire mass of the atom is concentrated at its centre in a very tiny region of the order of 10^{-15} m, called nucleus.

(ii) The negatively charged electrons revolve around the nucleus in different orbits.

(iii) The total positive charge on nucleus is equal to the total negative charge on electron. Therefore atom as a overall is neutral.

(iv) The centripetal force required by electron for revolution is provided by the electrostatic force of attraction between the electrons and the nucleus.

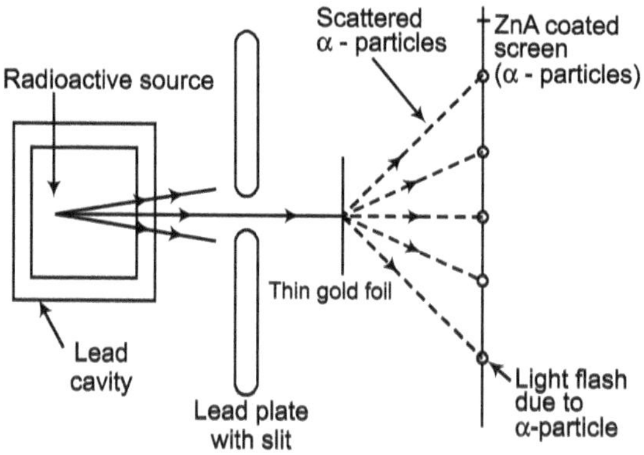

Distance of Closest Approach

$$r_0 = \frac{1}{4\pi\,\varepsilon_0} \cdot \frac{2Ze^2}{E_k}$$

where, E_k = kinetic energy of the α-particle.

Impact Parameter

The perpendicular distance of the velocity vector of α-particle from the central line of the nucleus, when the particle is far away from the nucleus is called impact parameter.

$$\text{Impact parameter, } b = \frac{1}{4\pi\,\varepsilon_0} \cdot \frac{Ze^2 \cot\left(\frac{\theta}{2}\right)}{E_k}$$

where, Z = atomic number of the nucleus, E_k = kinetic energy of the α-particle and θ = angle of scattering.

Rutherford's Scattering Formula

$$N(\theta) = \frac{N_i nt\, Z^2 e^4}{(8\pi\varepsilon_0)^2\, r^2 E^2 \sin^4\left(\frac{\theta}{2}\right)}$$

where, $N(\theta)$ = number of α-particles, N_i = total number of α-particles reach the screen, n = number of atoms per unit volume in the foil, Z = atoms number, E = kinetic energy of the alpha particles and t = foil thickness

$$\therefore \qquad N \propto \frac{1}{\sin^4\left(\frac{\theta}{2}\right)}$$

Limitations of Rutherford Atomic Model

(i) **About the Stability of Atom** According to Maxwell's electromagnetic wave theory electron should emit energy in the form of electromagnetic wave during its orbital motion. Therefore, radius of orbit of electron will decrease gradually and ultimately it will fall in the nucleus.

(ii) **About the Line Spectrum** Rutherford atomic model cannot explain atomic line spectrum.

Bohr's Atomic Model

Electron can revolve in certain non-radiating orbits called **stationary orbits** for which the angular momentum of electron is an integer multiple of $\left(\frac{h}{2\pi}\right)$.

$$mvr = \frac{nh}{2\pi}$$

where, $n = 1, 2, 3, \ldots$ called principle quantum number.

The radiation of energy occurs only when any electron jumps from one permitted orbit to another permitted orbit.

Energy of emitted photon, $h\nu = E_2 - E_1$

where, E_1 and E_2 are energies of electron in orbits.

Radius of orbit of electron is given by

$$r = \frac{n^2h^2}{4\pi^2\, mK\, Ze^2} \Rightarrow r \propto \frac{n^2}{Z}$$

where, n = principle quantum number, h = Planck's constant, m = mass of an electron, $K = \frac{1}{4\pi\,\varepsilon_0}$, Z = atomic number and e = electronic charge.

Velocity of electron in any orbit is given by

$$v = \frac{2\pi KZe^2}{nh} \quad \Rightarrow \quad v \propto \frac{Z}{n}$$

Frequency of electron in any orbit is given by

$$\nu = \frac{KZe^2}{nhr} = \frac{4\pi^2Z^2e^4mK^2}{n^3h^3}$$

$$\Rightarrow \qquad \nu \propto \frac{Z^2}{n^3}$$

Kinetic energy of electron in any orbit is given by

$$E_k = \frac{2\pi^2 me^4 Z^2 K^2}{n^2 h^2} = \frac{13.6\, Z^2}{n^2}\ \text{eV}$$

Potential energy of electron in any orbit is given by

$$E_P = \frac{-4\pi^2 me^4 Z^2 K^2}{n^2 h^2}$$

$$= -\frac{27.2\, Z^2}{n^2}\ \text{eV}$$

$$\Rightarrow \qquad E_P \propto \frac{Z^2}{n^2}$$

Total energy of electron in any orbit is given by

$$E = \frac{-2\pi^2 me^4 Z^2 k^2}{n^2 h^2} = -\frac{13.6\, Z^2}{n^2}\ \text{eV}$$

$$\Rightarrow \qquad E \propto \frac{Z^2}{n^2}$$

Wavelength of radiation emitted in the radiation from orbit n_2 to n_1 is given by

$$\frac{1}{\lambda} = \frac{2\pi^2 mK^2 e^4 Z^2}{ch^3}\left[\frac{1}{n_1^2} - \frac{1}{n_2^2}\right]$$

$$\frac{1}{\lambda} = R\left[\frac{1}{n_1^2} - \frac{1}{n_2^2}\right]$$

where,
$$R = \frac{2\pi^2 mK^2 e^4 Z^2}{ch^3}$$

$$= 1.097 \times 10^7\ \text{m}^{-1}$$

In quantum mechanics, the energies of a system are discreate or quantized. The energy of a particle of mass m is confined to a box of length L can have discrete values of energy given by the relation

$$E_n = \frac{n^2 h^2}{8mL^2};\ n < 1, 2, 3, \ldots$$

Hydrogen Spectrum Series

Each element emits a spectrum of radiation, which is characteristic of the element itself. The spectrum consists of a set of isolated parallel lines and is called the **line spectrum.**

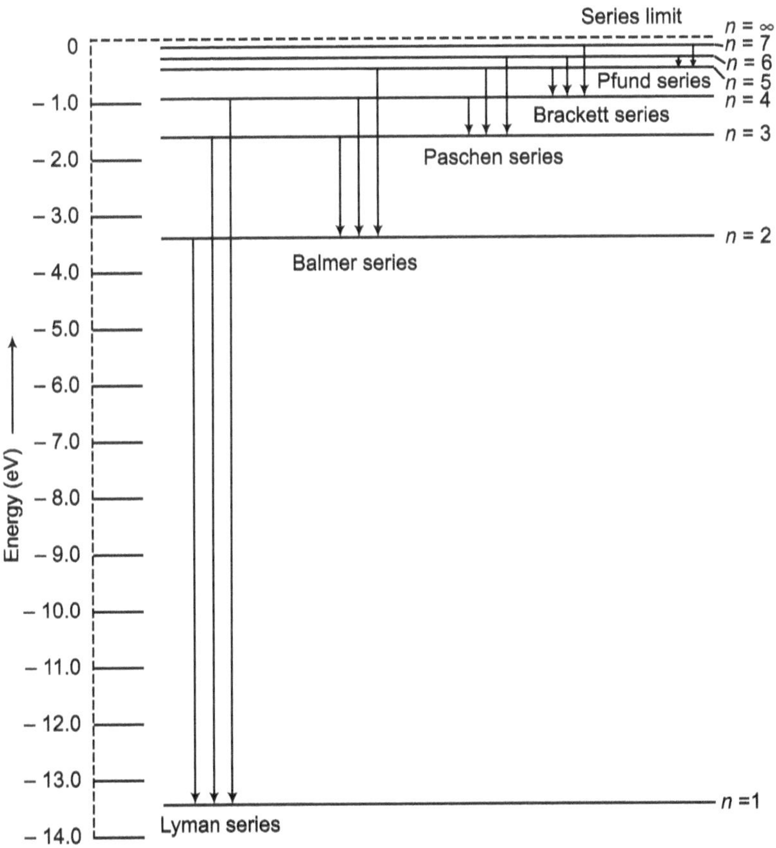

Hydrogen spectrum contains five series

(i) **Lyman Series** When electron jumps from $n = 2, 3, 4, \ldots$ orbit to $n = 1$ orbit, then a line of Lyman series is obtained.
This series lies in ultra **violet region.**

(ii) **Balmer Series** When electron jumps from $n = 3, 4, 5, \ldots$ orbit to $n = 2$ orbit, then a line of Balmer series is obtained.
This series lies in **visual region.**

(iii) **Paschen Series** When electron jumps from $n = 4, 5, 6, \ldots$ orbit to $n = 3$ orbit, then a line of Paschen series is obtained.
This series lies in **infrared region.**

(iv) **Brackett Series** When electron jumps from $n = 5, 6, 7, \ldots$ orbit to $n = 4$ orbit, then a line of Brackett series is obtained.
This series lies in **infrared region.**

(v) **Pfund Series** When electron jumps from $n = 6, 7, 8, \ldots$ orbit to $n = 5$ orbit, then a line of Pfund series is obtained.
This series lies in **infrared region.**

Wave Model

It is based on wave mechanics.

Quantum numbers are the numbers required to completely specify the state of the electrons.

In the presence of strong magnetic field, the four quantum number are

(i) Principle quantum number (n) can have value 1, 2, ...∞.

(ii) Orbital angular momentum quantum number l can have value $0, 1, 2, \ldots, (n-1)$.

(iii) Magnetic quantum number (m_e) which can have values $-l$ to l.

(iv) Magnetic spin angular momentum quantum number (m_s) which can have only two value $+\frac{1}{2}$.

30

Nuclear Physics

Nucleus

The entire positive charge and nearly the entire mass of atom is concentrated in a very small space called the nucleus of an atom.

The nucleus consists of protons and neutrons. They are called **nucleons.**

Terms Related to Nucleus

(i) **Atomic Number** The number of protons in the nucleus of an atom of the element is called atomic number (Z) of the element.

(ii) **Mass Number** The total number of protons and neutrons present inside the nucleus of an atom of the element is called mass number (A) of the element.

(iii) **Nuclear Size** The radius of the nucleus $R \propto A^{1/3}$

$$\Rightarrow \qquad R = R_0 A^{1/3}$$

where, $R_0 = 1.1 \times 10^{-15}$ m is an empirical constant.

(iv) **Nuclear Density** Nuclear density is independent of mass number and therefore same for all nuclei.

$$\rho = \frac{\text{mass of nucleus}}{\text{volume of nucleus}} \Rightarrow \rho = \frac{3m}{4\pi R_0^3}$$

where, m = average mass of a nucleon.

(v) **Atomic Mass Unit** It is defined as $\frac{1}{12}$th the mass of carbon nucleus.

It is abbreviated as amu and often denoted by u. Thus

$$1 \text{ amu} = \frac{1.992678 \times 10^{-26}}{12} \text{ kg}$$

$$= 1.6 \times 10^{-27} \text{ kg} = 931 \text{ MeV}$$

Isotopes

The atoms of an element having same atomic number but different mass numbers, are called isotopes. e.g. ${}_1H^1, {}_1H^2, {}_1H^3$ are isotopes of hydrogen.

Isobars

The atoms of different elements having same mass numbers but different atomic numbers, are called isobars. e.g. ${}_1H^3, {}_2He^3$ and ${}_{11}Na^{22}, {}_{10}Ne^{22}$ are isobars.

Isotones

The atoms of different elements having different atomic numbers and different mass numbers but having same number of neutrons, are called isotones. e.g., ${}_1H^3, {}_2He^4$ and ${}_6C^{14}, {}_8O^{16}$ are isotones.

Isomers

Atoms having the same mass number and the same atomic number but different radioactive properties are called isomers.

Nuclear Force

The force acting inside the nucleus or acting between nucleons is called nuclear force.
Nuclear forces are the strongest forces in nature.
It is a very short range attractive force.
It is non-central, non-conservative force.
It is neither gravitational nor electrostatic force.
It is independent of charge.
It is 100 times that of electrostatic force and 10^{38} times that of gravitational force.
According to the Yukawa, the nuclear force acts between the nucleons due to continuous exchange of meson particles.

Mass Defect

The difference between the sum of masses of all nucleons (M) and mass of the nucleus (m) is called mass defect.

$$\text{Mass defect } (\Delta m) = M - m = [Zm_p + (A - Z)m_n - m_N]$$

Mass-Energy

Einstein showed that mass is another form of energy and can convert mass energy into other forms of energy.

$$\text{Einstein-mass energy, } E = mc^2$$

Nuclear Binding Energy

The minimum energy required to separate the nucleons upto an infinite distance from the nucleus, is called nuclear binding energy.

$$\text{Nuclear binding energy per nucleon} = \frac{\text{Nuclear binding energy}}{\text{Total number of nucleons}}$$

Binding energy, $E_b = [Zm_p + (A - Z)\, m_n - m_N]c^2$

Packing Fraction (P)

$$P = \frac{(\text{Exact nuclear mass}) - (\text{Mass number})}{\text{Mass number}} = \frac{M - A}{M}$$

The larger the value of packing friction, greater is the stability of the nucleus.

The nuclei containing even number of protons and even number of neutrons are **most stable.**
The nuclei containing odd number of protons and odd number of neutrons are **most instable.**

Radioactivity

The phenomenon of disintegration of heavy elements into comparatively lighter elements by the emission of radiations is called radioactivity.
This phenomenon was discovered by Henry Becquerel in 1896.

Radiations Emitted by a Radioactive Element

Three types of radiations emitted by radioactive elements

(i) α-rays (ii) β-rays (iii) γ-rays

α-rays consists of α-particles, which are doubly ionised helium ion.
β-rays are consist of fast moving electrons.
γ-rays are electromagnetic rays.

When an α-particle is emitted by a nucleus its atomic number decreases by 2 and mass number decreases by 4.

$$_{Z}X^{A} \xrightarrow{\alpha\text{-particle}} {}_{Z-2}Y^{A-4}$$

When a β -particle is emitted by a nucleus its atomic number is increases by one and mass number remains unchanged.

$$_{Z}X^{A} \xrightarrow{\beta\text{-particle}} {}_{Z+1}Y^{A}$$

When a γ-particle is emitted by a nucleus its atomic number and mass number remain unchanged.

Radioactive Decay Law

The rate of disintegration of radioactive atoms at any instant is directly proportional to the number of radioactive atoms present in the sample at that instant.

$$\text{Rate of disintegration}\left(-\frac{dN}{dt}\right) \propto N - \frac{dN}{dt} = \lambda N$$

where λ is the decay constant.

The number of atoms present undecayed in the sample at any instant $N = N_0\, e^{-\lambda t}$

where, N_0 is number of atoms at time $t = 0$ and N is number of atoms at time t.

Half-life of a Radioactive Element

The time in which the half number of atoms present initially in any sample decays, is called half-life (T) of that radioactive element.

Relation between half-life and disintegration constant is given by

$$T = \frac{\log_e^2}{\lambda} = \frac{0.6931}{\lambda}$$

Average Life or Mean Life (τ)

Average life or mean life (τ) of a radioactive element is the ratio of total life time of all the atoms and total number of atoms present initially in the sample.

Relation between average life and decay constant $\tau = \dfrac{1}{\lambda}$

Relation between half-life and average life $\tau = 1.44\, T$

The number of atoms left undecayed after n half-lives is given by

$$N = N_0\left(\frac{1}{2}\right)^n = N_0\left(\frac{1}{2}\right)^{\frac{t}{T}}$$

where, $n = \dfrac{t}{T}$, here t = total time.

Activity of a Radioactive Element

The activity of a radioactive element is equal to its rate of disintegration.

$$\text{Activity } R = \left(-\frac{dN}{dt}\right)$$

Activity of the sample after time t, $R = R_0\, e^{-\lambda t}$

Its SI unit is Becquerel (Bq).

Its other units are Curie and Rutherford.

$$1 \text{ Curie} = 3.7 \times 10^{10} \text{ decay/s}$$

$$1 \text{ Rutherford} = 10^{6} \text{ decay/s}$$

Nuclear Fission

The process of the splitting of a heavy nucleus into two or more lighter nuclei is called nuclear fission.

When a slow moving neutron strikes with a uranium nucleus ($_{92}U^{235}$), it splits into $_{56}Ba^{141}$ and $_{36}Kr^{92}$ along with three neutrons and a lot of energy.

$$_{92}U^{235} + {}_0n^1 \longrightarrow {}_{56}Ba^{141} + {}_{36}Kr^{92} = 3\,{}_0n^1 + \text{energy}$$

Nuclear Chain Reaction

If the particle starting the nuclear fission reaction is produced as a product and further take part in the nuclear fission reaction, then a chain of fission reaction started, which is called nuclear chain reaction.

Nuclear chain reaction are of two types

(i) Controlled chain reaction (ii) Uncontrolled chain reaction

Nuclear Reactor

The main parts of a nuclear reactor are following

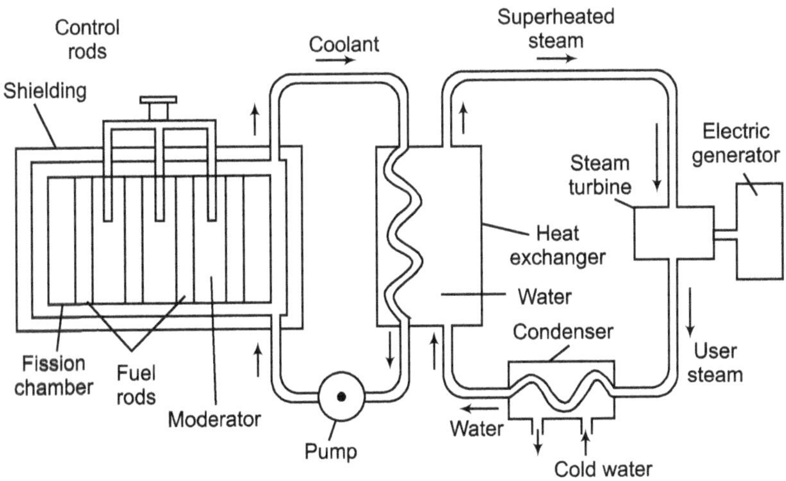

(i) **Fuel** Fissionable materials like ${}_{92}U^{235}$, ${}_{92}U^{238}$, ${}_{94}Pu^{239}$ are used as fuel.

(ii) **Moderator** Heavy water, graphite and baryllium oxide are used to slower down fast moving neutrons.

(iii) **Coolant** The cold water, liquid oxygen etc are used to remove heat generated in the fission process.

(iv) **Control rods** Cadmium or boron rods are good absorber of neutrons and therefore used to control the fission reaction.

Atom bomb working is based on uncontrolled chain reaction.

Nuclear Fusion

The process of combining of two lighter nuclei to form one heavy nucleus, is called nuclear fusion.

Three deuteron nuclei (${}_1H^2$) fuse, 21.6 MeV is energy released and nucleus of helium (${}_2He^4$) is formed.

$${}_1H^2 + {}_1H^2 + {}_1H^2 \longrightarrow {}_2He^4 + {}_1H^1 + {}_0n^1 + 21.6\,\text{MeV}$$

In this process, a large amount of energy is released.

Nuclear fusion takes place at very high temperature approximately about 10^7 K and at very high pressure 10^6 atmosphere.

Hydrogen bomb is based on nuclear fusion.

The source of Sun's energy is the nuclear fusion taking place at sun.

Thermonuclear Energy

The energy released during nuclear fusion is known as thermonuclear energy. Protons are needed for fusion while neutrons are needed for fission process.

31 Electronics

It is the branch of science which deals with the electron flow through a vacuum, gas or semiconductor.

Classification of substances on the basis of conduction of electricity.

Solid

We know that, each substance is composed of atoms. Substances are mainly classified into three categories namely solids, liquids and gases.

In each solid atoms are at a definite positions and the average distance between them is constant.

Depending upon the internal arrangement of atoms, solids are further divided into two groups.

1. Crystalline Solids

The solid in which the atoms are arranged in a regular order are called the crystalline solids. In other words, we can say that in a crystalline solid, there is periodicity and regularity of its component atoms in all the directions. For example sodium chloride (common salt), diamond, sugar, silver etc are the crystalline solids.

Their atoms are arranged in a definite geometrical shape.

They have a definite melting point.

They are anisotropic, i.e. their physical properties such as thermal conductivity, refractive index etc, are different in different directions.

They are the real solids.

2. Amorphous Solids

The solids in which the atoms do not have a definite arrangement are called the amorphous solids. They are also called the glassy solids. e.g. glass, rubber, plastic, power, etc are the amorphous solids.

They do not have a definite arrangement of its atoms, i.e. they do not have a characteristic geometrical shape.

They do not have a definite melting point.

They are isotropic, *i.e.*, their physical properties such as conductivity of heat refractive index etc, are same in all the directions.

They are not the real solids.

Monocrystal and Polycrystal

Monocrystal is a crystal in which the ordered arrangements of the atoms or molecules extends throughout the piece of solid, irrespective of its size.

Polycrystal is a crystalline solid in which each piece of the solid has a number of monocrystals with developed faces joined together.

The polycrystal ceramic made from PbO, ZnO_2 and TiO_2 are used in gas lighters and telephone receivers.

Liquid Crystals

Some organic crystalline solid, when heated acquire fluidity but retain their anisotropic properties. They are called liquid crystals.

Some liquid crystals like cyanobiphenyl can change the plane of polarization of light and such Liquid Crystal Displays (LCD) are used in watches and micro calculators.

Crystal Lattice

A crystal is made up of a three-dimensional array of points such that each point is surrounded by the height bouring points in an identical way. Such an array of points is known as bravais lattice or space lattice.

Unit cell is the smallest unit of the crystal lattice, repetition of which in three dimensions gives rise to crystal lattice.

The length of three sides of a unit cell are called primitives or lattice constant represented by a, b, c. The angle between three crystallographic axis are called **interfacial angles,** represented by α, β and γ. The primitives and interfacial angles constitute the lattice parameters of a unit cell.

The cubic crystal may be of the form, simple cubic (sc) lattice, the body centred cubic (bcc) lattice, the face centred cubic (fcc) lattice.

The coordination number is defined as the number of nearest neighbours around any lattice point (or atom) in the crystal lattice.

(a) For sc, coordination number is 6.

(b) For bcc, coordination number is 8.

(c) For fcc, coordination number is 12.

(d) For sc, atomic radius is $\frac{a}{2}$.

(e) For bcc, atomic radius is $\frac{a\sqrt{3}}{4}$.

(f) For fcc, atomic radius is $\frac{a}{2\sqrt{2}}$.

Classification of solids on the basis of conductivity

(i) **Conductor** Conductors are those substances through which electricity can pass easily, e.g. all metals are conductors.

(ii) **Insulator** Insulators are those substances through which electricity cannot pass, e.g. wood, rubber, mica etc.

(iii) **Semiconductor** Semiconductors are those substances whose conductivity lies between conductors and insulators, *e.g.*, germanium, silicon, carbon etc.

Energy Bands of Solids

1. Energy Band

In a crystal due to interatomic interaction valence electrons of one atom are shared by more than one atom in the crystal. Now splitting of energy levels takes place. The collection of these closely spaced energy levels is called an energy band.

2. Valence Band

This energy band contains valence electrons. This band may be partially or completely filled with electrons but never be empty.

The electrons in this band are not capable of gaining energy from external electric field to take part in conduction of current.

3. Conduction Band

This band contains conduction electrons. This band is either empty or partially filled with electrons.

Electrons present in this band take part in the conduction of current.

4. **Forbidden Band**

This band is completely empty. The minimum energy required to shift an electron from valence band to conduction band is called band gap (E_g).

Thermionic Emission

Thermionic emission occurs when a metal is heated to a high temperature, the free electrons in the metal gain kinetic energy sufficient to escape through the surface of the metal.

Thermionic Diode

The thermionic diode is a two electrode (cathode and plate) device based on thermionic emission.A diode allows unidirectional flow of electrons, i.e. only when the plate is positive with respect to cathode. Hence, it is also called a valve.

The triode value consists of three electrodes, e.g. cathode, plate and grid enclosed in an evacuated glass bulb.

Grid influences the space charge and controls the flow of plate current.

When the grid is given a negative potential with respect to cathode. It repels the electrons escaping from the cathode and increases the effect of space charge, at sufficiently negative grid potential is known as cut-off grid bias.
If the grid is given a positive potential with respect to cathode, it attracts the electrons and decreases the effect of space charge. The increasing the plate current. In this case, a current flow into the circuit, thus grid modifies the function of valve.

Grid is always kept at small negative potential with respect to cathode.

Triode can be used as an amplifier, oscillator modulator and demodulator. An oscillator is an electronic device which generates AC voltage from DC power. It is basically a positive feedback amplifier with infinite voltage gain.

Types of Semiconductor

(i) **Intrinsic Semiconductor** A semiconductor in its pure state is called intrinsic semiconductor.

(ii) **Extrinsic Semiconductor** A semiconductor doped with a suitable impurity to increase its impurity is called extrinsic semiconductor.

On the basis of doped impurity extrinsic semiconductors are of two types

(a) **n-type Semiconductor** Extrinsic semiconductor doped with pentavalent impurity like As, Sb, Bi, etc in which negatively charged electrons works as charge carrier, is called n-type semiconductor.

Every pentavalent impurity atom donate one electron in the crystal, therefore it is called a doner atom.

(b) **p-type Semiconductor** Extrinsic semiconductor doped with trivalent impurity like Al, B, etc, in which positively charged holes works as charge carriers, is called p-type semiconductor.

Every trivalent impurity atom have a tendency to accept one electron, therefore it is called an acceptor atom.

In a doped semiconductor $n_e n_h = n_i^2$, where n_e and n_h are the number density of electrons and holes and n_i is number density of intrinsic carriers, i.e. electrons or holes.

In n-type semiconductor, $n_e >> n_h$

In p-type semiconductor, $n_h >> n_e$

Electrical conductivity of extrinsic semiconductor is given by

$$\sigma = \frac{1}{\rho} = e\,(n_e \mu_e + n_h \mu_h)$$

where ρ is resistivity, μ_e and μ_h are mobility of electrons and holes respectively.

Note *Energy gap for Ge is* 0.72 eV *and for Si it is* 1.1 eV.

p-n Junction

An arrangement consisting a p-type semiconductor brought into a close contact with n-type semiconductor, is called a p-n junction.

The current in a p-n junction is given by

$$k_B = I_0\,(e^{eV/k_B T} - 1)$$

where I_0 is reverse saturation current, V is potential difference across the diode, and k_B is the Boltzmann constant.

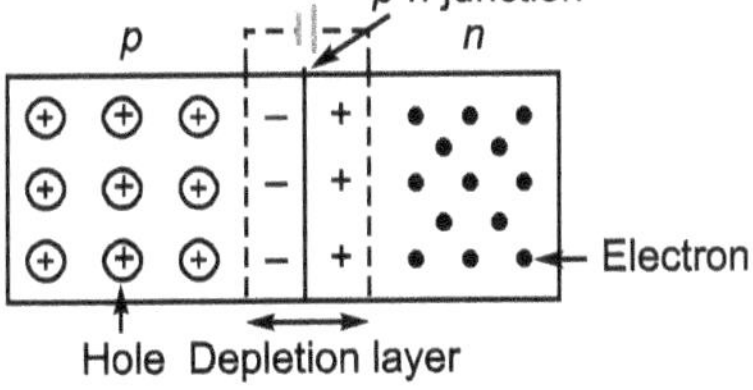

Terms Related to *p-n* Junction

(i) **Depletion Layer** At *p-n* junction a region is created, where there is no charge carriers. This region is called depletion layer.

The width of this region is of the order of 10^{-6} m.

(ii) **Potential Barrier** The potential difference across the deplation layer is called potential barrier.

Barrier potential for Ge is 0.3 V and for Si is 0.7 V.

(iii) **Forward Biasing** In this biasing, the *p* -side is connected to positive terminal and *n*-side to negative terminal of a battery.

In this biasing, forward current flows due to majority charge carriers.

The width of deplation layer decreases.

(iv) **Reverse Biasing** In this biasing, the *p*-side is connected to negative terminal and *n*-side to positive terminal of a battery.

In this biasing, reverse current flows due to minority charge carriers.

The width of depletion layer increases.

A *p-n* junction diode can be utilized as a rectifier.

Zener diode, photo-diode, light-emitting diode, etc are specially designed *p-n* junction diodes.

p-n Junction Diode

The current through *p-n* junction flow only from *p* toward *n* and not from *n* toward *p*.

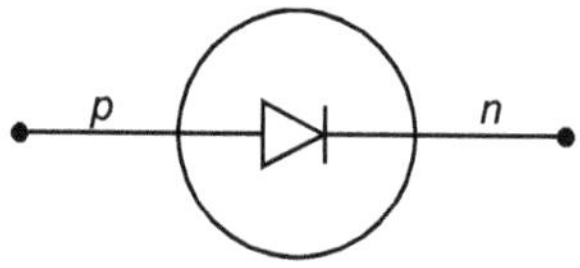

p-n junction diode

The maximum voltage that a junction diode can bear without break is called **zener voltage** and the junction diodes possessing this voltage is known as **zener diode**.

Resistance of diode $R = \dfrac{V}{I}$

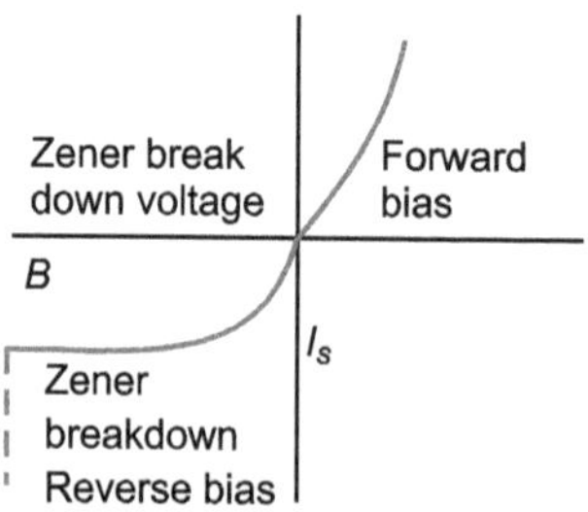

Characteristic curve of *p-n* junction diode

Light Emitting Diodes (LED)

It is forward biased *p-n* junction diode which emits light when recombination of electrons and holes takes place at the junction.

If the semiconducting material of *p-n* junction is transparent to light, the light is emitting and the junction becomes a light source, i.e. Light Emitting Diode (LED).

The colour of the light depends upon the types of material used in making the semiconductor diode.

(i) Gallium – Arsenide (Ga-As) – Infrared radiation

(ii) Gallium – Phosphide (GaP) – Red or green light

(iii) Gallium – Arsenide – Phosphide (GaAsP) – Red or yellow light

Photodiode

A photodiode is a special type of junction diode used for detecting optical signals. It is a reverse biased *p-n* junction made from a photosensitive material. In photodiode, current carries are generated by photons through photo excitation. Its symbol is

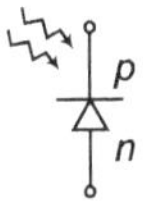

Solar Cell

Solar cell is a *p-n* junction diode which converts solar energy into electrical energy. Its symbol is

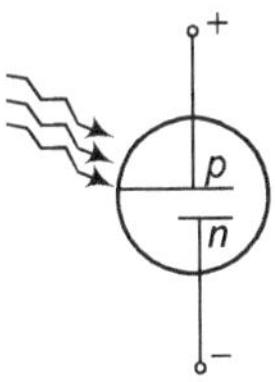

Rectifier

A device which converts alternating current or voltage into direct current or voltage is known as rectifier. The process of converting AC into DC is called **rectification**.

Half-Wave Rectifier

A half-wave rectifier converts the half cycle of applied AC signal into DC signal. Ordinary transformer may be used here.

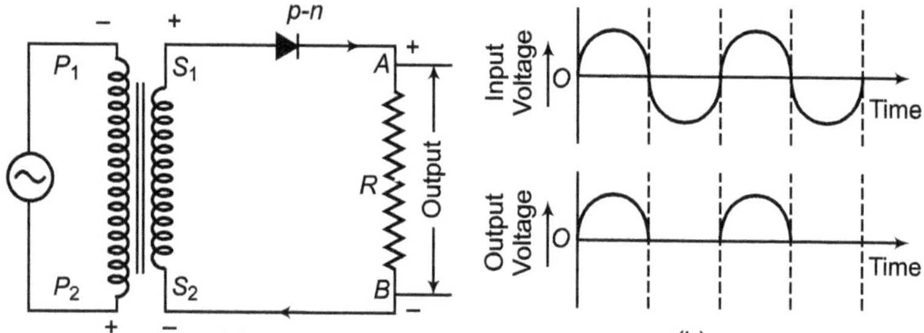

(a) (b)

Full-Wave Rectifier

A full-wave rectifier converts the whole cycle of applied AC signal into DC signal. Centre top-transformer is used here.

> Half-wave rectifier converts only one-half of AC into DC while full wave rectifier rectifies both halves of AC input.

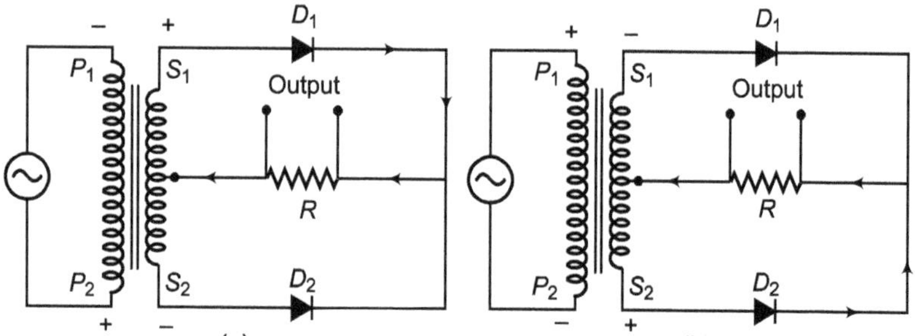

(a) (b)

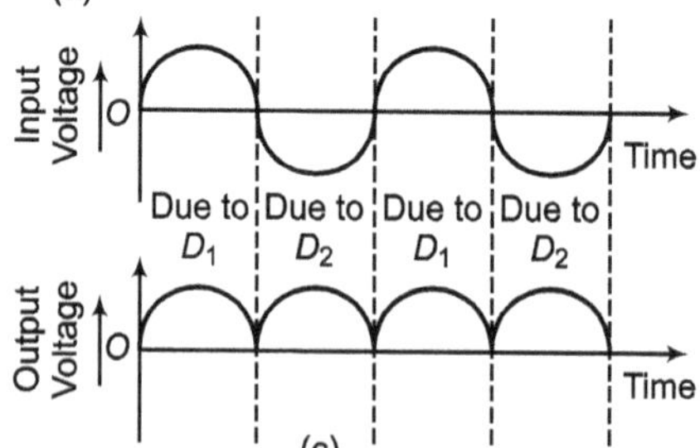

(c)

Transistor

A transistor is an arrangement obtained by growing a thin layer of one type of semiconductor between two thick layers of other similar type semiconductor.

Types of Transistors

(i) *p-n-p* transistor (ii) *n-p-n* transistor

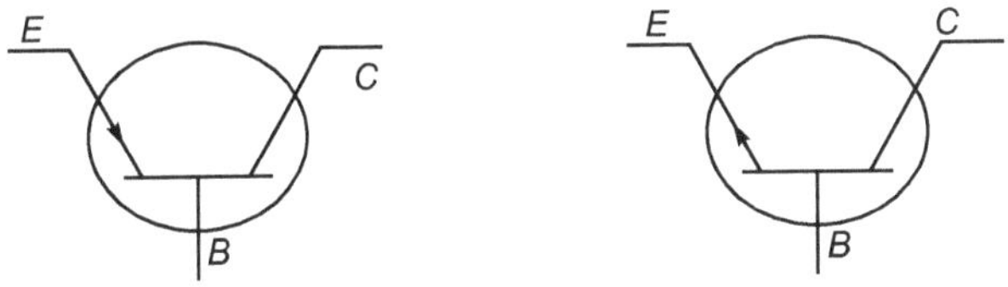

- The left side semiconductor is called **emitter,** the right side semiconductor is called **collector** and the thin middle layer is called **base**.
- Emitter is highly doped and base is feebly doped.
- A transistor can be utilized as an amplifier and oscillator but not a rectifier.
- Maximum amplification is obtained in common-emitter configuration.

Transistor as an Amplifier

An amplifier is a device which is used for increasing the amplitude of varition of alternating voltage or current or power.

The amplifier thus produces an enlarged version of the input signal.

The general concept of amplification is repesented in figure. There are two input terminals for the signal to be amplified and two output terminals for connecting the load; and a means of supplying power to the amplifier.

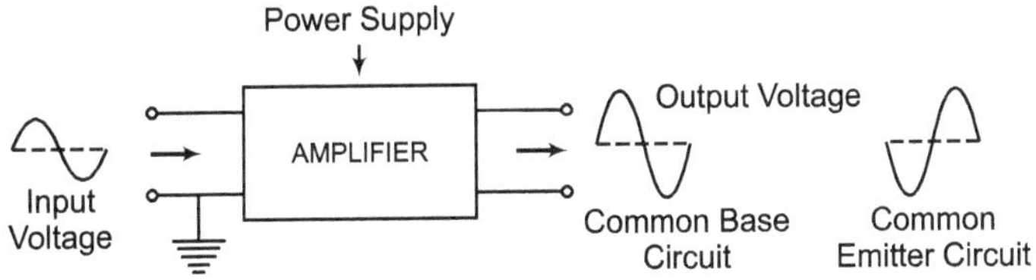

1. **In Common Base Amplifier,**

$$\text{AC current gain } (\alpha_{AC}) = \frac{\Delta I_c}{\Delta I_e}$$

where ΔI_c is change in collector current and ΔI_e change in emitter current.

$$\text{AC voltage gain } (A_V) = \frac{\text{Output voltage}}{\text{Input voltage}}$$

$$= \alpha_{AC} \times \text{Resistance gain}$$

$$= \alpha_{AC} \times \frac{R_o}{R_i}$$

where R_o is output resistance of the circuit and R_i is input resistance of the circuit.

$$\text{AC power gain} = \frac{\text{Change in output power}}{\text{Change in input power}}$$

$$= \text{AC voltage gain} \times \text{AC current gain}$$

$$= \alpha_{AC}^2 \times \text{Resistance gain.}$$

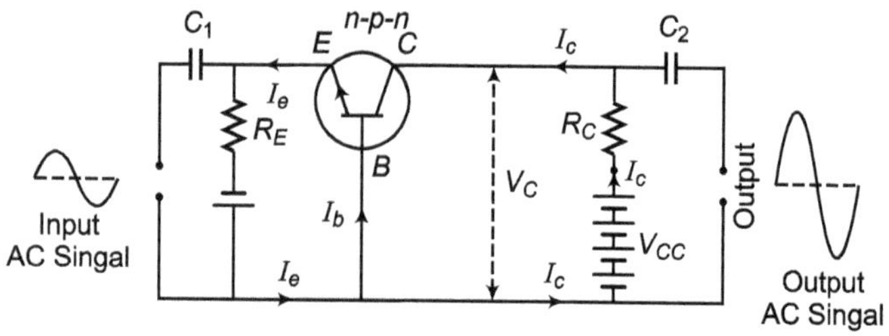

The input and output signals are in the same phase.

There is no amplification in current of a given signal.

There is an amplification in voltage and power of the given signal.

2. **In Common Emitter Amplifier**

$$\text{AC current gain } (\beta_{AC}) = \frac{\Delta I_c}{\Delta I_b}$$

where ΔI_c is change in collector current, and ΔI_b change in base current.

$$\text{AC voltage gain } (A_V) = \beta_{AC} \times \text{resistance gain}$$

$$\text{AC power gain} = \beta_{AC}^2 \times \text{resistance gain}$$

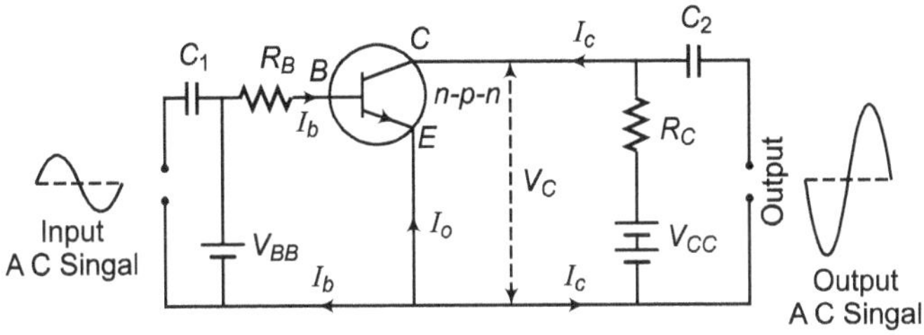

Relation between the current gain of common base and common emitter amplifier.

$$\beta = \frac{\alpha}{1-\alpha} = \frac{I_C}{I_B}$$

The input and output signals are out of phase by π or 180°.

There is amplification in current, voltage and power of the given signal.

Logic Gate

A digital circuit which allows a signal to pass through it, only when few logical relations are satisfied, is called a logic gate.

Truth Table

A table which shows all possible input and output combinations is called a truth table.

Basic Logic Gates

(i) **OR Gate** It is a two input and one output logic gate.

Symbol

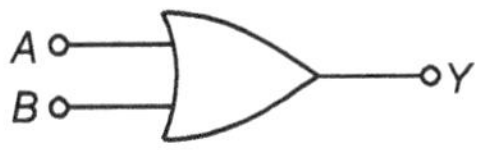

Truth table

A	*B*	*Y* = *A* + *B*
0	0	0
0	1	1
1	0	1
1	1	1

Boolean expression $Y = A + B$ (Y equals A OR B)

(ii) **AND Gate** It is a two input and one output logic gate.

Symbol

Truth table

A	B	$Y = A \cdot B$
0	0	0
0	1	0
1	0	0
1	1	1

Boolean expression $Y = A \cdot B$ (Y equals A AND B)

(iii) **NOT Gate** It is a one input and one output logic gate.

Symbol

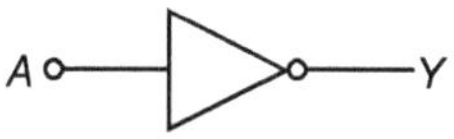

Truth table

A	$Y = \overline{A}$
0	1
1	0

Boolean expression $Y = \overline{A}$ (Y equals NOT A)

Combination of Gates

(i) **NAND Gate** When output of AND gate is applied as input to a NOT gate, then it is called a NAND gate.

Symbol

Truth table

A	B	$Y = \overline{A \cdot B}$
0	0	1
0	1	1
1	0	1
1	1	0

Boolean expression $Y = \overline{A \cdot B}$ (Y equals negated of A AND B)

(ii) **NOR Gate** When output of OR gate is applied as input to a NOT gate, then it is called a NOR gate.

Symbol

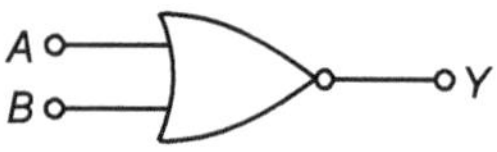

Truth table

A	B	Y
0	0	1
0	1	0
1	0	0
1	1	0

Boolean expression $Y = \overline{A + B}$ (Y equals negated of A OR B)

- The Boolean expression obey commutative law associative law as well as distributive law.

  (i) $A + B = B + A$ (ii) $A \cdot B = B \cdot A$ (iii) $A + (B + C) = (A + B) + C$

- De-Morgan's theorems

  (i) $\overline{A + B} = \overline{A} \cdot \overline{B}$ (ii) $\overline{A} \cdot \overline{B} = \overline{A} + \overline{B}$

32

Electromagnetic Waves and Communication

Displacement Current

It is a current which produces in the region in which the electric field and hence the electric flux changes with time.

Displacement current, $I_D = \varepsilon_0 \cdot \frac{d\phi_E}{dt}$

where, ϕ_E is the electric flux.

Ampere-Maxwell Law

$$\oint \mathbf{B} \cdot d\mathbf{l} = \mu_0 (I + I_D)$$

where, μ_0 = Permeability

$$= 4\pi \times 10^{-7} \text{ V/Am}$$

Maxwell's Equations

(i) $\oint_S \mathbf{E} \cdot d\mathbf{S} = \frac{q}{\varepsilon_0}$

This equation is Gauss's law in electrostatics.

(ii) $\oint_S \mathbf{E} \cdot d\mathbf{S} = 0$

This equation is Gauss's law in magnetostatics.

(iii) $\oint_S \mathbf{E} \cdot d\mathbf{l} = -\frac{d}{dt} \oint_S \mathbf{B} \cdot d\mathbf{S}$

This equation is Faraday's law of electromagnetic induction.

(iv) $\oint \mathbf{B} \cdot d\mathbf{l} = \mu_0 \left(I + \varepsilon_0 \frac{d\phi_E}{dt} \right)$

This equation is Ampere-Maxwell law.

Electromagnetic Waves

Electromagentic waves are those waves in which electric and magnetic field vectors changes sinusoidally and are perpendicular to each other as well as at right angles to the direction of propagation of wave.

The equation of plane progressive electromagnetic wave can be written as $E = E_0 \sin \omega \left(t - \frac{x}{c} \right)$ and $B = B_0 \sin \omega \left(t - \frac{x}{c} \right)$. Where, $\omega = 2\pi\nu$

Electromagnetic waves are produced by accelerated charge particles.

Properties of EM Waves

(i) These waves are transverse in nature.

(ii) These waves propagate through space with speed of light, *i.e.*, 3×10^8 m/s

(iii) The speed of electromagnetic wave, $c = \frac{1}{\sqrt{\mu_0 \varepsilon_0}}$

where, μ_0 is permeability of free space and ε_0 is permittivity of free space,

$$\therefore \qquad c = \frac{E_0}{B_0}$$

where E_0 and B_0 are maximum values of electric and magnetic field vectors.

According to Maxwell, when a charged particle is accelerated, it produces electromagnetic wave. The total radiant flux at any instant is given by, $P = \frac{q^2 a^2}{6\pi\varepsilon_0 c^2}$

(iv) The rate of flow of energy in an electromagnetic wave is described by the vector **S** called the poynting vector, which is defined by the expression,

$$\mathbf{S} = \frac{1}{\mu_0} \mathbf{E} \times \mathbf{B}$$

SI unit of **S** is watt/m^2.

(v) Its magnitude S is related to the rate at which energy is transported by a wave across a unit area at any instant.

(vi) The energy in electromagnetic waves is divided equally between electric field and magnetic field vectors.

(vii) The average electric energy density.

$$U_E = \frac{1}{2}\varepsilon_0 E^2 = \frac{1}{4}\varepsilon_0 E_0^2$$

(viii) The average magnetic energy density,

$$U_B = \frac{1}{2}\frac{B^2}{\mu_0} = \frac{1}{4}\frac{B_0^2}{\mu_0}$$

(ix) The electric vector is responsible for the optical effects of an electromagnetic wave.

(x) Intensity of electromagnetic wave is defined as energy crossing per unit area per unit time perpendicular to the directions of propagation of electromagnetic wave.

(xi) The intensity I is given by the relation,

$$I = <\mu> c = \frac{1}{2}\varepsilon_0 E_0^2 c$$

(xii) The existence of electromagnetic waves was confirmed by Hertz experimentally in 1888.

Propagation of Electromagnetic Waves

In radiowave communication between two places, the electromagnetic waves are radiated out by the transmitter antenna at one place which travel through the space and reach the receiving antenna at the other place.

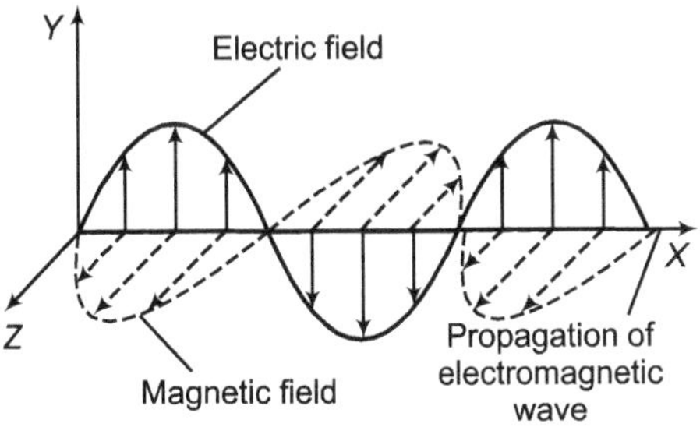

Electromagnetic Spectrum

The arranged array of electromagnetic radiations in the sequence of their wavelength or frequency is called electromagnetic spectrum.

Radio and microwaves are used in radio and TV communication.

Infrared rays are used to

(i) Treat muscular strain.
(ii) For taking photographs in fog or smoke.
(iii) In green house to keep plants warm.
(iv) In weather forecasting through infrared photography.

Visible light is the narrow region of electromagnetic spectrum, which is detected by the human eye. It is produced due to atomic excitation.

Ultraviolet rays are used

(i) In the study of molecular structure.
(ii) In sterilizing the surgical instruments.
(iii) In the detection of forged documents, finger prints.

X-rays are used

(i) In detecting faults, cracks, flaws and holes in metal products.
(ii) In the study of crystal structure.
(iii) For the detection of pearls in oysters.

γ-rays are used for the study of nuclear structure.

Earth's Atmosphere

The gaseous envelope surrounding the earth is called earth's atmosphere. It contain the following layers

(i) **Troposphere** This region extands upto a height of 12 km from earth's surface.
(ii) **Stratosphere** This region extands from 12 km to 50 km. In this region, most of the atmospheric ozone is concentrated from 30 to 50 km. This layer is called ozone layer.
(iii) **Mesosphere** The region extands from 50 km to 80 km.
(iv) **Ionosphere** This region extands from 80 km to 400 km.

In ionosphere the electron density is very large in a region beyond 110 km from earth's surface which extends vertically for a few kilometer. This layer is called **Kennelly Heaviside layer.**

In ionosphere a layer having large electron density is found at height 250 km from earth's surface, called **Appleton layer.**

There are four main layers in earth's atmosphere having high density of electrons and positive ions, produced due to ionisation by the high energy particles coming from sun, star or cosmos. These layers play their effective role in space communication. These layers are D, E, F_1 and F_2.

(i) ***D*-layer** is at a virtual height of 65 km from surface of earth and having electorn density $\approx 10^9 \text{ m}^{-3}$.

(ii) ***E*-layer** is at a virtual height of 100 km, from the surface of earth, having electron density $\approx 2 \times 10^{11} \text{m}^{-3}$.

(iii) ***F_1*-layer** is at a virtual height of 180 km from the surface of earth, having electron density $\approx 3 \times 10^{11} \text{m}^{-3}$.

(iv) ***F_2*-layer** is at a vertical height of about 300 km in night time and about 250 to 400 km in day time. The electron density of this layer is $\approx 8 \times 10^{11} \text{m}^{-3}$.

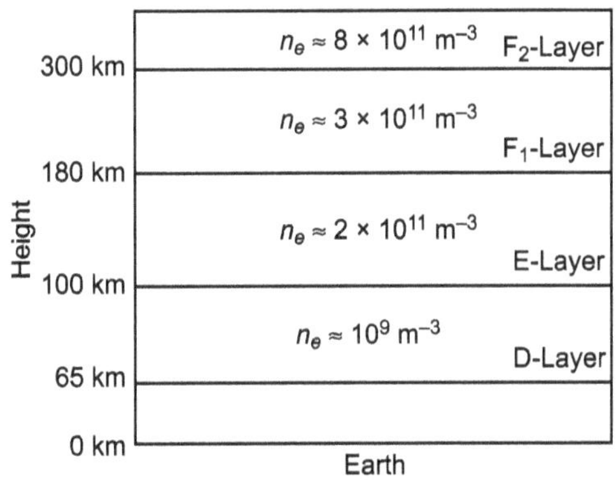

Communication

Faithful transmission of information from one place to another place is called communication.

Optical fibers are used in optical communication.

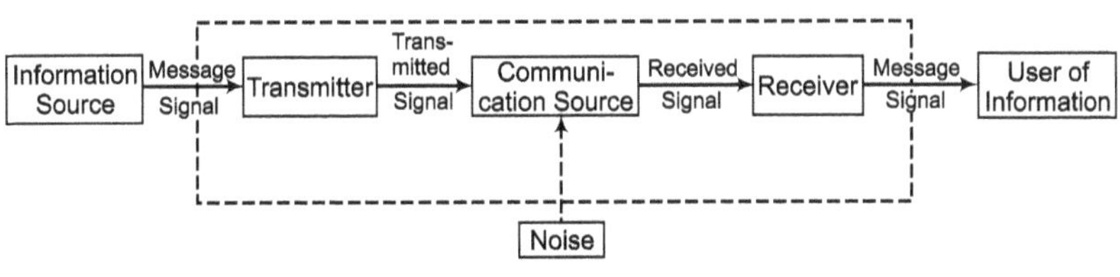

Communication System

A communication system contains three main parts

(i) **Transmitter** It process and encode the information and make it suitable for transmission.

The message signal for communication can be analog signals or digital signals.

An analog signal can be converted suitably into a digital signal and *vice-versa*.

An analog signal is that in which current or voltage value varies continuously with time.

A digital signal is a discontinuous function of time. Such a signal is usually in the form of pulses.

(ii) **Communication Channel** The medium through which information propagate from transmitter to receiver is called communication channel.

(iii) **Receiver** It receives and decode the signal.

Analog Signal

A signal in which current or voltage changes its magnitude continuously with time, is called an analog signal.

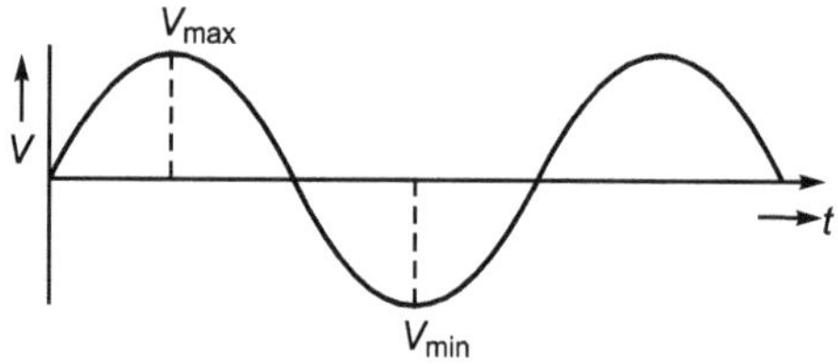

Digital Signal

A signal in which current or voltage have only two values, is called a digital signal.

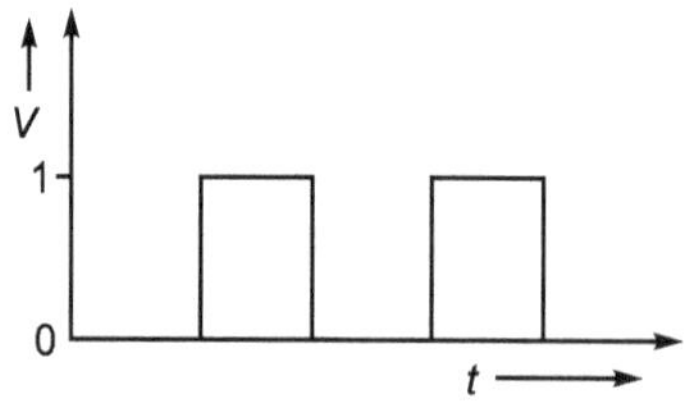

Note *An analog signal can be converted suitable into a digital signal and vice-versa.*

Modulation

The process of superimposing the audio signal over a high frequency carrier wave is called modulation.

In the process of modulation any one characteristic of carrier wave is varied in accordance with the instantaneous value of audio signal (modulating signal).

Need of Modulation

(i) Energy carried by low frequency audio waves (20 Hz to 20000 Hz) is very small.

(ii) For efficient radiation and reception of signal, the transmitting and receiving antennas should be very high approximately 5000 m.

(iii) The frequency range of audio signal is so small that overlapping of signals create a confusion.

Types of Modulation

(i) **Amplitude Modulation** In this type of modulation, the amplitude of high frequency carrier wave is varied in accordance to instantaneous amplitude of modulating signal.

Band width required for amplitude modulation

= twice the frequency of the modulating signal.

(ii) **Frequency Modulation** In this type of modulation, the frequency of high frequency carrier wave is varied in accordance to instantaneous frequency of modulating signal.

(iii) **Pulse Modulation** In this type of modulation, the continuous waveforms are sampled at regular intervals. Information is transmitted only at the sampling times.

Demodulation

The process of separating of audio signal from modulated signal is called demodulation.

Antenna

An antenna converts electrical energy into electromagnetic waves at transmitting end and pick up transmitted signal at receiving end and converts electromagnetic waves into electrical signal.

Modem

The term modem is cotraction of the term modulater and demodulater. Modem is a device which can modulate as well as demodulate the signal. It connect one computer to another through ordinary telephone lines.

The Internet

Internet permits communication and sharing of all types of information between any two or more computers connected through a large and complex network.

The exchange of information on internet is very fast (at the speed of light) as electronic signals (messages) of computers are communicated through electromagnetic waves.

Application of internet include: internet surfing, E-mails, file transfer, E-banking, E-commerce, E-booking and social networking.

Fax (Facsimile Telegraphy)

The electronic reproduction of a document at a distant place is called FAX.

Mobile Telephony

Mobile phone can wirelessly send and receive radio frequency signals. The central concept of this system is to divide the service area into a suitable number of cells centred an office called MTSO (Mobile Telephone Switching Office).

Mobile phones operate typically in UHF of frequencies (about 800-950 MHz).

Global Positioning System (GPS)

GPS is a space based satellite navigation system that provides accurate information about time and location anywhere on or near the earth.

GPS is a method of identifying location or position of any point (or a person) on the earth using a system of 24 satellites, which are continuously orbiting, monitoring and mapping the earth surface. Every such satellite orbits around the earth twice a day at a distance of about 20,000 km from it. The orbits of these satellites are aligned in such a way that atleast four of them always keep looking any given point on the earth surface.

Radio Waves

The radio waves are the electromagnetic waves of frequency ranging from 500 kHz to about 1000 MHz. These waves are used in the field of radio communication. With reference to the frequency range and wavelength range, the radio waves have been divided into various categories shown in table.

Frequency Range and Wavelength Range of Radio Waves

S.No.	Frequency band	Frequency range	Wavelength range	Main use
1.	Very-Low Frequency (VLF)	3 kHz to 30 kHz	10 km to 100 km	Long distance point to point communication
2.	Low Frequency (LF)	30 kHz to 300 kHz	1 km to 10 km	Marine and navigational purposes
3.	Medium Frequency (MF)	300 kHz to 3 MHz	100 m to 1 km	Marine and broadcasting purposes
4.	High Frequency (HF)	3 MHz to 30 MHz	10 m to 100 m	Communication of all types
5.	Very-High Frequency (VHF)	30 MHz to 300 MHz	1 m to 10 m	T V Radar and air navigation
6.	Ultra-High Frequency (UHF)	300 MHz to 3000 MHz	10 cm to 1 m	Radar and microwave communication
7.	Super-High-Frequency (SHF)	3 GHz to 30 GHz	1 cm to 10 cm	Radar, Radio relays and navigation purposes
8.	Extremely-High- Frequency (EHF)	30 GHz to 300 GHz	1 mm to 1 cm	Optical fibre communication

Propagation of Radio Waves

The three modes are discussed below

(i) **Ground Wave or Surface Wave Propagation** It is suitable for low and medium frequency upto 2 MHz. It is used for local broad casting.

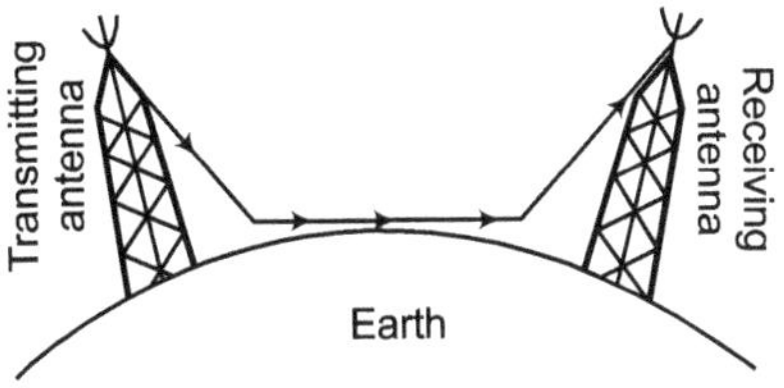

(ii) **Sky Wave Propagation** It is suitable for radiowaves of frequency between 2 MHz to 30 MHz. It is used for long distance radio communication.

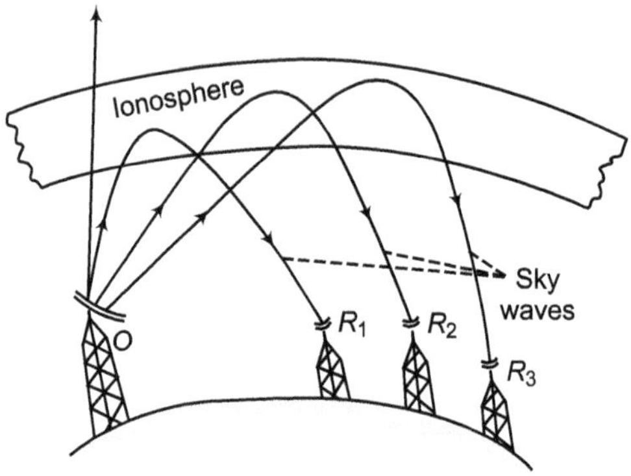

Critical Frequency The highest frequency of radio wave that can be reflected back by the ionosphere is called critical frequency.

$$\text{Critical frequency, } \nu_c = 9\,(N_{max})^{1/2}$$

where, N_{max} = number density of electrons/metre^3.

Skip Distance The minimum distance from the transmitter at which a sky wave of a frequency but not more than critical frequency, is sent back to the earth.

$$\text{Skip distance } (D_{skip}) = 2h\left(\frac{\nu_{max}}{\nu_c}\right)^2 - 1$$

where h is height of reflecting layer of atmosphere,
ν_{max} is maximum frequency of electromagnetic waves and
ν_c is critical frequency.

Fading The variation in the strength of a signal at receiver due to interference of waves, is called fading.

(iii) **Space Wave Propagation** It is suitable for 30 MHz to 300 MHz. It is used in television communication and radar communication. It is also called line of sight communication.

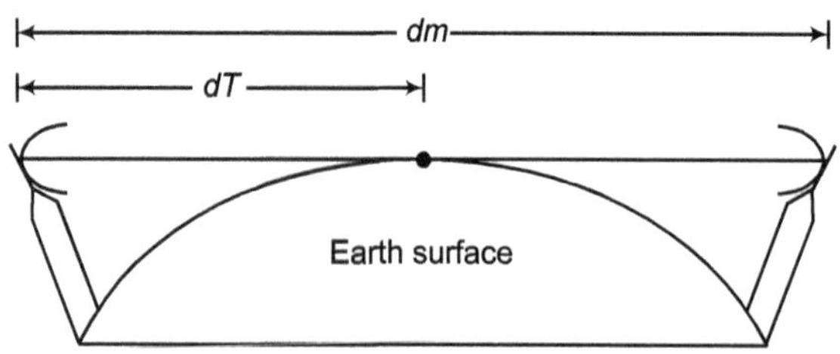

- Range is limited due to curvature of earth. If h be the height of the transmitting antenna, then signal can be received upto a maximum distance

$$d = \sqrt{2Rh}$$

- If height of transmitting and receiving antennas be h_T and h_R respectively. The effective range will

$$d = \sqrt{2Rh_T} + \sqrt{2Rh_R}$$

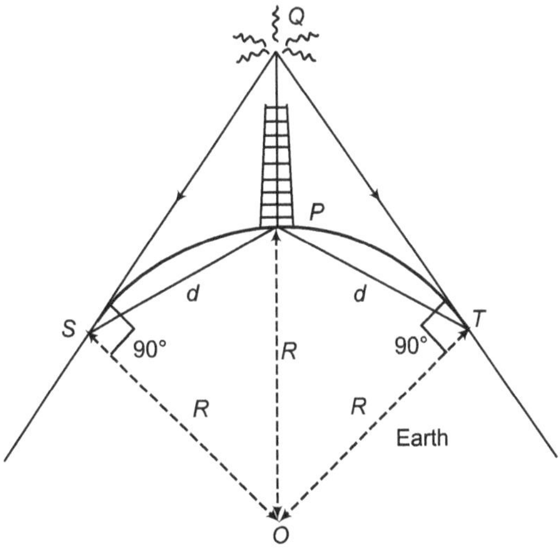

Microwave Propagation

- Microwaves are electromagnetic wave of frequency 1 to 300 GHz, greater than those of TV signals. The wavelength of microwaves is of the order of a few mm.
- Microwave communication is used in radar to locate the flying objects in space.
- These waves can be transmitted as beam signals in a particular direction, much better than radiowave.
- There is no diffraction of microwave around corners of an obstacle which happens to lie along its passage.

Satellite Communication

It is carried out between a transmitter and a receiver through a satellite. A geostationary satellite is utalized for this purpose, whose time period is 24 hours.

A communication satellite is a space craft, provided with microwave receiver and transmitter. It is placed in an orbit around the earth.

The India remote sensing satellites are

IRS-IA, IRS-IB and IRS-IC

The line-of-sight microwave communication through satellite is possible if the communication satellite is always at a fixed location with respect to the earth, *e.g.*, the satellite which is acting as a repeater must be at rest with respect to the earth. It is so far a satellite known as geostationary satellite.

The basic requirements for geostationary satellites are as follows:

1. The time period of revolution of the satellite around the earth is equal to the time period of rotation of earth about its polar axis i.e. 24 h.
2. The sense of revolution of the satellite around the earth is the same as that of the earth about its polar axis i.e. from west to east.
3. The orbital plane of revolution of satellite is concentric and coplanar with the equatorial plane of earth.
4. The height of geostationary satellite above the equator of earth is nearly 36000 km and its orbital velocity is nearly 3.1 km/s.

The orbit in which the geo-satellite above revolves around the earth is known as **geo-synchronous orbit**. As its angular speed is synchronised with the angular speed of the earth, therefore, the geostationary satellite is also known as geo-synchronous satellite.

Merits of Satellites Communication

1. The satellite communication covers wide area for broadcasting as compared to other communication systems i.e. it has wide coverage range.
2. The satellite communication is also used effectively in mobile communication.
3. The satellite communication is found to be much economical as compared to other communication systems on earth. Infact, the cost involved in satellite communication is independent of the distance.
4. The satellite communication is most cost effective in remote and hilly areas, such as Ladakh, Himachal Pardesh etc.
5. The satellite communication permits transmission of data at high rate.
6. The satellite communication is very accurate and economical for search, rescue and navigation purposes.

Demerits of Satellite Communication

1. If a system on the satellite goes out of order due to environmental stresses, it is almost impossible to repair it.
2. In satellite communication, there is a time delay between transmission and reception, due to extremely large communication path length (greater than 2×36000 km). This delay causes a time gap during talking, which proves quite annoying.

Remote Sensing

It is a technique of observing or measuring the characteristics of the object at a distance. A polar satellite is utalized for this purpose.

Distance upto which a signal can be obtained from an antenna is given by

$$d = \sqrt{2hR}$$

where, h is height of antenna and R is radius of earth.

LED and Diode Laser in Communication

Light Emitting Diode (LED) and diode laser are preferred sources for optical communication links to the following features.

(i) Each produces light of suitable power required in optical communication. Diode laser provides light which is monochromatic and coherent. This light is obtained as a parallel beam. It is used in very long distance transmission.

(ii) LED provides almost monochromatics light. This suitable for small distance transmission. It is infact, a low cost device as compared to diode lasers.

Line Communication

- Transmission lines are used to interconnect points separated from each other. For example interconnection between a transmitter and a receiver or a transmitter and antenna or an antenna and a receiver are achieved through transmission lines.
- Line communication may be in the form of electrical signal or optical signal.

Optical Fibres

An optical fibre is a long thread consisting of a central core of glass or plastic of uniform refractive index. It is surrounded by a cladding of material of refractive index less than that of the core and a protective jacket of insulating material.

There are three types of optical fibre configuration

Single mode step index fibre

Multi mode step index fibre

Multi mode graded index fibre.

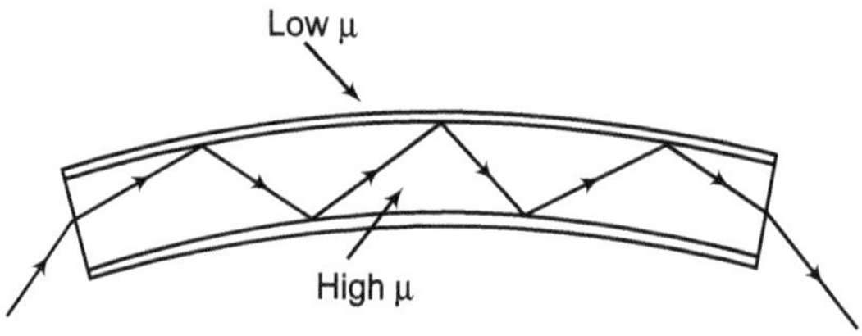

Applications of Optical Fibres

1. A bundle of optical fibres is called light pipe. This pipe can transmit as image. Since the pipe is flexible, it can be twisted in any desired manner. Hence it is used medical and optical examination of even the inaccessible parts of human body, *e.g.*, in endoscopy.
2. Optical fibres are used in transmission and reception of electrical signals by coverting them first into light signals.
3. Optical fibres are used in telephone and other transmitting cables. Each fibre can carry upto 2000 telephone messages without much loss of intensity.

33 Universe

Light Year

The distance travelled by light in vacuum in one year is called a light year.

It is a unit of length used to measure astronomical distances.

Its value is 9.46×10^{15} m.

Astronomical Unit (AU) The average distance between the earth and the sun is called astronomical unit.

Astronomy is the branch of science which deals with the study of heavenly bodies of universe.

Solar System

The sun and all the objects moving around it taken together is called solar system.

Solar system consists of the sun, the eight planets and other heavenly bodies like asteroids and comets etc.

Nearest star from the earth other than the sun is alpha centauri.

Sun

The sun is a typical example of a star.

Seventy per cent of sun's mass is hydrogen, twenty eight per cent helium and two per cent heavier elements from lithium to uranium.

The sun's mass is about 2×10^{30} kg which is more than 3×10^{5} times that of the earth and radius 1.4×10^{6} km.

The source of sun's energy is the process of nuclear fusion taking place in it.

The sun rotates about its axis and completes one rotation in 25 days.

Photosphere

It is the inner part of sun which appears as a bright disc. It is a denser mixture of gases and vapours. Its thickness is about 500 km and its temperature is about 6000 K.

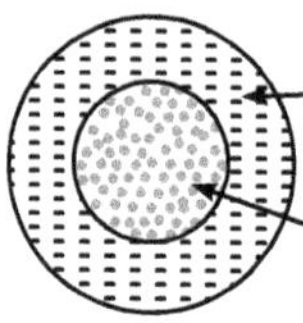

Chromosphere

It is the outer part of sun, just above the photosphere. It is a rarer mixture of gases and vapours.

Planets

The heavenly bodies which revolve around the sun in a fixed orbits, are called planets.

There are eight planets in solar system

(i) Mercury (ii) Venus
(iii) Earth (iv) Mars
(v) Jupiter (vi) Saturn
(vii) Uranus (viii) Neptune

Some important points regarding planets are as under

(i) **Mercury** Nearest, smallest and hottest planet of the solar system. Life is not possible on mercury. Mercury is the closest planet to the sun.

(ii) **Venus** Also called morning star and the evening star is brightest amongst all. Venus is the closest planet to the earth. It is the hottest planet of our solar system.

(iii) **Earth** We all know about earth. It is the only planet which contains suitable conditions for evolution and survival of life. It has only one natural satellite named **moon.**

(iv) **Mars** Nearest planet to earth. It is reddish. It has traces of O_2 but percentage of O_2 is not sufficient for evolution and survival of life.Mars is also called red planet. The temperature of mars varies from 21°C to 27°C. It has two satellites known as **phobos** and **deimos**.

(v) **Jupiter** Largest planet of the solar system having the maximum number of satellites. Jupiter is the largest planet of the solar system. Its temperature is about 140°C. It has 63 satellites and no life possible on the jupiter.

(vi) **Saturn** It has ring around it. Saturn is the second largest planet of the solar system. Its temperature is about 180°C. It has 61 satellites.

(vii) **Uranus** It is the only planet which rotates from east to west on it axis. Uranus appears green due to the large amount of methane and ammonia clouds in its atmosphere. Its temperature is about 127°C. It has 15 satellites.

(viii) **Neptune** It has no special characteristic. Neptune is very far away from the sun. It has two satellites.

Note *The ninth planet pluto is now taken as a dwarf planet.*

Terrestrial Planets

The four planets nearest the sun, Mercury, Venus, Earth and Mars are called **terrestrial planets.**

These planets have well defined surface of rocks and soil.

Jovian Planets

The four planets Jupiter, Saturn, Uranus and Neptune are called jovian planets.

These planets do not have a solid surface.

The Moon

It is the natural satellite of the earth.

It revolves around the earth once in 27.33 days.

Diameter is 3476 km.

Distance from the earth is 354000 km.

Mass of the moon is 0.0123 times the mass of the earth.

Maximum temperature at day is 117°C and at night is –171°C.

Asteroids (Minor Planets)

The small, rocky, irregularly shaped objects revolving around the sun, are called asteroids or minor planets.

There is a belt of asteroids between the orbits of mars and Jupiter.

The largest asteroid is 'Ceres'.

Asteroids can have moon too. The asteroid Ida has its moon. Dactyl.

Meteoroids

Small pieces of rock travelling through space are called meteroids. These are fragments from an asteroid, a comet, a moon, mars, etc.

Meteors

When a meteroid enter earth's atmosphere, it heats up due to friction. The hot vapours give off light before cooling down. A streak of light caused by a vaporizing meteoroid is called a meteor or a shooting star.

Meteorite

A meteoroid that hits the ground is called a meteorite.

Comets

Comets are objects which move in highly elongated orbits around the sun. They are made up of frozen gases and rocks. They have a long tail which always points away from the sun.

Inner Structure of the Earth

The earth is made up of three different layers.

(i) **Crust** It is the thinnest and outermost layer. It is about 10 km thick under the oceans and could be up to 45 km thick under the continents.

(ii) **Mantle** It lies below the crust. It extends up to a depth of about 3000 km. The temperature inside the mantle is about 1000°C.

(iii) **Core** It is the most inner part of earth. It is made up of iron mostly. The outer part of core is in molten state have a temperature about 4000°C and inner part is in solid state because of very high pressure.

Stars

(a) The heavenly objects which shine like the sun due to their own energy are called stars.

(b) The sun is the nearest star to the earth. After the sun, Alpha centauri is the next nearest star to the earth.

(c) A person can see nearly 5000 stars with the naked eye on a clear night.

(d) In addition to these stars, many groups of bright and faint stars called **constellations** can be see in the sky.

(e) The space between stars is almost empty with traces of interstellar gas and dust.

Brightness of Stars

(a) The brightness of stars is represented through a system of magnitudes.

(b) The Greek astronomer Hipparchus considered the brightest visible stars as first magnitude stars. Those about one-half as bright were called second magnitude stars and so on; the sixth magnitude stars which were the faintest stars in the sky.

(c) The magnitude scale is now defined so that a magnitude difference of 5 corresponds exactly to a factor of 100 in the amount of light energy arriving at the earth.

(d) It now follows that if l_1 and l_2 are the brightness values of two stars of magnitudes m_1 and m_2 respectively, then

$$\frac{l_1}{l_2} = 100^{(m_2 - m_1)/5}$$

Taking log of both sides, we get

$$\log\left(\frac{l_1}{l_2}\right) = \left(\frac{m_1 - m_2}{5}\right)\log 100 = \left(\frac{m_2 - m_1}{5}\right)2$$

or
$$(m_2 - m_1) = \frac{5}{2}\log\left(\frac{l_1}{l_2}\right) = -2.5\log\left(\frac{l_2}{l_1}\right)$$

If $l_1 = l_0$, the brightness of a star of zero magnitude, i.e. $m_1 = 0$,

then
$$m_2 = -2.5\log\left(\frac{l_2}{l_0}\right)$$

So, in general;
$$m = -2.5\log\left(\frac{l}{l_0}\right)$$

(e) This shows that the stars which are brighter than the standard one have negative magnitude values. Thus, a star having magnitude – 5 will be 100 times more brighter than a star of zero magnitude. The value of m for the sun is nearly – 26.5

Life Cycle of a Star

Birth of a Star

(a) Interstellar dust and gas present in outer space come together under gravitational force to from a cloud.

(b) When the cloud of dust and gas grows to a large size (of the order of 1000 solar masses) it starts contracting under the efffect of gravitational force. As a result of contraction, the cloud heats up and starts radiating. This in turn causes further contraction and results in further heating up of the cloud. This process repeats again and again till the temperature of the central core rises to about 10^7 K.

(c) When this temperature is attained, the fusion of hydrogen atoms into helium atoms is initiated with the release of energy. This energy keeps the star shining. The dwarf stars which from more than 90% of the whole population are found to be in this stage of evolution.

Death of a Star

Phase I

(i) When the hydrogen in the core of the star is exhausted it heads towards its death.

(a) With complete conversion of hydrogen atom into helium atoms, the core of the star starts contracting again with a consequent rise of temperature.

(b) With the rises in temperature, the outer layers of the star expand and the size of the star increases.

(c) With the successive contraction of the core and the expansion of the outer layers, the temperature of the outer layers falls to make the star appear red. At this state, it is called **red giant**.

(d) At this stage, the release of energy in the star is so large and rapid that a violent explosion called **nova** or **super nova** occurs and throws out a large portion of outer layers back into outer space leaving behind the core of the star.

Phase II

(ii) This decay of the star is called the **death of a star**. The core that remains behind may end up as one of the following corpses (a) white dwarf, (b) neutron star, (c) black hole.

(a) When the original mass of the star ≤ 1.4 solar mass, the core of the star tends to die as white dwarf which just cools off slowly, changing its colour from white through yellow to red and finally becomes black. This mass limit of 1.4 solar mass is called **Chandra Shekher limit**.

(b) When the mass of the star is between 1.4 to 5 solar masses, the core of the star tends to die as neutron star.

(c) When the mass of the star is more than 5 solar masses, the core continues to contract; continuously as a result of explosion, the core tends to die as a black hole.

Black Hole

It is a region in space in which the gravitational force is so great that no object can escape from it.

The magnitude of a star is the measure of its brightness, when observed from the earth.

Galaxy

A huge group of stars and other celestial bodies bound together by gravity is called a galaxy.

There are over 100 billion (10^{11}) galaxies in the universe.

Our galaxy is called **Milky Way Galaxy**.

Milky Way Galaxy is a spiral galaxy.

There are 250 billion stars in our milky way galaxy.

Constellation

A group of stars forming a recognizable pattern is called a constellation.

Few constellations are Ursa Major (great bear), Leo (lion), Pisces (fish), Taurus (bull), Hercules (a hero), Orion (a hunter), etc.

Ursa Major have seven brightest stars forms a shape of great bear.

The Big-Bang Theory

About 15 billion years ago, the whole matter of the universe was concentrated in a highly dense small region. A sudden explosion broke this region into pieces which start to move away from each other. The instant of this explosion is called the big bang. Therefore, the universe is expanding continuously.

Eclipses

It is the total or partial obscuration of light from a celestial body as it passes through the shadow of another body.

Eclipse occurs when the sun, moon and earth lie in straight line.

(i) **Solar eclipse** When moon passes between the earth and the sun, cutting off the light of the sun from the earth.

Solar eclipse may be partial, total or annular.

(ii) **Lunar eclipse** When the earth comes between the sun and the moon a full moon during rotation in their respective orbits.

A ring shaped lunar eclipse is not possible because the size of the moon is less than that of the earth.

34

Basic Formulae of Physics

1. Absolute error $(a_m) = \dfrac{a_1 + a_2 + a_3 + \dots a_n}{n}$

2. Mean absolute error $(\Delta\overline{a}) = \dfrac{|\Delta a_1| + |\Delta a_2| + \dots + |\Delta a_n|}{n}$

3. Relative error $= \dfrac{\text{Mean absolute error } (\Delta\overline{a})}{\text{Absolute error}(a_m)}$

4. Percentage error $= \dfrac{\Delta\overline{a}}{a_m} \times 100\%$

5. Error in addition or subtraction

$$\Delta x = \pm(\Delta a + \Delta b)$$

6. Error in multiplication or division

$$\frac{\Delta x}{x} = \pm\left(\frac{\Delta a}{a} + \frac{\Delta b}{b}\right)$$

7. Scalar product of orthogonal unit vectors

$$\hat{\mathbf{i}} \cdot \hat{\mathbf{i}} = \hat{\mathbf{j}} \cdot \hat{\mathbf{j}} = \hat{\mathbf{k}} \cdot \hat{\mathbf{k}} = 1$$

and $\hat{\mathbf{i}} \cdot \hat{\mathbf{j}} = \hat{\mathbf{j}} \cdot \hat{\mathbf{k}} = \hat{\mathbf{k}} \cdot \hat{\mathbf{i}} = 0$

8. Scalar product in cartesian coordinates

$$\mathbf{A} \cdot \mathbf{B} = A_x B_x + A_y B_y + A_z B_z$$

9. Vector product of orthogonal unit vectors

$$\hat{\mathbf{i}} \times \hat{\mathbf{i}} = \hat{\mathbf{j}} \times \hat{\mathbf{j}} = \hat{\mathbf{k}} \times \hat{\mathbf{k}} = 0$$

and $\hat{\mathbf{i}} \times \hat{\mathbf{j}} = \hat{\mathbf{k}}, \hat{\mathbf{j}} \times \hat{\mathbf{k}} = \hat{\mathbf{i}}, \hat{\mathbf{k}} \times \hat{\mathbf{i}} = \hat{\mathbf{j}}$

10. Vector product in cartesian coordinates

$$\mathbf{A} \times \mathbf{B} = (A_y B_z - A_z B_y)\,\hat{\mathbf{i}} - (A_x\, B_z - B_x A_z)\,\hat{\mathbf{j}} + (A_x B_y - A_y B_x)\,\hat{\mathbf{k}}$$

11. Area = Length × Breadth

12. Volume = Length × Breadth × Height

13. Density $= \dfrac{\text{Mass}}{\text{Volume}}$

14. Speed $= \dfrac{\text{Distance}}{\text{Time}}$

15. Instantaneous speed $= \lim\limits_{\Delta t \to 0} \dfrac{\Delta s}{\Delta t}$

16. Velocity $= \dfrac{\text{Displacement}}{\text{Time taken}}$

17. Acceleration $= \dfrac{\text{Change in velocity } (\Delta v)}{\text{Time interval } (\Delta t)}$

18. Equations of Motion

    (i) $v = u + at$

    (ii) $s = ut + \dfrac{1}{2} at^2$

    (iii) $v^2 = u^2 + 2as$

    (iv) Distance travelled in nth second

$$S_n = u + \frac{a}{2}(2n - 1)$$

19. Equation of the path of projectile

$$y = x \tan\theta - \frac{g}{2u^2 \cos^2\theta} \cdot x^2$$

20. Time of flight $T = \dfrac{2u \sin\theta}{g}$

21. Maximum height $H = \dfrac{u^2 \sin^2\theta}{2g}$

22. Horizontal range $R = \dfrac{u^2 \sin 2\theta}{g} \quad \Rightarrow \quad R_{\text{max}} = \dfrac{u^2}{g}$

23. Angular displacement $(\Delta\theta) = \dfrac{\text{Linear displacement}}{\text{Radius}}$

24. Angular velocity $(\omega) = \dfrac{\text{Angular displacement}}{\text{Time interval}}$

25. Angular acceleration $(\alpha) = \dfrac{\text{Change in angular velocity}}{\text{Time taken}}$

26. Angular momentum $L = mvr = m\omega r^2$

27. Relation between angular velocity and linear velocity; $v = r\omega$

28. Relation between angular acceleration and linear acceleration; $a = r\alpha$

29. Centripetal acceleration $a = \dfrac{v^2}{r} = r\omega^2$

30. Centripetal force $F = \dfrac{mv^2}{r} = mr\omega^2$

31. Kinematical equations in circular motion
    (i) $\omega = \omega_0 + \alpha t$
    (ii) $\theta = \omega_0 t + \frac{1}{2}\alpha t^2$
    (iii) $\omega^2 = \omega_0^2 + 2\alpha\theta$

32. Time period of conical pendulum,
$$T = 2\pi\sqrt{\frac{l\cos\theta}{g}}$$

33. Linear momentum $p = mv$

34. Impulse = Force × Time = Change in momentum

35. Force = Mass × Acceleration

36. Weight = Mass × Acceleration due to gravity

37. Thrust on the rocket at any instant
$$F = -u\frac{dM}{dt}$$

38. Velocity of rocket at any instant
$$v = v_o + u\log_e\left(\frac{M_o}{M}\right)$$

39. Limiting friction $(f_s) = \mu_s \times R$ (normal reaction)

40. Kinetic friction $(f_K) = \mu_K \times R$

41. Normal reaction of plane

$$R = mg\cos\theta$$

42. Maximum speed of a car on a level road;

$$v_{\max} = \sqrt{\mu_s R_g}$$

43. Maximum speed of a car on a banked road;

$$v_{\max} = \left(R_g \frac{\mu_s + \tan\theta}{1 - \mu_s \tan\theta} \right)^{1/2}$$

For $\mu_s = 0$, $\quad v = \left(\dfrac{R_g}{\tan\theta} \right)^{1/2}$

44. Work = Force × Displacement × $\cos\theta$

45. Power = Rate of doing work = $\dfrac{\text{Work done}}{\text{Time taken}}$

46. Kinetic energy $= \dfrac{1}{2} mv^2 = \dfrac{p^2}{2m}$

47. Gravitational potential energy $= mgh$

48. Elastic potential energy $= \dfrac{1}{2} kx^2$

49. Electric potential energy $= \dfrac{1}{4\pi\varepsilon_o} \cdot \dfrac{q_1 q_2}{r}$

50. Work-Energy theorem

$$W = \int_{v_1}^{v_2} F \cdot ds$$

$$= \frac{1}{2} mv_2^2 - \frac{1}{2} mv_1^2 = K_f - K_i$$

$$= \Delta \text{KE}$$

51. Mass-Energy Equivalence

$$E = \Delta mc^2$$

52. Centre of mass of system of n-particles

$$\mathbf{r}_{\text{CM}} = \frac{m_1 \mathbf{r}_1 + m_2 \mathbf{r}_2 + \ldots + m_n \mathbf{r}_n}{m_1 + m_2 + \ldots + m_n} = \sum_{i=1}^{n} \frac{m_i \mathbf{r}_i}{\Sigma m_i}$$

53. Moment of force = Force × Perpendicular distance

54. Moment of inertia

$$I = MK^2$$

55. Radius of gyration

$$K = \sqrt{\frac{r_1^2 + r_2^2 + \ldots + r_n^2}{n}}$$

56. Torque = Moment of inertia × Angular acceleration

57. Rotational kinetic energy $(K) = \frac{1}{2} I\omega^2$

58. Gravitational force, $F = \frac{Gm_1m_2}{r^2}$

59. Relation between universal gravitational constant (G) and acceleration due to gravity (g);

$$g = \frac{GM}{R^2}$$

60. Gravitational field $E = \frac{F}{m} = \frac{GM}{r^2}$

61. Gravitational potential $V = \frac{W}{m} = -\frac{GM}{r}$

62. Gravitational potential energy $U = -\frac{GMm}{r}$

63. Time period of satellite $T = 2\pi\sqrt{\frac{r^3}{GM}}$

64. Orbital velocity, $v_o = \sqrt{\frac{GM}{r}} = R\sqrt{\frac{g}{R+h}}$

65. Energy of a satellite in orbit $E = -\frac{GMm}{2r}$

66. Binding energy $= \frac{+GMm}{2r}$

67. Escape velocity, $v_e = \sqrt{\frac{2GM}{R}} = \sqrt{2gR}$

$$= \sqrt{\frac{8\pi\rho GR^2}{3}}$$

68. Relation between escape velocity and orbital velocity;

$$v_e = \sqrt{2}\, v_o$$

69. Inertial mass $= \dfrac{\text{Force}}{\text{Acceleration}}$

70. Gravitational mass $= \dfrac{\text{Weight of body}}{\text{Acceleration due to gravity}}$

71. Stress $= \dfrac{\text{Restoring force}}{\text{Area}}$

72. Strain $= \dfrac{\text{Change in configuration}}{\text{Original configuration}}$

73. Hooke's Law

    Stress = Modulus of elasticity × Strain

74. Young's Modulus of elasticity

$$Y = \frac{\text{Normal stress}}{\text{Longitudinal strain}}$$

75. Bulk modulus of elasticity

$$K = \frac{\text{Normal stress}}{\text{Volumetric strain}}$$

76. Modulus of rigidity

$$\eta = \frac{\text{Tangential stress}}{\text{Shearing strain}}$$

77. Compressibility $= \dfrac{1}{\text{Bulk modulus}}$

78. Safety factor $= \dfrac{\text{Breaking stress}}{\text{Working stress}}$

79. Elastic potential energy in a stretched wire

$$= \frac{1}{2}\,(\text{stress}) \times \text{strain} \times \text{volume of the wire}$$

80. Elastic potential energy per unit volume

$$= \frac{1}{2} \times \text{stress} \times \text{strain}$$

$$= \frac{1}{2} \times \text{Young's modulus} \times (\text{strain})^2$$

81. Elastic potential energy of a stretched spring $= \frac{1}{2} kx^2$

82. Thermal stress $= \frac{F}{A} = Y\alpha\Delta\theta$

83. Interatomic force constant, $K = Yr_o$

84. Poisson's ratio $(\sigma) = \frac{\text{Lateral strain}}{\text{Longitudinal strain}}$

85. Depression of a cantilever $\delta = \frac{wl^3}{3YI_G}$

86. Pressure exerted by the liquid $p = h\rho g$

87. Archimedes' principle $W = \frac{T}{\left(1 - \frac{\rho}{\sigma}\right)}$

88. Density of a mixture of substances

$$\rho = \frac{m_1 + m_2}{\left(\frac{m_1}{\rho_1}\right) + \left(\frac{m_2}{\rho_2}\right)}$$

89. Viscosity $(\eta) = -\frac{F}{A\left(\frac{dv}{dx}\right)}$

90. Viscous force $(F) = -\eta A \frac{dv}{dx}$

91. Variation of viscosity $\eta_t = \frac{\eta_o}{(1 + \alpha t + \beta t^2)}$

92. Poiseuille's formula $v = \frac{\pi}{8} \frac{pr^4}{\eta l}$

93. Rate of flow of liquid through a tube

$$v = \frac{\text{Liquid pressure}}{\text{Liquid resistance}}$$

94. Stoke' law, $F = 6\pi\eta rv$

95. Terminal velocity

$$v = \frac{2}{9} \cdot \frac{r^2 (\rho - \sigma)g}{\eta}$$

96. Critical velocity $v_c = \frac{k\eta}{r\rho}$

97. Reynold's number $K = \frac{v_c \rho r}{\eta}$

98. Equation of continuity

$$a_1 v_1 = a_2 v_2 \quad \Rightarrow \quad av = \text{constant}$$

99. Bernoulli's theorem

$$p + \frac{1}{2}\rho v^2 + \rho g h = \text{constant}$$

100. Rate of flow of liquid in venturimeter;

$$v = a_1 a_2 \sqrt{\frac{2gh}{a_1^2 - a_2^2}}$$

101. Surface tension $= \frac{\text{Force}}{\text{Length}}$

$$= \frac{\text{Work done}}{\text{Change in area}}$$

102. Surface energy = Surface tension × Increase in surface area.

103. Ascent of a liquid column in a capillary tube:

$$h = \frac{2s\cos\theta}{r\rho g} - \frac{r}{3}$$

104. Zurin's law

$$Rh = \text{constant}$$

$$\Rightarrow \quad R_1 h_1 = R_2 h_2$$

105. Relation between different scales of temperatures

$$\frac{C}{100} = \frac{F - 32}{180} = \frac{K - 273}{100} = \frac{R}{80}$$

106. Thermo emf, $E = at + bt^2$

107. Thermal (heat) capacity = mc.

108. Water equivalent; $W = ms$ = heat capacity of a body

109. Latent heat, $Q = mL$

110. Principle of calorimetry

$$\text{Heat lost} = \text{Heat gained}$$

111. Coefficient of linear expansion

$$\alpha = \frac{\Delta l}{l \times \Delta t}$$

112. Coefficient of superficial expansion

$$\beta = \frac{\Delta A}{A \times \Delta t}$$

113. Coefficient of cubical expansion

$$\gamma = \frac{\Delta V}{V \times \Delta t}$$

114. Relation between coefficient of linear, superficial and cubical expansions

$$\beta = 2\alpha \text{ and } \gamma = 3\alpha$$

$$\therefore \qquad \alpha : \beta : \gamma = 1 : 2 : 3$$

115. Coefficient of apparent expansion of a liquid

$$(\gamma_a) = \frac{\text{Apparent increase in volume}}{\text{Original volume} \times \text{rise in temperature}}$$

116. Coefficient of real expansion of a liquid

$$(\gamma_r) = \frac{\text{Real increase in volume}}{\text{Original volume} \times \text{rise in temperature}}$$

117. Boyle's law, $pV = \text{constant}$

$$p_1V_1 = p_2V_2$$

118. Charles' law

$$\frac{V}{T} = \text{constant}$$

$$\frac{V_1}{T_1} = \frac{V_2}{T_2}$$

119. Gay Lussac's or Regnault's law

$$\frac{p}{T} = \text{constant}$$

$$\Rightarrow \qquad \frac{p_1}{T_1} = \frac{p_2}{T_2}$$

120. Standard gas equation

$$pV = nRT$$

121. van der Waals' gas equation

$$\left(p + \frac{a}{V^2}\right)(V - b) = RT$$

122. Pressure due to an ideal gas

$$p = \frac{1}{3}\frac{mn}{V}c^2 = \frac{1}{3}\rho c^2$$

123. Average speed of molecules of a gas

$$v = \sqrt{\frac{8kT}{\pi m}} = \sqrt{\frac{8RT}{\pi M}}$$

124. The most probable speed of molecules of a gas

$$v_{mp} = \sqrt{\frac{2kT}{m}} = \sqrt{\frac{2RT}{M}}$$

125. Degree of freedom

f or $N = 3$ (Number of particles) – Number of independent relation.

126. Specific heat of a gas at constant volume

$$C_V = \frac{f}{2}R$$

127. Specific heat of a gas at constant pressure

$$C_p = \left(\frac{f}{2} + 1\right)R$$

128. Ratio of specific heats of a gas at constant pressure and at constant volume

$$\gamma = \left(1 + \frac{2}{f}\right)$$

129. Mean free path $\lambda = \dfrac{kT}{\sqrt{2}\,\pi\sigma^2 p}$

130. Work done by a thermodynamic system

$$W = p \times \Delta V$$

131. First law of thermodynamics

$$\Delta Q = \Delta U + \Delta W$$

132. Efficiency of the cycle

$$\eta = \frac{\text{Work done}}{\text{Heat supplied}}$$

133. Slope of the adiabatic curve

$$= \gamma \times \text{slope of the isothermal curve}$$

134. Isothermal modulus of elasticity $E_s = p$

135. Adiabatic modulus of elasticity $E_T = \gamma p$

136. Ratio between isothermal and adiabatic modulus

$$\frac{E_S}{E_T} = \gamma = \frac{C_p}{C_V}$$

137. Change in entropy $= \dfrac{\text{Heat supplied to the system}}{\text{Absolute temperature}}$

138. Thermal efficiency of a heat engine

$$\eta = \frac{\text{Work done / cycle}}{\text{Total amount of heat absorbed/ cycle}}$$

$$= 1 - \frac{Q_2}{Q_1} = 1 - \frac{T_2}{T_1}$$

139. Coefficient of performance of refrigerator

$$\beta = \frac{Q_2}{W}$$

$$= \frac{Q_2}{Q_1 - Q_2} = \frac{T_2}{T_1 - T_2}$$

140. Relation between efficiency (η) and coefficient of performance (β)

$$\beta = \frac{1-\eta}{\eta}$$

141. Temperature gradient $= \dfrac{\text{Change in temperature}}{\text{Perpendicular distance}}$

$$= -\frac{\Delta\theta}{\Delta x}$$

142. The amount of heat flow in a conducting rod $Q = \dfrac{KA\Delta\theta t}{l}$

143. Thermal resistance $R = \frac{\Delta\theta}{H} = \frac{l}{KA}$

144. Ingen-Hausz experiment $\frac{k_1}{k_2} = \frac{l_1^2}{l_2^2}$

145. Emissivity (e) $= \frac{\text{Emissive power of the body } (e_\lambda)}{\text{Emissive power of a perfectly black body} (E_\lambda)}$

146. Kirchhoff 's law

$$\frac{e_\lambda}{(a_\lambda)} = \text{constant} \, (E_\lambda)$$

147. Stefan's law

$$E \propto T^4 \quad \Rightarrow \quad E = \sigma T^4$$

148. Newton's law of cooling

$$\frac{dT}{dt} = E\alpha \, (T - T_o)$$

149. Wien's Displacement law

$$\lambda_m T = \text{constant} \, (b)$$

150. Solar constant $S = \left(\frac{r}{R}\right)^2 \sigma T^4$

151. Frequency $= \frac{1}{\text{Time period}}$

152. Angular frequency

$$\omega = 2\pi\nu$$

153. Velocity of a particle executing SHM

$$v = \omega \sqrt{(a^2 - y^2)}$$

154. Acceleration of a particle executing SHM

$$\alpha = \omega^2 y$$

155. Time period in SHM

$$T = 2\pi \sqrt{\frac{\text{Displacement}}{\text{Acceleration}}}$$

156. Time period of the simple pendulum

$$T = 2\pi\sqrt{\frac{l}{g}}$$

157. Time period of the conical pendulum

$$T = 2\pi\sqrt{\frac{mr}{T\sin\theta}}$$

158. Time period of the compound pendulum

$$T = 2\pi\sqrt{\frac{I}{mgl}}$$

159. Time period of the torsional pendulum

$$T = 2\pi\sqrt{\frac{I}{C}}$$

160. Time period of the oscillation

$$T = 2\pi\sqrt{\frac{h}{g}}$$

161. Restoring force (F) = – Force constant of spring × distance

$$= -ky$$

162. Hooke's law, $mg = kl$

163. Time period of a loaded spring

$$T = 2\pi\sqrt{\frac{m}{k}}$$

164. Displacement of the damped oscillator

$$x = x_o e^{-bt/2m}\cos(\omega' t + \phi)$$

165. Mechanical energy E of the damped oscillator

$$E = \frac{1}{2}Kx_0^2 e^{-bt/m}$$

166. Wave velocity

$$v = f\lambda$$

167. Particle velocity $= \dfrac{dy}{dt}$

168. Velocity of longitudinal (sound) waves $v = \sqrt{\dfrac{E}{\rho}}$

169. Newton's formula for isothermal process

$$v = \sqrt{\frac{E_T}{\rho}} = \sqrt{\frac{p}{\rho}}$$

170. Laplace's correction for adiabatic process

$$v = \sqrt{\frac{E_s}{\rho}} = \sqrt{\frac{\gamma p}{\rho}}$$

171. Speed of transverse motion

$$v = \sqrt{\frac{T}{m}}$$

172. Plane progressive simple harmonic wave

$$y = a \sin 2\pi \left(\frac{t}{T} - \frac{x}{\lambda}\right)$$

$$= a \sin \frac{2\pi}{\lambda}(vt - x)$$

173. Intensity of constructive interference = (amplitude)^2

174. Equation of a stationary wave

$$y = 2a \cdot \sin \frac{2\pi t}{T} \cos \frac{2\pi x}{\lambda}$$

175. Melde's experiment

$$\text{Frequency of vibration of string} = \frac{\text{Frequency of tunning fork}}{2}$$

176. Quantization of charge

$$Q = \pm\, ne$$

177. Coulomb's law of electrostatics

$$F = \frac{1}{4\pi\varepsilon_0} \cdot \frac{q_1 q_2}{r^2}$$

178. Electric field intensity

$$\mathbf{E} = \lim_{q_0 \to 0} \frac{\mathbf{F}}{q_0}$$

179. Electric field intensity due to a point charge q

$$E = \frac{1}{4\pi\varepsilon_0} \cdot \frac{q}{r^2}$$

180. Electric potential $V = \frac{W}{q}$

181. Electric potential due to a point change q

$$V = \frac{1}{4\pi\varepsilon_0} \cdot \frac{q}{r}$$

182. Potential gradient $= \frac{dV}{dr}$

183. Relation between potential gradient and electric field intensity

$$E = -\left(\frac{dV}{dr}\right)$$

184. Electric flux $(\phi_E) = \mathbf{E} \cdot \mathbf{ds}$

185. Gauss' theorem

$$\phi_E = \oint_{\mathbf{s}} \mathbf{E} \cdot \mathbf{ds} = \frac{1}{\varepsilon_0} \Sigma q$$

186. Electric dipole moment

$$\mathbf{p} = q \times 2\mathbf{l}$$

187. Torque

$$\tau = Ep \sin\theta = \mathbf{p} \times \mathbf{E}$$

188. Work done of an electric dipole

$$W = pE\,(\cos\theta_1 - \cos\theta_2)$$

189. Potential energy of an electric dipole

$$U = -\,pE\cos\theta$$

190. Potential energy of charge system

$$U = \frac{1}{4\pi\varepsilon_0} \cdot \frac{q_1 q_2}{r}$$

191. Capacitance of a conductor $C = \frac{q}{V}$

192. Capacitance of an isolated spherical conductor

$$C = 4\pi\varepsilon_0 R$$

193. Electric current $I = \frac{q}{t}$

194. Current density $J = \frac{I}{A}$

195. Drift velocity, $v_d = \frac{eE\tau}{m} = \frac{eV\tau}{ml}$

196. Relation between electric current and drift velocity $v_d = \frac{I}{Ane}$

197. Mobility of electron $\mu = \frac{v_d}{E}$

198. Ohm's law; $I \propto V$

$$\Rightarrow \quad V = IR$$

199. Electrical resistance $R = \frac{Ane^2\tau}{ml} = \frac{V}{I}$

200. Resistivity $\rho = \frac{m}{ne^2\tau}$

201. Electrical conductivity

$$\sigma = \frac{1}{\rho} = \frac{l}{RA} = \frac{ne^2\tau}{m}$$

202. Relation between current density (J) and electrical conductivity

$$J = \sigma E$$

203. Electromotive force (emf) of a cell

$$E = \frac{W}{q}$$

204. Terminal potential difference

$$V = \frac{W}{q}$$

205. Relation between E, V and internal resistance of a cell

$$E = V + Ir$$

206. Principle of Wheatstone bridge

$$\frac{P}{Q} = \frac{R}{S}$$

207. Principle of Meter bridge

$$\frac{R}{S} = \frac{l_1 \text{ (length of wire)}}{100 - l_1}$$

208. Principle of potentiometer

$$K = \frac{V}{L} = \frac{IR}{L} = \frac{E_o R}{(R_o + R)L}$$

209. Joule's law

$$H = I^2 Rt = \frac{V^2 \cdot t}{R} \text{ joule}$$

or $$H = \frac{I^2 Rt}{4.18} = \frac{V^2 t}{4.18R} \text{ calories}$$

210. Electric power $P = VI = I^2 R = \dfrac{V^2}{R}$

211. Electric energy

$$W = V.q = VIt = I^2 Rt = \frac{V^2 t}{R}$$

212. Chemical equivalent $= \dfrac{\text{Atomic weight}}{\text{Valency}}$

213. Faraday's constant F = Avogadro's number × electric charge

214. Relation between neutral temperature (T_n) and temperature of inversion (T_i)

$$T_n = \frac{T_i + T_o}{2}$$

215. Thermoelectric power $S = \dfrac{dE}{dt} = \alpha + \beta T$

216. Peltier coefficient $\pi = \dfrac{\text{Peltier heat}}{\text{Charge flowing}}$

217. Thomson's coefficient $\sigma = \dfrac{dV}{dT}$

218. Biot Savart's law

$$\mathbf{dB} = \frac{\mu_o}{4\pi} \frac{\mathbf{Idl} \times \mathbf{r}}{r^3} \quad \Rightarrow \quad dB = \frac{\mu_o}{4\pi} \cdot \frac{Idl \sin\theta}{r^2}$$

219. Magnetic dipole

$$|M| = NiA$$

220. Ampere's circuital law

$$\oint \mathbf{B} \cdot \mathbf{d}l = \mu_0 I$$

221. Magnetic field inside the turns of toroid

$$B = \mu_0 n I$$

222. Radius of circular path

$$r = \frac{mv}{Bq}$$

223. Cyclotron frequency $\upsilon = \dfrac{Bq}{2\pi m}$

224. Torque acting on a current carrying coil placed inside a uniform magnetic field

$$\tau = NBIA \sin\theta$$

225. Current sensitivity

$$I_s = \frac{\theta}{I} = \frac{NBA}{K}$$

226. Voltage sensitivity

$$V_s = \frac{\theta}{V} = \frac{NBA}{KR}$$

227. Principle of ammeter $S = \left(\dfrac{I_g}{I - I_g}\right) G$

228. Principle of voltmeter

$$V = I_g (G + R)$$

229. Coulomb's law

$$F = \frac{\mu_0}{4\pi} \cdot \frac{m_1 m_2}{r^2}$$

230. Magnetic dipole moment

$$\mathbf{M} = \mathbf{m}\,(2l)$$

231. Torque acting on a magnetic dipole

$$\tau = \mathbf{M} \times \mathbf{B} = MB \sin\theta$$

232. Potential energy of a magnetic dipole

$$U = W = -MB\cos\theta = -\mathbf{M}\cdot\mathbf{B}$$

233. Magnetic moment of an atom

$$M = \frac{1}{2}e\omega r^2 = n\frac{eh}{4\pi m}$$

234. Total intensity of earth's magnetic field

$$I = I_0\sqrt{1 + 3\sin^2\lambda}$$

235. Tangent law

$$B = H\tan\theta$$

236. Time period of vibrations

$$T = 2\pi\sqrt{\frac{I}{MH}}$$

237. Magnetic flux $\phi = \mathbf{B}\cdot\mathbf{A} = BA\cos\theta$

238. Magnetic induction $B = \dfrac{\phi}{A} = \mu_0(H + I)$

239. Magnetic permeability $\mu = \dfrac{B}{H}$

240. Magnetising force or Magnetic intensity, $H = \dfrac{B}{\mu}$

241. Intensity of magnetisation

$$\mathbf{I} = \frac{\mathbf{M}}{\mathrm{V}} = \frac{m}{A}$$

242. Magnetic susceptibility $\chi_m = \dfrac{I}{H}$

243. Relation between magnetic permeability and susceptibility

$$\mu = \mu_0(1 + \chi_m)$$

244. Curie law in magnetism

$$\chi_m \propto \frac{1}{T}$$

$$\chi_m T = \text{constant}$$

245. Motional emf $E = \mathbf{B}\cdot\mathbf{v}\times l = Bvl$

246. Coefficient of self-induction

$$L = \frac{\phi}{I} = \frac{\mu_0 N^2 A}{l} = \mu n^2 Al$$

247. Coefficient of mutual induction

$$M = \frac{\phi}{I}$$

248. Induced emf in the secondary coil

$$E = -M \frac{dI}{dt}$$

249. Coefficient of coupling

$$K = \sqrt{\frac{M}{L_1 L_2}}$$

250. Mutual induction of two long coaxial solenoids

$$M = \frac{\mu_0 N_1 N_2 A}{l} = \mu_0 n_1 n_2 Al$$

251. Growth of current in an inductor

$$I = I_0 (1 - e^{-Rt/L})$$

252. Decay of current in an inductor

$$I = I_0 e^{-Rt/T}$$

253. The instantaneous charge on a capacitor on charging

$$q = q_0 [1 - e^{-t/RC}]$$

254. The instantaneous charge on a capacitor in discharging

$$q = q_0 e^{-t/RC}$$

255. Instantaneous value of alternating current

$$I = I_0 \sin \omega t$$

256. Inductive reactance

$$X_L = L\omega = L \cdot 2\pi f = \frac{L \cdot 2\pi}{T}$$

257. Capacitive reactance

$$X_C = \frac{1}{C\omega} = \frac{1}{C \cdot 2\pi f} = \frac{T}{C \cdot 2\pi}$$

258. Impedance of an AC circuit

$$Z = \sqrt{R^2 + (X_L - X_C)^2}$$

259. Average power in AC circuit

$$P_{av} = \frac{Vi\cos\theta}{2}$$

260. Resonant frequency $f = \frac{1}{2\pi\sqrt{LC}}$

261. Q-factor or sharpness at resonance

$$Q = \frac{1}{R}\sqrt{\frac{L}{C}}$$

262. Induced emf produced by the AC generator

$$e = NBA\omega \sin\omega t = e_o \sin\omega t$$

263. Torque acting on a current carrying coil

$$\tau = NBIA\sin\theta$$

264. Efficiency of a motor

$$\eta = \frac{\text{Back emf}}{\text{Applied emf}} = \frac{E}{V}$$

265. Transformation ratio

$$K = \frac{N_S}{N_P} = \frac{E_S}{E_P} = \frac{I_P}{I_S}$$

266. Mirror formula, $\frac{1}{f} = \frac{1}{v} + \frac{1}{u}$

267. Relation between focal length and radius of curvature $f = \frac{R}{2}$

268. The power of a mirror $P = \frac{1}{f(\text{metre})} = \frac{100}{f(\text{cm})}$

269. Newton's formula for a concave mirror

$$f = \sqrt{x_1 x_2}$$

270. Linear magnification $m = \frac{I}{O} = -\frac{v}{u}$

271. Areal magnification $= m^2 = \dfrac{\text{Area of image}}{\text{Area of object}}$

$$= \frac{-dv}{du} = \left(\frac{v}{u}\right)^2 = \left(\frac{f}{f-u}\right)^2 = \left(\frac{f-v}{u}\right)^2$$

272. Snell's law, $\dfrac{\sin i}{\sin r} = \text{constant } ({}_1\mu_2)$

273. Refractive index of a medium $(\mu) = \dfrac{c}{v}$

274. Relative refractive index

$${}_1\mu_2 = \frac{v_1}{v_2}$$

275. Cauchy's formula, $\mu = A + \dfrac{B}{\lambda^2} + \dfrac{C}{\lambda^4}$

276. Refractive index of denser medium

$$\mu = \frac{1}{\sin C}$$

277. Lens formula, $\dfrac{1}{f} = \dfrac{1}{v} - \dfrac{1}{u}$

278. Lens Maker's formula, $\dfrac{1}{f} = (\mu - 1)\left(\dfrac{1}{R_1} - \dfrac{1}{R_2}\right)$

279. Linear magnification for lens

$$m = \frac{I}{O} = \frac{v}{u}$$

280. Focal length of a convex lens by displacement method

$$f = \frac{a^2 - d^2}{4a}$$

281. Height of the object, $O = \sqrt{I_1 I_2}$

282. Prism formula

$$\mu = \frac{\sin\left(\dfrac{A + \delta_m}{2}\right)}{\sin\left(\dfrac{A}{2}\right)}$$

283. Angular dispersion

$$\theta = \delta_V - \delta_R = (\mu_V - \mu_R)A$$

284. Dispersive power

$$W = \frac{\theta}{\delta_Y} = \frac{(\mu_V - \mu_R)}{(\mu_Y - 1)}$$

285. f- number for a camera $= \dfrac{\text{Focal length of the lens } (F)}{\text{Diameter of the lens } (d)}$

286. Velocity of electromagnetic wave in vacuum

$$c = \frac{1}{\sqrt{\mu_o \varepsilon_o}}$$

287. Energy of a photon

$$E = h\nu = \frac{hc}{\lambda}$$

288. de-Broglie wave equation

$$\lambda = \frac{h}{p} = \frac{h}{mv}$$

289. Interference fringe width

$$\beta = \frac{D\lambda}{d}$$

290. Law of Malus, $I \propto \cos^2 \theta$

291. Brewster's law, $\mu = \tan i_p$

292. Specific charge of electron

$$\frac{e}{m} = \frac{E^2}{2VB^2}$$

293. Energy of a photon $E = h\nu$

294. Momentum of a photon

$$E = \frac{h\nu}{c} = \frac{h}{\lambda}$$

295. Dynamic or kinetic mass of photon

$$m = \frac{h\nu}{c^2} = \frac{h}{c\lambda}$$

296. Relation between work function, threshold frequency and threshold wavelength

$$\phi = h\nu_o = \frac{hc}{\lambda_{max}}$$

297. Einstein's photoelectric equation

$$(E_K)_{max} = h\nu - \phi = h(\nu - \nu_o)$$

298. Maximum kinetic energy of photo electrons

$$(E_k)_{max} = \frac{1}{2} mv_{max}^2 = eV_o$$

299. Compton effect $\Delta\lambda = \frac{h}{m_0 c}(1 - \cos\phi)$

300. Kinetic energy of recoil electron

$$E_K = \frac{hc}{\lambda} - \frac{hc}{\lambda'}$$

301. de-Broglie wavelength

$$\lambda = \frac{h}{p} = \frac{h}{mv} = \frac{h}{\sqrt{2meV}}$$

302. Bragg's law,

$$2d \sin\theta = n\lambda$$

303. Moseley's law,

$$\nu = a(z - b)^2$$

304. Distance of closest approach

$$r_0 = \frac{1}{4\pi\varepsilon_0} \cdot \frac{2ze^2}{E_K}$$

305. Impact parameter

$$b = \frac{1}{4\pi\varepsilon_0} \cdot \frac{Ze^2 \cot\left(\frac{\theta}{2}\right)}{E_K}$$

306. Rutherford's scattering formula

$$N(\theta) = \frac{N_i ntZ^2 e^4}{(8\pi\varepsilon_0)^2 r^2 E^2 \sin^4\left(\frac{\theta}{2}\right)}$$

307. Radius of orbit of electron

$$r = \frac{n^2h^2}{4\pi^2 mkZe^2}$$

308. Velocity of electron in any orbit

$$v = \frac{2\pi\, kZe^2}{nh}$$

309. Frequency of electron in any orbit

$$\nu = \frac{kZe^2}{nhr} = \frac{4\pi^2 Z^2 e^4 mk^2}{n^3h^3}$$

310. Kinetic energy of electron in any orbit

$$E_k = \frac{2\pi^2 me^4 Z^2 k^2}{n^2h^2} = \frac{13.6Z^2}{n^2} \text{ eV}$$

311. Potential energy of electron in any orbit

$$E_p = -\frac{4\pi^2 me^4 Z^2 k^2}{n^2h^2} = -\frac{27.2Z^2}{n^2} \text{ eV}$$

312. Total energy of electron in any orbit

$$E = \frac{-2\pi^2 me^4 Z^2 k^2}{n^2h^2} = \frac{-13.6Z^2}{n^2} \text{ eV}$$

313. Radius of the nucleus $R \propto A^{1/3}$

$\Rightarrow$ $$R = R_0 A^{1/3}$$

314. Nuclear density

$$\rho = \frac{\text{Mass of nucleus}}{\text{Volume of nucleus}} = \frac{3m}{4\pi R_0^3}$$

315. Mass defect $\Delta m = M - m = [Zm_p + (A - Z)m_n - m_N]$

316. Nuclear binding energy $= (\Delta m)c^2$

$$= [Zm_p + (A - Z)m_n - m_N]c^2$$

317. Nuclear binding energy per nucleon

$$= \frac{\text{Nuclear binding energy}}{\text{Total number of nucleons}}$$

$$= \frac{[Zm_p + (A - Z)m_n - m_N]c^2}{A}$$

318. Packing fraction P

$$= \frac{\text{(Exact nuclear mass)} - \text{(Mass number)}}{\text{Mass number}}$$

319. Rate of disintegration

$$\frac{-dN}{dt} \propto N \Rightarrow \frac{-dN}{dt} = \lambda N$$

320. Relation between half-life and disintegration constant

$$T = \frac{\log_e 2}{\lambda} = \frac{0.6931}{\lambda}$$

321. Activity of a radioactive element

$$R = \left(\frac{-dN}{dt}\right)$$

322. Electrical conductivity of extrinsic semiconductor

$$\sigma = \frac{1}{\rho} = e\,(n_e\mu_e + n_h\mu_h)$$

323. Resistance of diode $R = \frac{V}{I}$

324. AC current gain $(\alpha_{AC}) = \frac{\Delta I_c}{\Delta I_e}$

325. AC voltage gain $(A_V) = \frac{\text{Output voltage}}{\text{Input voltage}} = \alpha_{AC} \times$ resistance gain

$$= \alpha_{AC} \times \frac{R_o}{R_i}$$

326. AC power gain $= \frac{\text{Change in output power}}{\text{Change in input power}}$

327. Relation between the current gain of common base and common emitter amplifier

$$\beta = \frac{\alpha}{1-\alpha} = \frac{I_C}{I_B}$$

328. Displacement current,

$$I_D = \varepsilon_0 \cdot \frac{d\phi_E}{dt}$$

329. Ampere-Maxwell law

$$\oint \mathbf{B} \cdot d\mathbf{l} = \mu_0 (I + I_D)$$

330. Poynting vector

$$\mathbf{S} = \frac{1}{\mu_0} \mathbf{E} \times \mathbf{B}$$

331. Average electric energy density

$$U_E = \frac{1}{2} \varepsilon_0 E^2 = \frac{1}{4} \varepsilon_0 E_0^2$$

332. Average magnetic energy density

$$U_B = \frac{1}{2} \frac{B^2}{\mu_0} = \frac{1}{4} \frac{B_0^2}{\mu_0}$$

333. Critical frequency $\nu_c = 9 (N_{max})^{1/2}$

334. Skip distance $(D_{skip}) = 2h \left(\frac{v_{max}}{v_c}\right)^2 - 1$

335. Effective range in space wave propagation

$$d = \sqrt{2Rh_T} + \sqrt{2Rh_R}$$

336. Principle of meter bridge or slide wire bridge

$$\frac{P}{Q} = \frac{l}{100 - l} = \frac{R}{S}$$

35

Nobel Laureates in Physics

Year	Laureate	Country	Rationale
2014	Shuji Nakamura, Isamu Akasaki Hiroshi Amano	Japan United States Japan Japan	For the invention of efficient blue light emitting diodes which has enabled bright and energy - saving white light sources.
2013	Francois Englert and Peter W. Higgs	Belgium United Kingdom	For the theoretical discovery of a mechanism that contributes to our understanding of the origin of mass of sub-atomic particles and which recently was confirmed through the discovery of the predicted fundamental particle by the ATLAS and CMS experiments at CERN's large Hadron Collider.
2012	Serge Haroche David J. Wineland	France United States	For ground-breaking experimental methods that enable measuring and manipulation of individual quantum systems.
2011	Saul Perlmutter Brian P. Schmidt Adam G. Riess	United States Australia United States United States	For the discovery of the accelerating expansion of the Universe through observations of distant supernovae
2010	Andre Geim Konstantin Novoselov	Russia Netherlands Russia United Kingdom	For ground-breaking experiments regarding the two-dimensional material graphene

Year	Laureate	Country	Rationale
2009	Charles K. Kao	Hong Kong United Kingdom United States	For ground-breaking achievements concerning the transmission of light in fibers for optical communication
	Willard S. Boyle	Canada United States	For the invention of an imaging semiconductor circuit -the CCD sensor
	George E. Smith	United States	
2008	Makoto Kobayashi Toshihide Maskawa	Japan Japan	For the discovery of the origin of the broken symmetry which predicts the existence of at least three families of quarks in nature
	Yoichiro Nambu	Japan United States	For the discovery of the mechanism of spontaneous broken symmetry in subatomic physics
2007	Albert Fert Peter Grunberg	France Germany	For the discovery of giant magnetoresistance
2006	John C. Mather George F. Smoot	United States United States	For their discovery of the blackbody form and anisotropy of the cosmic microwave background radiation
2005	Roy J. Glauber	United States	For his contribution to the quantum theory of optical coherence
	John L. Hall Theodor W. Hansch	United States Germany	For their contributions to the development of laser based precision spectroscopy, including the optical frequency comb technique
2004	David J. Gross Hugh David Politzer Frank Wilczek	United States United States United States	For the discovery of asymptotic freedom in the theory of the strong interaction
2003	Alexei Alexeyevich Abrikosov	Russia	For pioneering contributions to the theory of superconductors and superfluids
	Vitaly Lazarevich Ginzburg	Russia	
	Anthony James Leggett	United Kingdom United States	

Year	Laureate	Country	Rationale
2002	Raymond Davis, Jr.	United States	For pioneering contributions to astrophysics, in particular for the detection of cosmic neutrinos
	Masatoshi Koshiba	Japan	
	Riccardo Giacconi	Italy United States	For pioneering contributions to astrophysics, which have led to the discovery of cosmic X-ray sources
2001	Eric Allin Cornell	United States	For the achievement of Bose-Einstein condensation in dilute gases of alkali atoms, and for early fundamental studies of the properties of the condensates
	Carl Edwin Wieman	United States	
	Wolfgang Ketterle	Germany	
2000	Zhores Ivanovich Alferov	Russia	For developing semiconductor hetero structures used in high speed and optoelectronics
	Herbert Kroemer	Germany	
	Jack St. Clair Kilby	United States	For his part in the invention of the integrated circuit
1999	Gerard't Hooft	Netherlands	For elucidating the quantum structure of electroweak interactions in physics
	Martinus J. G. Veltman	Netherlands	
1998	Robert B. Laughlin	United States	For their discovery of a new form of quantum fluid with fractionally charged excitations
	Horst Ludwing	Germany	
	Daniel Chee Tsui	United States	
1997	Steven Chu	United States	For development of methods to cool and trap atoms with laser light.
	Claude Cohen-Tannoudji	France	
	William Daniel Phillips	United States	
1996	David Morris Lee	United States	For their discovery of superfluidity in helium-3
	Douglas D. Osheroff	United States	
	Robert Coleman Richardson	United States	
1995	Martin Lewis Perl	United States	For the discovery of the tau lepton and for pioneering experimental contributions to lepton physics
	Frederick Reines	United States	For the detection of the neutrino and for pioneering experimental contributions to lepton physics

Year	Laureate	Country	Rationale
1994	Bertram N. Brockhouse	Canada	For the development of neutron spectroscopy and for pioneering contributions to the development of neutron scattering techniques for studies of condensed matter
	Clifford Glenwood Shull	United States	For the development of the neutron diffraction technique and for pioneering contributions to the development of neutron scattering techniques for studies of condensed matter
1993	Russell Alan Hulse	United States	For the discovery of a new type of pulsar, a discovery that has opened up new possibilities for the study of gravitation
	Joseph Hooton taylor, Jr.	United States	
1992	Georges Charpak	France	For his invention and development of particle detectors, in particular the multiwire proportional chamber
1991	Pierre-Gilles de Gennes	France	For discovering that methods developed for studying order phenomena in simple systems can be generalised to more complex forms of matter, in particular to liquid crystals and polymers
1990	Jerome I. Friedman	United States	For their pioneering investigations concerning deep inelastic scattering of electrons on protons and bound neutrons which have been of essential importance for the development of the quark model in particle physics
	Henry Way Kendall	United States	
	Richard E. Taylor	Canada	
1989	Norman Foster Ramsey	United States	For the invention of the separated oscillatory fields method and its use in the hydrogen maser and other atomic clocks
	Hans Georg Dehmelt	United States	For the development of the ion trap technique
	Wolfgang Paul	West Germany	
1988	Leon Max Lederman	Unites States	For the neutrino beam method and the demonstration of the doublet structure of the leptons through the discovery of the muon neutrino
	Melvin Schwartz	United States	
	Jack Steinberger	United States	
1987	Johannes Georg Bednorz	West Germany	For their important break through in the discovery of superconductivity in ceramic materials
	Karl Alexander	Switzerland	

36

Famous Physicists & their Contributions

1. Andre Marie Ampere
   Father of electrodynamics.
2. Amadeo Avogadro
   Developed hypothesis that all gases at same volume, pressure, and temperature contain same number of atoms.
3. Augustin-Jean Fresnel
   Studied transverse nature of light waves.
4. Armand-Hippolyte - Louis Fizeau
   Made the first terrestrial measurement of the speed of light, invented one of the first interferometers, took the first pictures of the sun on daguerreotypes, argued that the Doppler effect with respect to sound should also apply to any wave motion, particularly that of light.
5. Antoine Henri Becquerel
   Discovered natural radioactivity.
6. Albert A. Michelson
   Devised an interferometer and used it to try to measure earth's absolute motion, precisely measured speed of light.
7. Albert Einstein
   Explained Brownian motion and photoelectric effect, contributed to theory of atomic spectra, formulated theories of special and general relativity.
8. Arthur Compton
   Discovered the increase in wavelength of X-rays when scattered by an electron.

9. **Alfred Kastler**
Discovered and developed optical methods for studying the Hertzian resonances that are produced when atoms interact with radiowaves or microwaves.

10. **Andrei Sakharov**
Father of the Soviet hydrogen bomb, awarded the Nobel Peace Prize for his struggle for human rights, for disarmament, and for cooperation between all nations.

11. **Arthur L. Schawlow**
Contributed to the development of laser spectroscopy.

12. **Aage Bohr**
Contributed to theoretical understanding of collective motion in nuclei.

13. **Abdus Salam**
Co-developed gauge field theory of the electroweak interaction, suggested that the proton might be unstable.

14. **Arno A. Penzias**
Co-discovered the cosmic microwave background radiation.

15. **Albert Fert**
Co-discovered Giant Magnetoresistance, which brought about a breakthrough in gigabyte hard disks.

16. **Andre Geim**
Co-discovered a simple method for isolating single atomic layer of graphite, known as graphene.

17. **Arnold Sommerfeld**
Generalized the circular orbits of the atomic Bohr model to elliptical orbits, introduced the magnetic quantum number, used statistical mechanics to explain the electronic properties of metals.

18. **Arthur Jeffrey Dempster**
Discovered the isotope uranium-235.

19. **Albert V. Crewe**
Developed the first practical scanning electron microscope.

20. **Akito Arima**
Co-developed the Interacting Boson Model of the atomic nucleus.

21. **Blaise Pascal**
Discovered that pressure applied to an enclosed fluid is transmitted undiminished to every part of the fluid and to the walls of its container (Pascal's principle).

22. **Benjamin Franklin**
The first American physicist, characterized two kinds of electric charge, which he named "positive" and "negative".

23. **Ben Mottelson**
Contributed to theoretical understanding of collective motion in nuclei.

24. **Burton Richter**
Carried out an experiment leading to the discovery of charmonium.

25. **Brian Josephson**
Contributed to theoretical predictions of the properties of a supercurrent through a tunnel barrier.

26. **Benoit Mandelbrot**
Developed theory of fractals.

27. **Christiaan Huygens**
Proposed a simple geometrical wave theory of light, now known as "Huygen's principle"; pioneered use of the pendulum in clocks.

28. **Charles Augustin de Coulomb**
Experiments on elasticity, electricity, and magnetism, established experimentally nature of the force between two charges.

29. **Count Alessandro Volta**
Pioneer in study of electricity, invented the first electric battery.

30. **Christian Doppler**
Experimented with sound waves; derived an expression for the apparent change in wavelength of a wave due to relative motion between the source and observer.

31. **Charles Wilson**
Invented the cloud chamber.

32. **Charles Glover Barkla**
Discovered that every chemical element, when irradiated by X-rays, can emit an X-ray spectrum of two line-groups, which he named the *K*-series and L-series, that are of fundamental importance to understanding atomic structure.

33. **Clinton Joseph Davisson**
Co-discovered electron diffraction.

34. **Cecil F. Powell**
Developed the photographic emulsion method of studying nuclear processes, discovered the charged pion.

35. **Carl David Anderson**
Discovered the positron and the muon.

36. **Clifford G. Shull**
Developed a neutron scattering technique in which a neutron diffraction pattern is produced that may be used to determine the atomic structure of a material.

37. **Charles H. Townes**
Created first maser using ammonia to produce coherent microwave radiation.

38. **Chen Ning Yang**
Co-proposed parity violation in weak interactions.

39. **Claude Cohen-Tannoudji**
Developed methods, with his colleagues, of using laser light to cool helium atoms to a temperature of about 0.18 μK and capturing the chilled atoms in a trap.

40. **Charles K. Kao**
Pioneer in the development and use of fibre optics in telecommunications.

41. **Carlo Rubbia**
Contributed to experiments that led to the discovery of the carriers ($W\pm$ and $Z°$) of the weak interaction.

42. **Charles Francis Richter**
Established the Richter scale for the measurement of earthquake intensity.

43. **Chien-Shiung Wu**
Experimentally proved that parity is not conserved in nuclear beta decay.

44. **Calvin F. Quate**
Made pioneering contributions to nanoscale measurement science through the development and application of scanning probe microscopes.

45. **Chris Quigg**
Contributed to theoretical understanding of high energy collisions and the fundamental interactions of elementary particles.

46. **Daniel Bernoulli**
Developed the fundamental relationship of fluid flow now known as Bernoulli's principle.

47. **Dennis Gabor**
Invented and developed the holographic method whereby it is possible to record and display a three-dimensional display of an object.

48. **Donald A. Glaser**
Invented the bubble chamber.

49. **David M. Lee**
Co-discovered that the isotope Helium-3 becomes a quantum superfluid near absolute zero.

50. **David J. Thouless**
Contributed to condensed matter theory, especially vortices in superfluids, the quantum Hall effect, and topological quantum numbers.

51. **Douglas D. Osheroff**
Co-discovered that the isotope Helium-3 becomes a quantum superfluid near absolute zero.

52. **D. Allan Bromley**
Served as Science Advisor to the President of the United States, carried out pioneering studies of nuclear structure and dynamics, considered the father of modern heavy-ion science.

53. Daniel Kleppner
Co-invented the hydrogen maser, explores quantum chaos by optical spectroscopy of Rydberg atoms.

54. Ernst Mach
Studied conditions that occur when an object moves through a fluid at high speed (the "Mach number" gives the ratio of the speed of the object to the speed of sound in the fluid); proposed "Mach's principle," which states that the inertia of an object is due to the interaction between the object and the rest of the universe.

55. Edwin H. Hall
Discovered the "Hall effect," which occurs when charge carriers moving through a material are deflected because of an applied magnetic field - the deflection results in a potential difference across the side of the material that is transverse to both the magnetic field and the current direction.

56. Erwin Schrödinger
Contributed to creation of quantum mechanics, formulated the Schrodinger wave equation.

57. Enrico Fermi
Performed experiments leading to first self-sustaining nuclear chain reaction, developed a theory of beta decay that introduced the weak interaction, derived the statistical properties of gases that obey the Pauli exclusion principle.

58. Ernest Orlando Lawrence
Invented the cyclotron.

59. Eugene Wigner
Contributed to theoretical atomic and nuclear physics; introduced concept of the nuclear cross-section.

60. Ernest Walton
Co-invented the first particle accelerator.

61. Emilio Segre
Co-discovered the antiproton, discovered technetium.

62. Ernst Ruska
Designed the first electron microscope.

63. Edwin M. McMillan
Made discoveries concerning the transuranium elements.

64. Edward Mills Purcell
Developed method of nuclear resonance absorption that permitted the absolute determination of nuclear magnetic moments, co-discovered a line in the galactic radiospectrum caused by atomic hydrogen.

65. Ernst Ising
Developed the Ising model of ferromagnetism.

66. Edward Teller
Helped develop atomic and hydrogen bombs.

67. **Ernest M. Henley**
Contributed to the theoretical understanding of how symmetries place restrictions on theories and models, the connection of quarks and gluons to nucleon-meson degrees of freedom, the changes that occur when hadrons are placed in a nuclear medium.

68. **Edward Witten**
Made fundamental contributions to manifold theory, string theory, and the theory of supersymmetric quantum mechanics.

69. **Felix Savart**
Co-discovered that intensity of magnetic field set up by a current flowing through a wire varies inversely with the distance from the wire.

70. **Frits Zernike**
Invented the phase-contrast microscope, a type of microscope widely used for examining specimens such as biological cells and tissues.

71. **Frederic Joliot-Curie**
Co-discovered artificial radioactivity.

72. **Felix Bloch**
Contributed to development of the NMR technique, measured the magnetic moment of the neutron, contributed to the theory of metals.

73. **Francis Crick**
Co-proposed the double-helix structure of DNA.

74. **Frederick Reines**
Established, together with Clyde L. Cowan, Jr., the existence of the electron antineutrino by detecting them using a reactor experiment.

75. **Frank Wilczek**
Co-discovered "asymptotic freedom" in non-Abelian gauge theories; contributed to the study of "anyons" (particle-like excitations in two-dimensional systems that obey "fractional statistics").

76. **Fritz London**
Co-developed the phenomenological theory of superconductivity, co-developed the first quantum-mechanical treatment of the hydrogen molecule, determined that the electromagnetic gauge is the phase of the Schrödinger wave function.

77. **Freeman J. Dyson**
Made many important contributions to quantum field theory, including the demonstration that the Feynman rules are direct and rigorous consequences of quantum field theory, advocated exploration of the solar system by humans; speculated on the possibility of extraterrestrial civilizations.

78. Felix Hans Boehm
Pioneered the use of nuclear-physics techniques for exploring fundamental questions concerning the weak interactions and the nature of neutrinos.

79. Francesco Iachello
Co-developed the Interacting Boson Model of the atomic nucleus; introduced supersymmetry in nuclei (1980); developed the Vibron Model of molecules (1981).

80. Galileo Galilei
Performed fundamental observations, experiments and mathematical analysis in astronomy and physics; discovered mountains and craters on the moon, the phases of Venus, and the four largest satellites of Jupiter: Io, Europa, Callisto, and Ganymede.

81. Georg Ohm
Discovered that current flow is proportional to potential difference and inversely proportional to resistance (Ohm's law).

82. Guglielmo Marconi
Invented the first practical system of wireless telegraphy.

83. George Francis FitzGerald
Hypothesized foreshortening of moving bodies (Lorentz-FitzGerald contraction) to explain the result of the Michelson-Morley experiment.

84. Glenn T. Seaborg
Co-discovered plutonium and all further transuranium elements through element 102.

85. Georges Charpak
Invented the multiwire proportional chamber.

86. George E. Smith
Co-invented the CCD (charge-coupled device).

87. Gerard t' Hooft
Contributed to theoretical understanding of gauge theories in elementary particle physics, quantum gravity and black holes, and fundamental aspects of quantum physics.

88. Gerd Binnig
Co-designed the scanning tunneling microscope (STM), a type of microscope in which a fine conducting probe is held close the surface of a sample.

89. George E. Uhlenbeck
Co-discovered that the electron has an intrinsic spin.

90. George Gamow
First suggested hydrogen fusion as source of solar energy.

91. Gordon A. Baym
Contributed to several areas of theoretical physics, including condensed matter, low-temperature physics including superfluidity, statistical physics, nuclear

physics, and astrophysics; made advances in quantum statistical mechanics and the study of neutron stars.

92. **Gabriele Veneziano**
First introduced string theory to describe the strong force without using quantum fields.

93. **Henry Cavendish**
Discovered and studied hydrogen; first to measure Newton's gravitational constant; calculated mass and mean density of Earth.

94. **Hans Christian Oersted**
Discovered that a current in a wire can produce magnetic effects.

95. **Hermann von Helmholtz**
Developed first law of thermodynamics, a statement of conservation of energy.

96. **Henri Poincaré**
Founded qualitative dynamics (the mathematical theory of dynamical systems); created topology; contributed to solution of the three-body problem; first described many properties of deterministic chaos; contributed to the development of special relativity.

97. **Heinrich Hertz**
Worked on electromagnetic phenomena; discovered radio waves and the photoelectric effect.

98. **Hendrik Antoon Lorentz**
Introduced Lorentz transformation equations of special relativity; advanced ideas of relativistic length contraction and relativistic mass increase; contributed to theory of electromagnetism.

99. **Heike Kamerlingh-Onnes**
Liquified helium; discovered superconductivity.

100. **Harold Clayton Urey**
Discovered deuterium.

101. **Hans Bethe**
Contributed to theoretical nuclear physics, especially concerning the mechanism for energy production in stars.

102. **Hideki Yukawa**
Predicted existence of the pion.

103. **Henry W. Kendall**
Co-discovered, through investigations of deep-inelastic electron scattering, clear signs that there exists an inner structure (quarks and gluons) in the protons and neutrons of the atomic nucleus.

104. **Heinrich Rohrer**
Co-designed the scanning tunneling microscope (STM), a type of microscope in which a fine conducting probe is held close the surface of a sample.

105. **H. David Politzer**
Co-discovered "asymptotic freedom" in non-Abelian gauge theories; co-predicted the existence of charmonium - the bound state of a charm quark and its anti-particle.

106. **Hans Geiger**
Helped measure charge-to-mass ratio for alpha particles; invented Geiger counter for detecting ionizing particles.

107. **Hermann Weyl**
Attempted to incorporate electromagnetism into general relativity; evolved the concept of continuous groups using matrix representations and applied group theory to quantum mechanics.

108. **Henry Moseley**
Developed the modern form of the period table of elements based on their atomic numbers.

109. **Homi Jehangir Bhabha**
Initiated nuclear research programs in India; carried out experiments in cosmic rays; calculated cross-section for elastic electron-positron scattering.

110. **Henry Primakoff**
Co-developed the theory of spin waves; first described the process that became known as the "Primakoff effect" (the coherent photoproduction of neutral mesons in the electric field of an atomic nucleus); contributed to understanding of various manifestations of the weak interaction, including muon capture, double-beta decay, and the interaction of neutrinos with nuclei.

111. **Haim Harari**
Predicted the existence of the top quark, which he named; also named the bottom quark.

112. **Igor Y. Tamm**
Co-developed the theoretical interpretation of the radiation of electrons moving through matter faster than the speed of light (the "Cerenkov effect"), and developed the theory of showers in cosmic rays.

113. **Irene Joliot-Curie**
Co-discovered artificial radioactivity.

114. **Isador Isaac Rabi**
Developed the resonance technique for measuring the magnetic properties of atomic nuclei.

115. **Il'ja M. Frank**
Co-developed the theoretical interpretation of the radiation of electrons moving through matter faster than the speed of light (the "Cerenkov effect"), and carried out experimental investigations of pair creation by gamma rays.

116. **Igor Vasilievich Kurchatov**
Headed the Soviet atomic and hydrogen bomb programs.

117. **Joseph-Louis Lagrange**
Developed new methods of analytical mechanics.

118. **James Watt**
Invented the modern condensing steam engine and a centrifugal governor.

119. **Joseph Fourier**
Established the differential equation governing heat diffusion and solved it by devising an infinite series of sines and cosines capable of approximating a wide variety of functions.

120. **Jean-Babtiste Biot**
Studied polarization of light; co- discovered that intensity of magnetic field set up by a current flowing through a wire varies inversely with the distance from the wire.

121. **Johann Carl Friedrich Gauss**
Formulated separate electrostatic and electrodynamical laws, including "Gauss' law"; contributed to development of number theory, differential geometry, potential theory, theory of terrestrial magnetism, and methods of calculating planetary orbits.

122. **Joseph Henry**
Performed extensive fundamental studies of electromagnetic phenomena; devised first practical electric motor.

123. **James Prescott Joule**
Discovered mechanical equivalent of heat.

124. **Jean-Bernard-Léon Foucault**
Accurately measured speed of light; invented the gyroscope; demonstrated the Earth's rotation.

125. **Johann Balmer**
Developed empirical formula to describe hydrogen spectrum.

126. **James Clerk Maxwell**
Propounded the theory of electromagnetism; developed the kinetic theory of gases.

127. **Josef Stefan**
Studied blackbody radiation.

128. **Josiah Gibbs**
Developed chemical thermodynamics; introduced concepts of free energy and chemical potential.

129. **James Dewar**
Liquified nitrogen and invented the Dewar flask, which is critical for low-temperature work.

130. **John Henry Poynting**
Demonstrated that the energy flow of electromagnetic waves could be calculated by an equation (now called Poynting's vector).

131. **Janne Rydberg**
Analyzed the spectra of many elements; discovered many line series were described by a formula that depended on a universal constant (the Rydberg constant).

132. **Johannes van der Waals**
Worked on equations of state for gases and liquids.

131. **Jean Baptiste Perrin**
Experimentally proved that cathode rays were streams of negatively charged particles; experimentally confirmed the correctness of Einstein's theory of Brownian motion, and through his measurements obtained a new determination of Avogadro's number.

134. **Johannes Stark**
Discovered splitting of spectral lines in a strong electric field.

135. **James Franck**
Experimentally confirmed that atomic energy states are quantized.

136. **J. Hans D. Jensen**
Advanced shell model of nuclear structure.

137. **Jack Steinberger**
Made many important discoveries in particle physics; co-discovered the neutral pion *via* photoproduction; co-discovered the muon neutrino.

138. **Jack S. Kilby**
Invented the monolithic integrated circuit- the microchip which laid the foundation for the field of microelectronics, co-invented the hand held calculator.

139. **Jerome I. Friedman**
Co-discovered, through investigations of deep inelastic electron scattering, clear signs that there exists an inner structure (quarks and gluons) in the protons and neutrons of the atomic nucleus.

140. **James W. Cronin**
Co-discovered that decays of neutral kaons sometime violate CP conservation.

141. **J. Robert Schrieffer**
Contributed to condensed matter theory on phenomena of superconductivity.

142. **J. Georg Bednorz**
Co-discovered the first ceramic superconductors.

143. **John von Neumann**
Formulated a fully quantum mechanical generalization of statistical mechanics.

144. **J. Robert Oppenheimer**
Headed Manhattan Project to develop the nuclear fission bomb.

145. **James E. Zimmerman**
Co-invented the radio frequency superconducting quantum interference

device (SQUID), a practical magnetometer/amplifier with extreme sensitivity limited only by the uncertainty principle.

146. **John Stewart Bell**
Proved the inherent non-locality of quantum mechanics.

147. **Joel Lebowitz**
Contributed to condensed matter theory, especially involving statistical mechanics: phase transitions; derivation of hydrodynamical equations from microscopic kinetics; statistical mechanics of plasmas.

148. **John P. Schiffer**
Studied nuclear structure, pion absorption in nuclei, ion traps and crystalline beams, heavy-ion physics, and the Mössbauer effect.

149. **John Dirk Walecka**
Contributed to the theoretical understanding of the atomic nucleus as a relativistic quantum many body system; provided theoretical guidance in exploiting electromagnetic and weak probes of the nucleus.

150. **Jeffrey Goldstone**
Contributed to understanding the role of massless particles in spontaneous symmetry breaking (Goldstone bosons).

151. **John N. Bahcall**
Made important theoretical contributions to understanding solar neutrinos and quasars.

152. **James D. Bjorken**
Formulated the scaling law for deep inelastic processes and made other outstanding contributions to particle physics and quantum field theory.

153. **Kai M. Siegbahn**
Contributed to the development of high resolution electron spectroscopy.

154. **K. Alexander Müller**
Co-discovered the first ceramic superconductors.

155. **Kenneth Wilson**
Invented renormalization group methods to develop a theory for critical phenomena in connection with phase transitions; contributed to solving QCD using lattice gauge theory.

156. **Klaus von Klitzing**
Discovered the quantized Hall effect.

157. **Konstantin Novoselov**
Co-discovered a simple method for isolating single atomic layers of graphite, known as graphene.

158. **Kip S. Thorne**
Contributed to theoretical understanding of black holes and gravitational radiation; co-founded the Laser Interferometer

Gravitational Wave Observatory Project (LIGO).

159. **Lev Landau**
Contributed to condensed matter theory on phenomena of superfluidity and superconductivity.

160. **Leonard Euler**
Made fundamental contributions to fluid dynamics, lunar orbit theory (tides), and mechanics; also contributed prolifically to all areas of classical mathematics.

161. **Lord Kelvin** (born William Thomson)
Proposed absolute temperature scale, of essence to development of thermodynamics.

162. **Lord Rayleigh** (born John William Strutt)
Discovered argon; explained how light scattering is responsible for red colour of sunset and blue colour of sky.

163. **Lord Ernest Rutherford**
Theorized existence of the atomic nucleus based on results of the alpha-scattering experiment performed by Hans Geiger and Ernest Marsden; developed theory of Rutherford scattering (scattering of spinless, pointlike particles from a Coulomb potential).

164. **Leo Szilard**
First suggested possibility of a nuclear chain reaction.

165. **Lincoln Wolfenstein**
Contributed to theory of weak interactions, especially concerning neutrino masses, the origin of CP violation, lepton number violation, the solar neutrino problem, and Higgs boson properties.

166. **Ludvig Faddeev**
Made many theoretical contributions in quantum field theory and mathematical physics; developed the Faddeev equation in connection with the three body system; co-developed the Faddeev-Popov covariant prescription for quantizing non-abelian gauge theories; contributed to the quantum inverse scattering method and the quantum theory of solitons.

167. **Ludwig Boltzmann**
Developed statistical mechanics and applied it to kinetic theory of gases.

168. **Michael Faraday**
Discovered electromagnetic induction and devised first electrical transformer.

169. **Max Planck**
Formulated the quantum theory; explained wavelength distribution of blackbody radiation.

170. **Marie Curie**
Discovered radioactivity of thorium; co-discovered radium and polonium.

171. **Max von Laue**
Discovered diffraction of X-rays by crystals.

172. **Max Born**
Contributed to creation of quantum mechanics; pioneer in the theory of crystals.

173. **Maria Goeppert-Mayer**
Advanced shell model of nuclear structure.

174. **Maurice Wilkins**
Investigated the structure of DNA.

175. **Martin L. Perl**
Discovered the tau lepton.

176. **Murray Gell-Mann**
Advanced an explanation of strange particles; predicted the existence of the Omega particle; postulated existence of quarks; founded the study of QCD.

177. **Melvin Schwartz**
Proposed that it should be possible to produce and use a beam of neutrinos; co-discovered the muon neutrino.

178. **Makato Kobayashi**
Contributed to theoretical understanding of CP violation; co-discovered the origin of the broken symmetry that predicts the existence of at least three families of quarks.

179. **Maurice Goldhaber**
First measured (with James Chadwick) an accurate mass for the neutron; participated in experiments proving that beta rays are identical to atomic electrons; developed (with Edward Teller) the concept of coherent oscillations of protons and neutrons in nuclei leading to the giant dipole resonance; performed an experiment showing that neutrinos are created with negative helicity, which provided conclusive evidence for the V-A theory of weak interactions; participated in experiments that obtained an upper limit on the rate of proton decay and that provided evidence for neutrino oscillations.

180. **Mildred S. Dresselhaus**
Contributed to the advance of solid-state physics, especially involving carbon-based materials, including fullerenes and nanotubes (a.k.a., buckyballs and buckytubes).

181. **Nikola Tesla**
Created alternating current.

182. **Niels Bohr**
Contributed to quantum theory and to theory of nuclear reactions and nuclear fission.

183. **Norman F. Ramsey, Jr.**
Developed the separated oscillatory fields method, which is the basis of the caesium atomic clock (our present time standard); co-invented the hydrogen maser.

184. **Nicolaas Bloembergen**
Contributed to the development of laser spectroscopy.

185. **Nikolai Basov**
Worked in quantum electronics; independently worked out theoretical basis of the maser.

186. **Nikolai N. Bogolubov**
Theoretical physicist and mathematician who contributed to the microscopic theory of superfluidity; also contributed to theory of elementary particles, including the S-matrix and dispersion relations, and to non-linear mechanics and the general theory of dynamical systems.

187. **Nathan Isgur**
Contributed to understanding the quark structure of baryon resonances; discovered a new symmetry of nature that describes the behavior of heavy quarks.

188. **Nathan Seiberg**
Contributed to the development of supersymmetric field theories and string theories in various dimensions.

189. **Otto Hahn**
Discovered the fission of heavy nuclei.

190. **Otto Stern**
Contributed to development of the molecular beam method; discovered the magnetic moment of the proton.

191. **Owen Chamberlain**
Co-discovered the antiproton.

192. **Oskar Klein**
Introduced the physical notion of extra dimensions that helped develop the Kaluza-Klein theory; co-developed the Klein-Gordon equation describing the relativistic behaviour of spinless particles; co-developed the Klein-Nishina formula describing relativistic electron-photon scattering.

193. **Oscar Wallace Greenberg**
Introduced colour as a quantum number to resolve the quark statistics paradox.

194. **Osborne Reynolds**
Contributed to the fields of hydraulics and hydrodynamics; developed mathematical framework for turbulence and introduced the "Reynolds number," which provides a criterion for dynamic similarity and correct modeling in many fluid flow experiments.

195. **Oliver Heaviside**
Contributed to the development of electromagnetism; introduced operational calculus and invented the modern notation for vector calculus; predicted existence of the Heaviside layer (a layer of the earth's ionosphere).

196. Philipp von Lenard
Studied cathode rays and the photoelectric effect.

197. Pieter Zeeman
Discovered splitting of spectral lines in a strong magnetic field.

198. Percy Williams Bridgman
Invented an apparatus to produce extremely high pressures; made many discoveries in high pressure physics.

199. Peter Debye
Used methods of statistical mechanics to calculate equilibrium properties of solids; contributed to knowledge of molecular structure.

200. Prince Louis-Victor de-Broglie
Predicted wave properties of the electron.

201. Pjotr Leonidovich Kapitsa
Heralded a new era of low temperature physics by inventing a device for producing liquid helium without previous cooling with liquid hydrogen; demonstrated that helium-II is a quantum superfluid.

202. Paul Adrien Maurice Dirac
Helped found quantum electrodynamics; predicted the existence of anti-matter by combining quantum mechanics with special relativity.

203. Pavel A. Cherenkov
Discovered the "Cerenkov effect" whereby light is emitted by a particle passing through a medium at a speed greater than that of light in the medium.

204. Polykarp Kusch
Experimentally established that the electron has an anomalous magnetic moment and made a precision determination of its magnitude.

205. Pierre-Gilles de Gennes
Developed theories in condensed matter physics applicable to liquid crystals and polymers.

206. Samuel C. C. Ting
Carried out an experiment leading to the discovery of charmonium.

207. Paul Ehrenfest
Applied quantum mechanics to rotating bodies; helped develop the modern statistical theory of non-equilibrium thermodynamics.

208. Pierre Auger
Discovered the Auger effect whereby an electron is ejected from an atom without the emission of an X-ray or gamma-ray photon as the result of the de-excitation of an excited electron within the atom; discovered cosmic-ray air showers.

209. **Peter Higgs**
Proposed with others the Higgs mechanism by which particles are endowed with mass by interacting with the Higgs field, which is carried by Higgs bosons.

210. **Peter A. Carruthers**
Contributed to several areas of theoretical physics, including condensed matter, quantum optics, elementary particle physics, and field theory; statistics and dynamics of galaxy distributions.

211. **Pierre Curie**
Studied radioactivity with wife, Marie Curie; discovered piezoelectricity.

212. **Robert Hooke**
Discovered Hooke's law of elasticity.

213. **Rudolf Clausius**
Developed second law of thermodynamics, a statement that the entropy of the Universe always increases.

214. **Roland Eötvös**
Demonstrated equivalence of gravitational and inertial mass.

215. **Robert Millikan**
Measured the charge of an electron; introduced term "cosmic rays" for the radiation coming from outer space; studied the photoelectric effect.

216. **Robert S. Mulliken**
Introduced the theoretical concept of the molecular orbital, which led to a new understanding of the chemical bond and the electronic structure of molecules.

217. **Robert Hofstadter**
Measured charge distributions in atomic nuclei with high-energy electron scattering; measured the charge and magnetic moment distributions in the proton and neutron.

218. **Richard P. Feynman**
Co-developed quantum electrodynamics; created a new formalism for practical calculations by introducing a graphical method called Feynman diagrams.

219. **Roy J. Glauber**
Made important contributions to the theoretical understanding of quantum optics and high energy collisions.

220. **Rudolf Ludwig Mössbauer**
Experimented with resonance absorption of gamma radiation; discovered "Mössbauer effect," the recoilless emission of gamma rays by nuclei.

221. **Richard E. Taylor**
Co-discovered, through investigations of deep inelastic electron scattering, clear signs that there exists an inner structure (quarks and gluons) in the protons and neutrons of the atomic nucleus.

222. **Robert W. Wilson**
Co-discovered the cosmic microwave background radiation.

223. **Robert C. Richardson**
Co-discovered that the isotope helium-III becomes a quantum superfluid near absolute zero.

224. **Robert Laughlin**
Developed a theory of quantum fluids that explained the fractional quantum Hall effect.

225. **Robert J. Van de Graaf**
Invented the Van de Graaf electrostatic generator.

226. **Robert Rathbun Wilson**
Driving force behind creation of Fermilab and Cornell University's Laboratory of Nuclear Studies; a leader in the formation of the Federation of Atomic Scientists; did extensive measurements of kaon and pion photoproduction in which he made the first observation of a new state of the nucleon, N(1440).

227. **Robert E. Marshak**
Contributed to theoretical particle physics; independently proposed (with George Sudarshan) the V-A theory of weak interactions; developed explanation of how shock waves behave under conditions of extremely high temperatures.

228. **Robert V. Pound**
Used the Mössbauer effect to measure (with Glen A. Rebka, Jr.) the gravitational redshift predicted by Einstein's theory of general relativity.

229. **Ralph Charles Merkle**
Leading theorist of molecular nanotechnology; invented the encryption technology that allows secure translations over the internet.

230. **Sir Isaac Newton**
Developed theories of gravitation and mechanics, and invented differential calculus.

231. **Sir David Brewster**
Deduced "Brewster's law" giving the angle of incidence that produces reflected light which is completely polarized; invented the kaleidoscope and the stereoscope, and improved the spectroscope.

232. **Sadi Carnot**
Founded the science of thermodynamics.

233. **Sir William Hamilton**
Developed the principle of least action and the Hamiltonian form of classical mechanics.

234. **Sir George Gabriel Stokes**
Described the motion of viscous fluids by independently discovering the Navier-Stoke's equations of fluid

mechanics (or hydrodynamics); developed Stoke's theorem by which certain surface integrals may be reduced to line integrals; discovered fluorescence.

235. **Sir Joseph Wilson Swan**
Developed a carbon filament incandescent light; patented the carbon process for printing photographs in permanent pigment.

236. **Sir Joseph John Thomson**
Demonstrated existence of the electron.

237. **Sir William Henry Bragg**
Worked on X-ray spectrometry.

238. **Sir Owen Richardson**
Discovered the basic law of thermionic emission, now called the Richardson (or Richardson-Dushman) equation, which describes the emission of electrons from a heated conductor.

239. **Sir Chandrasekhara Raman**
Studied light scattering and discovered the Raman effect.

240. **Sir William Lawrence Bragg**
Worked on crystal structure and X-rays.

241. **Sir James Chadwick**
Discovered the neutron.

242. **Sir Edward Appleton**
Discovered the layer of the earth's atmosphere, called the Appleton layer, which is the part of the ionosphere having the highest concentration of free electrons and is the most useful for radio transmission.

243. **Sir George Paget Thomson**
Co-discovered electron diffraction.

244. **Sir John Cockcroft**
Co-invented the first particle accelerator.

245. **Sir Nevill F. Mott**
Contributed to theoretical condensed matter physics by applying quantum theory to complex phenomena in solids; calculated cross section for relativistic Coulomb scattering.

246. **Shin-Ichiro Tomonaga**
Co-developed quantum electrodynamics.

247. **Subramanyan Chandrasekhar**
Made important theoretical contributions concerning the structure and evolution of stars, especially white dwarfs.

248. **Sheldon Glashow**
Co-developed gauge field theory of the electroweak interaction.

249. **Steven Weinberg**
Co-developed gauge field theory of the electroweak interaction.

250. **Steven Chu**
Developed the Doppler cooling method of using laser light (optical molasses) to cool gases and capturing the chilled atoms in a Magneto-Optical Trap (MOT).

251. **Sir Robert Watson-Watt**
Developed radar.

252. **Satyendra Bose**
Worked out statistical method of handling bosons (a group of particles named in his honor).

253. **Samuel Abraham Goudsmit**
Co-discovered that the electron has an intrinsic spin.

254. **Sir Rudolf Peierls**
Many contributions in theoretical physics, including an improved calculation of the critical mass needed to make a fission bomb.

255. **Sidney D. Drell**
Made important theoretical contributions to particle physics and quantum electrodynamics; specialist in arms control and national security.

256. **Stanley Mandelstam**
Contributed to the modern understanding of relativistic particle scattering through his representation of the analytic properties of scattering amplitudes in the form of double dispersion relations (Mandelstam representation); applied path integral quantization methods to string theory.

257. **Stanley J. Brodsky**
Contributed to theoretical understanding of high energy physics, especially the quark gluon structure of hadrons in quantum chromodynamics.

258. **Stephen Wolfram**
Created Mathematica, the first modern computer algebra system; contributed to development of complexity theory.

259. **William Shockley**
Co-discovered the transistor effect.

260. **Tsung-Dao Lee**
Co-proposed parity violation in weak interactions.

261. **Toshihide Maskawa**
Contributed to theoretical understanding of CP violation; co-discovered the origin of the broken symmetry that predicts the existence of at least three families of quarks.

262. **Theodor von Karman**
Provided major contributions to our understanding of fluid mechanics, turbulence theory, and supersonic flight.

263. **T. Kenneth Fowler**
Contributed to the theory of plasma physics and magnetic fusion.

264. **Tullio Regge**
Developed the theory of Regge trajectories by investigating the asymptotic behaviour of potential scattering processes through the analytic continuation of the angular momentum to the complex plane.

265. **Thomas A. Witten**
Contributed to theory of soft condensed matter; structured fluids.

266. **Victor Franz Hess**
Discovered cosmic radiation.

267. **Val Logsdon Fitch**
Co-discovered that decays of neutral kaons sometime violate CP conservation.

268. **Vladimir A. Fock**
Made fundamental contributions to quantum theory; invented the Hartree-Fock approximation method and the notion of Fock space.

269. **Victor F. Weisskopf**
Made theoretical contributions to quantum electrodynamics, nuclear structure, and elementary particle physics.

270. **Vitaly L. Ginzburg**
Contributed to theory of superconductivity and theory of high energy processes in astrophysics; co-discovered transition radiation, emitted when charged particles traverse interface between two different media.

271. **Vernon W. Hughes**
Participated in experiments to test the fundamental QED interaction using the muonium atom.

272. **William Gilbert**
Hypothesized that the earth is a giant magnet.

273. **Willebrod Snell**
Discovered law of refraction (Snell's law).

274. **Wilhelm E. Weber**
Developed sensitive magnetometers; worked in electrodynamics and the electrical structure of matter.

275. **Wilhelm Wien**
Discovered laws governing radiation of heat.

276. **Walther Bothe**
Devised a coincidence counter for studying cosmic rays; demonstrated validity of energy momentum conservation at the atomic scale.

277. **Wolfgang Pauli**
Discovered the exclusion principle; suggested the existence of the neutrino.

278. **Werner Heisenberg**
Contributed to creation of quantum mechanics; introduced the "uncertainty principle" and the concept of exchange forces.

279. **William Fowler**
Studied nuclear reactions of astrophysical significance; developed, with others, a theory of the formation of

chemical elements in the universe.

280. **Willis E. Lamb, Jr.**
Made discoveries concerning fine structure of hydrogen.

281. **Willard S. Boyle**
Co-invented the CCD (charge-coupled device).

282. **William D. Phillips**
Developed, with his colleagues, a device called a Zeeman slower, with which he could slow down and capture atoms in a purely magnetic trap.

283. **Wallace Clement Sabine**
Founded the science of architectural acoustics.

284. **Walther Meissner**
Co-discovered the "Meissner effect", whereby a superconductor expells a magnetic field.

285. **Wolfgang K. H. Panofsky**
Co-discovered the neutral pion via photoproduction; studied gamma rays from pi-captured in hydrogen and first measured the "Panofsky ratio".

286. **Yoichiro Nambu**
Contributed to elementary particle theory; recognized the role played by spontaneous symmetry breaking in analogy with superconductivity theory; formulated QCD (quantum chromodynamics), the gauge theory of colour.

Appendix 1

Conversion Factors

Length

- 1 km = 0.6215 mi
- 1 mi = 1.609 km
- 1 m = 1.0936 yd = 3.281 ft = 39.37 in
- 1 in = 2.54 cm
- 1 ft = 12 in = 30.48 cm
- 1 yd = 3 ft = 91.44 cm = 36 inch
- 1 light year = 1 ly = 9.461×10^{15} m
- 1 Å = 0.1 nm

Area

- $1\ m^2 = 10^4\ cm^2$
- $1\ km^2 = 0.3861\ mi^2 = 247.1$ acres
- $1\ in^2 = 6.4516\ cm^2$
- $1\ ft^2 = 9.29 \times 10^{-2}\ m^2$
- $1\ m^2 = 10.76\ ft^2$
- 1 acre = $43560\ ft^2$
- $1\ mi^2 = 460$ acres $= 2.590\ km^2$

Volume

- $1\ m^3 = 10^6\ cm^3$
- $1\ L = 1000\ cm^3 = 10^{-3}\ m^3$
- 1 gal = 3.786 L
- 1 gal = 4 qt = 8 pt = 128 oz = $231\ in^3$
- $1\ in^3 = 16.39\ cm^3$
- $1\ ft^3 = 1728\ in^3$
  $= 28.32\ L = 2.832 \times 10^4\ cm^3$

Speed

- $1\ km\ h^{-1} = 0.2778\ m\ s^{-1} = 0.6215\ mi\ h^{-1}$
- $1\ mi\ h^{-1} = 0.4470\ m\ s^{-1} = 1.609\ km\ h^{-1}$
- $1\ mi\ h^{-1} = 1.467\ ft\ s^{-1}$

Magnetic Field

- $1\ G = 10^{-4}\ T$
- $1\ T = 1\ Wb\ m^{-2} = 10^4\ G$

Angle and Angular Speed

- π rad = 180°
- 1 rad = 57.30°
- $1° = 1.745 \times 10^{-2}$ rad
- $1\ rev\ min^{-1} = 0.1047\ rad\ s^{-1}$
- $1\ rad\ s^{-1} = 9.549\ rev\ min^{-1}$

Mass

- 1 kg = 1000 g
- 1 tonne = 1000 kg = 1 Mg
- $1\ u = 1.6606 \times 10^{-27}$ kg
- $1\ kg = 6.022 \times 10^{26}$ u
- 1 slug = 14.59 kg
- $1\ kg = 6.852 \times 10^{-2}$ slug
- $1\ u = 931.50\ MeV/c^2$

Density

- $1\ g\ cm^{-3} = 1000\ kg\ m^{-3} = 1\ kg\ L^{-1}$

Force

- 1 N = 0.2248 lbf = 10^5 dyne
- 1 lbf = 4.4482 N
- 1 kgf = 2.2046 lbf

Time

- 1 h = 60 min = 3.6 ks
- 1 d = 24 h = 1440 min = 86.4 ks
- 1 y = 365.24 d = 31.56 Ms

Pressure

- $1\ Pa = 1\ Nm^{-2}$
- 1 bar = 100 kPa
- 1 atm = 101.325 kPa = 1.01325 bar
- $1\ atm = 14.7\ lbf/in^2 = 760$ mm Hg
  = 29.9 in Hg = 33.8 ft H_2O
- $1\ lbf\ in^{-2} = 6.895$ kPa
- 1 torr = 1 mm Hg = 133.32 Pa

Energy

- 1 kWh = 3.6 MJ
- 1 cal = 4.186 J
- 1 ft lbf = 1.356 J = 1.286×10^{-3} Btu
- 1 L atm = 101.325 J
- 1 L atm = 24.217 cal
- 1 Btu = 778 ft lb = 252 cal = 1054.35 J
- 1 eV = 1.602×10^{-19} J
- 1 u c^2 = 931.50 MeV
- 1 erg = 10^{-7} J

Power

- 1 horse power, hp = 550 ft lbf/s
  = 745.7 W
- 1 Btu min^{-1} = 17.58 W
- 1 W = 1.341×10^{-3} hp
  = 0.7376 ft lbf/s

Thermal Conductivity

- 1 W $\text{m}^{-1}\text{K}^{-1}$ = 6.938 Btu in/hft^{2}°F
- 1 Btu in/hft^{2}°F = 0.1441 W/m-K

Note *For simplicity conversion factors are written as equations.*

Appendix 2

The Greek Alphabet

Alpha	Α	α
Beta	Β	β
Gamma	Γ	γ
Delta	Δ	δ
Epsilon	Ε	ε
Zeta	Ζ	ζ
Eta	Η	η
Theta	Θ	θ
Iota	Ι	ι
Kappa	Κ	κ
Lambda	Λ	λ
Mu	Μ	μ
Nu	Ν	ν
Xi	Ξ	ξ
Omicron	Ο	ο
Pi	Π	π
Rho	Ρ	ρ
Sigma	Σ	σ
Tau	Τ	τ
Upsilon	Υ	υ
Phi	Φ	ϕ
Chi	Χ	χ
Psi	Ψ	ψ
Omega	Ω	ω

Appendix 3

Common SI Prefixes and Symbols for Multiples and Sub-multiples

Multiple

Factor	Prefix	Symbol
10^{18}	Exa	E
10^{15}	Peta	P
10^{12}	Tera	T
10^{9}	Giga	G
10^{6}	Mega	M
10^{3}	kilo	k
10^{2}	Hecto	h
10^{1}	Deca	da

Sub-multiple

Factor	Prefix	Symbol
10^{-18}	atto	a
10^{-15}	femto	f
10^{-12}	pico	p
10^{-9}	nano	n
10^{-6}	micro	µ
10^{-3}	milli	m
10^{-2}	centi	c
10^{-1}	deci	d

Appendix 4

Some Important Constants

Name	Symbol	Value
Speed of light in vacuum	c	2.9979×10^{8} ms^{-1}
Charge of electron	e	1.602×10^{-19} C
Gravitational constant	G	6.673×10^{-11} Nm^{2} kg^{-2}
Planck constant	h	6.626×10^{-34} J s
Boltzmann constant	k	1.381×10^{-23} JK^{-1}
Avogadro number	N_A	6.022×10^{23} mol^{-1}
Universal gas constant	R	8.314 J mol^{-1} K^{-1}
Mass of electron	m_e	9.110×10^{-31} kg
Mass of neutron	m_n	1.675×10^{-27} kg
Mass of proton	m_p	1.673×10^{-27} kg
Electron-charge to mass ratio	e/m_e	1.759×10^{11} C/kg
Faraday constant	F	9.648×10^{4} C/mol
Rydberg constant	R	1.097×10^{7} m^{-1}

Name	Symbol	Value
Bohr radius	a_0	5.292×10^{-11} m
Stefan-Boltzmann constant	σ	5.670×10^{-8} W m^{-2} K^{-4}
Wien's constant	b	2.898×10^{-3} mK
Permittivity of free space	ε_0	8.854×10^{-12} C^{2}N^{-1}m^{-2}
Permeability of free space	μ_0	$4\pi \times 10^{-7}$ Tm A^{-1} $\cong 1.257 \times 10^{-6}$ Wb A^{-1}m^{-1}
Mechanical equivalent of heat	J	4.186 J cal^{-1}
Standard atmospheric pressure	1 atm	1.013×10^5 Pa
Absolute zero	0 K	– 273.15°C
Electron-volt	1 eV	1.602×10^{-19} J
Unified atomic mass unit	1 u	1.661×10^{-27} kg
Electron rest energy	mc^2	0.511 MeV
Energy equivalent of 1 u	uc^2	931.5 MeV
Volume of ideal gas (0°C and 1 atm)	V	22.4 L mol^{-1}
Acceleration due to gravity (sea level, at equator)	g	9.78049 ms^{-2}

Appendix 5

SI Derived Units

SI Derived Units Expressed in SI Base Units

Physical Quantity	SI Unit	
	Name	Symbol
Area	square metre	m^2
Volume	cubic metre	m^3
Speed, velocity	metre per second	m/s or ms^{-1}
Angular velocity	radian per second	rad/s or rad s^{-1}
Acceleration	metre per second square	m/s^2 or ms^{-2}
Angular acceleration	radian per second square	rad/s^2 or rad s^{-2}
Wave number	per metre	m^{-1}
Density, mass density	kilogram per cubic metre	kg/m^3 or kg m^{-3}

Physical Quantity	SI Unit	
	Name	Symbol
Current density	ampere per square metre	A/m^2 or $A\ m^{-2}$
Magnetic field strength, magnetic intensity, magnetic moment density	ampere per metre	A/m or $A\ m^{-1}$
Concentration (an amount of substance)	mole per cubic metre	mol/m^3 or $mol\ m^{-3}$
Specific volume	cubic metre per kilogram	m^3/kg or $m^3 kg^{-1}$
Luminance, intensity of illumination	candela per square metre	cd/m^2 or $cd\ m^{-2}$
Kinematic viscosity	square metre per second	m^2/s or $m^2 s^{-1}$
Momentum	kilogram metre per second	$kg\ m\ s^{-1}$
Moment of inertia	kilogram square metre	$kg\ m^2$
Radius of gyration	metre	m
Linear/superficial/volume expansivities	per kelvin	K^{-1}
Flow rate	cubic metre per second	$m^3 s^{-1}$

SI Derived Units with Special Names

Physical Quantity	SI Unit			
	Name	Symbol	Expression in Terms of Other Units	Expression in Terms of SI Base Units
Frequency	hertz	Hz	—	s^{-1}
Force	newton	N	—	$kg\ m\ s^{-2}$
Pressure, stress	pascal	Pa	N/m^2 or $N\ m^{-2}$	$kg\ m^{-1} s^{-2}$
Energy, work, quantity of heat	joule	J	N m	$kg\ m^2\ s^{-2}$ or $kg\ m^2/s^2$
Power, radiant flux	watt	W	J/s or $J\ s^{-1}$	$kg\ m^2\ s^{-3}$
Quantity of electricity, electric charge	coulomb	C	—	A s
Electric potential, potential difference, electromotive force	volt	V	W/A or $W\ A^{-1}$	$kg\ m^2\ s^{-3} A^{-1}$ or $kg\ m^2/s^3 A$

Physical Quantity	SI Unit			
	Name	Symbol	Expression in Terms of Other Units	Expression in Terms of SI Base Units
Capacitance	farad	F	C/V	$A^2s^4 kg^{-1} m^{-2}$
Electric resistance	ohm	Ω	V/A	$kg\ m^2 s^{-3}A^{-2}$
Conductance	siemens	S	A/V	$m^{-2}kg^{-1} s^3A^2$
Magnetic flux	weber	Wb	V s or J/A	$kg\ m^2 s^{-2}A^{-1}$
Magnetic field, magnetic flux density	tesla	T	Wb/m^2	$kg\ s^{-2}A^{-1}$
Inductance	henry	H	Wb/A	$kg\ m^2 s^{-2}A^{-2}$
Luminous flux, luminous power	lumen	lm	—	cd/sr
Illuminance	lux	lx	lm/m^2	m^{-2} cd sr^{-1}
Activity (of a radio nuclide/radioactive source)	becquerel	Bq	—	s^{-1}
Absorbed dose, absorbed dose index	gray	Gy	J/kg	m^2/s^2 or m^2s^{-2}

SI Derived Unit Represented in terms of SI Unit with Special Names

Physical Quantity	SI Unit	
	Symbol	SI
Magnetic moment	$J\ T^{-1}$	$m^2\ A$
Dipole moment	C m	m A s
Kinetic viscosity	Pl or Pa s or N s m^{-2}	$kg\ m^{-1}s^{-1}$
Couple, Moment of force	Nm	$kg\ m^2 s^{-2}$
Surface tension	N/m or N m^{-1}	$kg\ s^{-2}$
Power density, Thermal flux density	W/m^2	$kg\ s^{-3}$
Heat-capacity, Entropy	J/K	$kg\ m^2 s^{-2}K^{-1}$
Specific heat capacity, Specific entropy	J/kg K	$m^2 s^{-2} K^{-1}$
Specific energy, Latent heat	J/kg or J kg^{-1}	$m^2 s^{-2}$
Radiation/radiant intensity	W/sr or W sr^{-1}	$kg\ m^2 s^{-3} sr^{-1}$
Thermal conductivity	W/m K or W $m^{-1}K^{-1}$	$kg\ m\ s^{-3} K^{-1}$
Energy-density	J/m^3 or J m^{-3}	$kg\ m^{-1} s^{-2}$

Physical Quantity	SI Unit	
	Symbol	SI
Electric field intensity	V/m or V m^{-1}	kg m $s^{-3}A^{-1}$
Electric charge density	C/m^2 or C m^{-2}	m^{-3} s A
Permittivity	F/m or F m^{-1}	$kg^{-1}m^{-3}s^4A^2$
Permeability	H/m or H m^{-1}	kg m $s^{-2}A^{-2}$
Molar energy	J/mol or J mol^{-1}	kg $m^2s^{-2}mol^{-1}$
Angular momentum, Planck's constant	J-s	kg m^2s^{-1}
Molar entropy, Molar heat capacity	J/mol K or J mol^{-1} K^{-1}	kg m^2 $s^{-2}K^{-1}mol^{-1}$
Exposure time (X- and γ-rays)	C/kg or C kg^{-1}	kg^{-1} s A
Absorbed lose rate	Gy/s or Gy s^{-1}	m^2 s^{-3}
Compressibility	Pa^{-1}	kg^{-1} m s^2
Coefficient of Elasticity	N/m^2 or N m^{-2}	kg $m^{-1}s^{-2}$
Pressure gradient	Pa/m or N m^{-3}	kg $m^{-2}s^{-2}$
Surface potential	J/kg or J kg^{-1}; N m/kg or N m kg^{-1}	m^2 s^{-2}
Pressure energy	Pa m^3 or N m	kg m^2 s^{-2}
Impulse	N s	kg m s^{-1}
Angular impulse	Nm s	kg m^2s^{-1}
Specific resistance	Ω m	kg $m^3s^{-3}A^{-2}$
Surface energy	J/m^2 or J m^{-2}; N/m or N m^{-1}	kg s^{-2}

Appendix 6

Timeline of Indian Space Programme

1962

Indian Space Research Committee was established.

1963

First sounding rocket launched from the Thumb Equatorial Rocket Launching Station (TERLS) (November 21).

1965

Space Science and Technology Centre set-up as Thumba.

1967

Satellite Telecommunication Earth Station set-up at Ahmedabad.

1968

TERLS dedicated to United Nations (February 12).

1969

ISRO formed under Department of Atomic Energy (August 15, 1962).

1972

Space Commission and Department of Space set-up. Remote sensing testing committee established.

1975

First Indian Satellite Aryabhatta launched from the Soviet Union (April 19). This satellite survived in orbit beyond its life of six months.

1979

Experimental satellite for earth observation Bhaskara launched from the Soviet Union (June 7). This satellite was designed and fabricated by ISRO. First experimental launch of satellite launch vehicle SLV-3 from Sriharikota (SHAR) (August 10).

1980

Second experimental launch of SLV-3 with the Rohini Satellite on board (July 18).

1981

First developmental launch of SLV-3 with the Rohini Satellite (RS-D1) on board (May 31) "Apple", an experimental geostationary communication satellite, launched by the Ariane Vehicle of the Europe Space Agency from Kourou in French Guyana (June 19). With this, India entered the domestic satellite communication era. Bhaskara-II was launched.

1982

Multipurpose Indian national satellite built by the Ford Aerospace and Communication Corporation of U.S.A. and launched by NASA's Delta rocket from Cape Canaveral, Florida, (April 10). It was the first of the first generation Indian national satellites, which combined the services of telecommunication, meteorology, TV relay and radio broadcasting. It was parked 37000 km out in space over Indonesia but ended its life prematurely (September 6, 1982).

1983

Second development launch of SLV-3 with the RS-D2 on board (April 17). INSAT 1-B launched by U.S. space shuttle "Challenger" from Cape Canaveral, Florida (August 30). It remained in orbit for more than life span of seven years.

1984

The first joint Indo-Soviet manned space flight mission. Sq. Ldr. Rakesh Sharma became the first Indian to undertake a space voyage (April 3).

1988

First operational Indian remote- sensing satellite (IRS-1A) launched from Balkanour in the Soviet Union (March 17).
Second developmental launch of ASLV fails (July 13).
INSAT-1C launched by Ariane vehicle from "Kourou" in French Guyana (July 21). Like INSAT-1B, INSAT-1C was a multipurpose satellite in geostationary orbit for domestic long-distance telephonic communication, continuous meteorological earth observation, disaster warning direct TV broadcasting and programme distribution. It was declared non-usable in November 1989.

1990

INSAT-1D launched (June 12) by US Delta rocket from Cape Canaveral Florida (June 12). After becoming operational (July 17), it became the kingpin of the country's satellite based system providing telecommunications radio and television networking and round the clock weather monitoring.

1991

Second remote sensing satellite IRS-1B launched from the Soviet Union (August 29).

1992

Augmented Satellite Launch Vehicle (ASLV-D3) launched from SHAR and placed SROSS-C in low-earth orbit (May 20).
First indigenously-built second generation satellite, INSAT-2A, launched by the European Space Agency's Ariane rocket at Kourou, French Guyana in South America (July 10) and became fully operational (August 15).

1993

INSAT-2B launched by the Ariane rocket from Kourou, French Guyana (July 23) and was successfully injected into Geosynchronous Transfer Orbit.
First developmental launch of PSLV from SHAR fails (September 20). Its purpose is to help expand regional television services and usher in an era of satellite news gathering in India.

1994

Augmented Satellite Launch Vehicle (ASLV-D4) launched with Stretched Rohini Satellite System (SROSS-C2) as payload from SHAR (May 4). SROSS-C2 is successfully put in its final intended orbit with a perigee of 429 km and an apogee of 628 km at the same inclination of 46 degrees after using manoeuvring operations (July 9).

1995

INSAT 2C, the heaviest and the first exclusive, third indigenous satellite was launched by the Ariane rocket from Kourou island in French Guyana (December 7) and was successfully put into its orbit.

1996

PSLV-D3 was successfully launched from Sriharikota, placing the 922-kg Indian remote sensing satellite, IRS-P-3, into orbit. With this, India has entered the US-dominated global market for satellite launch vehicle.

1997

INSAT-2D is launched (June 4).

1999

INSAT-2E launched by Ariane from Kourou French Guyana. (April 3) Indian remote sensing satellite, IRS-P4 (OCEANS) launched by Polar Satellite Launch Vehicle (PSLV-C2) with Korean KITSAT-2 and German DLR-TUBSAT from Sriharikota.

2000

INSAT-3B launched by Ariane from Kourou French Guyana (March 22).

2001

PSLV-C3 successfully launched three satellite TES, BIRD and PROBA.

2002

PSLV-C4 successfully launched KALPANA-1 satellite from Sriharikota.

2003

INSAT-3A launched by Ariane from Kourou French Guyana (April 10)

- Successfully launched of INSAT-3E (September)
- ISRO's PSLV-C5, successfully launched RESOURCESAT-1 (IRS-P6) satellite from Sriharikota (October-17)

2004

Maiden operational flight of GSLV (GSLV-F01) launched EDUSAT from SDSC, SHAR, Sriharikota
(20 September).

2005

PSLV-C6 carries CARTSAT-1 and HAMSAT satellite from Sriharikota on May 5, 2005 into orbit. Launch of INSAT-4A by Ariane from Kourou French Guyana (December 22).

2007

ISRO launches India's CARTSAT-2 and SRE-I and Indonesia's LAPAN-TUBSAT and Argentina's PEHUENSAT-1 (January 10)

- Successful recovery of SRE-1 from Bay of Bengal (January 22)
- Successful launch of INSAT-4B by Ariane-5 from Kourou French Guyana (March 12)
- PSLV-C8 successfully launched Italian astronomical satellite AGILE from Sriharikota on April 23.

- Successful launch of GSLV with INSAT-4CR on board from SDSC SHAR on September-2.

2008

PSLV-C10 successfully launches TECSAR satellite on January 21.

- PSLV-C9 successfully launches CARTSAT-2A, IMS-1 and 8 foreign satellite from Sriharikota on April 28.
- Chandrayan-I launched on October 22 from Satish Dhawan Space Centre, Sriharikota.
- Moon Impact probe lands on moon's south pole on November 14.

2009

- On 20-04-2009 PSLV-C12 successfully launched RISAT-2 and ANUSAT. There are earth observation satellite and experimental satellite respective.
- On 23-09-2009 PSLV-C14 launches Oceansat-2, it is Earth observation satellite.

2010

- On 15-04-2010 GSLV-D3 launched GSAT-4, it is geostationary satellite.
- On 12-07-2010 PSLV-C15 launched CATTOSAT-2B and STUDSAT respectively, there are earth observation satellite and experimental satellite respectively.
- On 25-12-2010 PSLV-C15 launched GSAT-5P which is geostationary satellite.

2011

- On 20-04-2011 PSLV-C16 launched YOUTHSAT and RESOURCESAT-2 there are experimental satellite and earth observation satellite respectively.
- On 21-05-2011 Araine-5VA-202 launched GSAT-8 which is a geostationary satellite.
- On 15-07-2011 PSLV-C17 launched GSAT-12 which is a geostationary satellite.

2012

- Discussion of the impact of budget austerity on the global space economy and space programs around the world.
- Demographic information for the European Space Agency (ESA) and the Japan Aerospace Exploration Agency (JAXA).
- Outlook for human space flight and global positioning systems in major space faring countries.

2013

- On 25-02-2013 PSLV – C 20 launched SARAL and six commercial payloads from Sriharikota.
- On 01-07-2013 PSLV – C 22 Launched IRNSS – 1 A from Sriharikota.
- On 26-07-2013 Ariane – 5VA – 214 launched INSAT – 3D from Kourou French Guiana.
- On 30-08-2013 Ariane – 5VA–215 launched GSAT – 7 from Kourou French Guiana.

- On 05-11-2013 PSLV – C 25 launched Mars oribiter Mission Spacecraft from Sriharikota.

2014

- On 05-01-2014 GSLV–D5 launched GSAT–14 from Sriharikota.
- On 04-04-2014 PSLV–C24 launched IRNSS–1B from Sriharikota.
- On 30-06-2014 PSLV–C23 launched SPOT 7 and four co-passenger satellites AISAT, NLS 7.1, NLS 7.2 and VELOX–1 from Sriharikota.
- On 16-10-2014 PSLV–C26 launched IRNSS–1C from Sriharikota.
- On 07-12-2014 Ariane–5VA–221 launched GSAT–16 from Kourou French Guiana.
- On 18-12-2014 Experimental suborbital flight of India's next generation launch vehicle that tested the vehicle for crucial atmospheric phase of its flight. The payload CARE Module – reentered the atmosphere and was slowed using parachutes before splash down in Andaman Sea.

Appendix 7

Important Scientific Instruments & their use

Altimeter	It measures altitudes and is used in aircrafts.
Ammeter	It measures strength of electric current (in ampere).
Audiometer	It measures intensity of sound.
Barometer	It measures atmospheric pressure.
Binocular	It is used to view distant objects.
Burette	It is used to deliver any required volume of a liquid upto its maximum capacity
Calorimeter	It measures quanitity of heat.
Cardiogram	It traces movements of the heart, recorded on a cardiograph.
Chronometer	It determines longitude of a place kept on board ship.
Cinematography	It is an instrument used in Cinema making to throw on screen and enlarged image of photograph.
Dynamo	It converts mechanical energy into electrical energy.
Dynamometer	It measures electrical power.
Electrometer	It measures electricity.
Electroscope	It detects presence of an electric charge.
Endoscope	It examines internal parts of the body.
Fathometer	It measures the depth of the ocean.
Galvanometer	It measures the electric current of low magnitude.
Hydrometer	It measures the specific gravity of liquids.
Hygrometer	It measures humidity in air.
Hydrophone	It measures sound under water.

Lactometer	It determines the purity of milk.
Manometer	It measures the pressures of gases.
Mariner's compass	It is an instrument used by the sailors to determine the direction.
Microphone	It converts the sound waves into electrical vibration and to magnify the sound.
Microscope	It is used to obtain magnified view of small objects.
Odometer	An instrument by which the distance covered by wheeled vehicles is measured.
Phonograph	It is used for producing sound.
Photometer	It compares the luminous intensity of the source of light.
Periscope	It is used to view objects above sea level (Used in sub-marines).
Radar	It is used for detecting the direction and range of an approaching plane by means or radio microwaves.
Radiometer	It mesures the emission of radiant energy.
Screw gauge	It is used to measure thickness of a thin glass plate and diameter of a thinwire or a small sphere.
Seismograph	It measures the intensity of earthquake shocks.
Salinometer	It determines salinity of solution.
Sonometer	To measure frequency of a tunning fork.
Spectrometer	It is an instrument for measuring the energy distribution of a particular type of radiation.
Speedometer	It is an instrument placed in a vehicle to record its speed.
Sphygmomanometer	It measures blood pressure.
Spherometer	It measures the curvaturs of surfaces.
Stereoscope	It is used to view two dimensional pictures.
Stethoscope	An instrument which is used by the doctors to hear and analyse heart and lung sounds.
Straboscope	It is used to view rapidly moving objects.
Tachnometer	An instrument used in measuring speeds of aeroplanes and motor boats.
Telescope	It views distant objects in space.
Thermometer	This instrument is used for the measurement of temperatures.
Thermostat	It regulates the temperature at a particular point.
Voltmeter	It measures the electric potential difference between two points.
Vernier callipers	To measure lengths accurately.

Appendix 8

Inventions and Discoveries

Invention	Year	Inventor	Country
Adding machine	1642	Pascal	France
Aeroplane	1903	Orville and Wilbur Wright	USA
Air conditioning	1902	Carrier	USA
Airplane (Jet engine)	1939	Ohain	Germany
Airship (Non-rigid)	1852	Henri Giffard	France
Atomic bomb	1945	J Robert Oppenheimer	USA
Ball-point pen	1888	John J Loud	USA
Barometer	1644	Evangelista Torricelli	Italy
Battery (Electric)	1800	Alessandro Volta	Italy
Bicycle	1839-40	Kirkpatrick Macmillan	Britain
Bicycle tyres (Pneumatic)	1888	Johan Boyd Dunlop	Britain
Bifocal lens	1780	Benjamin Franklin	USA
Bleaching powder	1798	Tennant	Britain
Busnen burner	1855	R Willhelm von Bunsen	Germany
Burglar alarm	1858	Edwin T Holmes	USA
Camera (Kodak)	1888	Walker Eastman	USA
Car (Steam)	1769	Nicolas Cugnot	France
Car (Petrol)	1888	Karl Benz	Germany
Carburetor	1876	Gottlieb Daimler	Germany
Cassette (Videotape)	1969	Sony	Japan
Cement (Portland)	1824	Joseph Aspdin	Britain
Cinema	1895	Nicolas and Jean Lumiere	France
Clock (Mechanical)	1725	I-Hsing and Liang Ling-Tsan	China
Clock (Pendulum)	1656	Christian Huygens	Netherlands
Compact disc	1972	RCA	USA
Compact disc player	1979	Sony, Philips	Japan, Netherlands
Computer (Laptop)	1987	Sinclair	Britain
Computer (Mini)	1960	Digital Corp	USA
Diesel engine	1895	Rudolf Diesel	Germany
Dynamo	1832	Hypolite Pixii	France
Electric flat iron	1882	H W seeley	USA
Electric lamp	1879	Thomas Alva Edison	USA
Electric motor (DC)	1873	Zenobe Gramme	Belgium
Electri motor (AC)	1888	Nikola Tesla	USA
Electric iron	1882	Henry W Seeley	USA
Electric washing machine	1906	Alva J Fisher	USA

Invention	Year	Inventor	Country
Electro-magnet	1824	William Sturgeon	Britain
Electroplating	1805	Luigi Brugnatelli	Italy
Electronic computer	1824	Dr Alan M Turing	Britain
Facsimile machine	1843	Alexander Bain	Britain
Fibre optics	1955	Kepany	Britain
Film (Moving outlines)	1885	Louis Prince	France
Film (Talking)	1922	J Engl, J Mussolle and H Vogt	Germany
Galvanometer	1834	Andre-Marie Ampere	France
Gramophone	1878	Thomos Alva Edison	USA
Helicopter	1924	Etienne Oehmichen	France
Hydrogen Bomb	1952	Edward Teller	USA
Intelligence testing	1905	Simon Binet	France
Jet engine	1937	Sir Frank Whittle	Britain
Laser	1960	Theodore Maiman	USA
Launderette	1934	J F Cantrell	USA
Lift (Mechanical)	1852	Elisha G Otis	USA
Lighting conductor	1752	Benjamin Franklin	USA
Loudspeaker	1900	Horace Short	Britain
Machine gun	1918	Richard Gatling	Britain
Magnetic recording tape	1928	Fritz Pfleumer	Germany
Microphone	1876	Alexander Graham Bell	USA
Microscope (Comp.)	1590	Z Jansseen	Netherlands
Microscope (Elect.)	1931	Ruska Knoll	Germany
Microwave oven	1947	Percy Le Baron Spencer	USA
Motor cycle	1885	G Daimler	Germany
Movie projector	1893	Thomas Edison	USA
Neon lamp	1910	Georges Claude	France
Neutron bomb	1958	Samuel Cohen	USA
Optical fibre	1955	Narinder Kapany	Germany
Pacemaker	1952	Zoll	USA
Photoelectric cell	1893	Julius Elster, Hans F Geitel	Germany
Photography (On metal)	1826	J N Niepce	France
Photography (On paper)	1835	WH Fox Talbot	Britain
Photography (On film)	1888	John Carbutt	USA
Piano	1709	Cristofori	Italy
Pistol, revolver	1836	Colt	USA
Radar	1922	A H Laylor and Leo C Young	USA
Radiocarbon dating	1947	Libby	USA
Radio telegraphy	1864	Dr Mohlon Lommis	USA

Invention	Year	Inventor	Country
Radio telegraphy (Trans Atlantic)	1901	G Marconi	Italy
Rayon	1883	Sir Joseph Swan	Britain
Razor (Electric)	1931	Col Jacob Scick	USA
Razor (Safety)	1895	King C Gillatte	USA
Refrigerator	1850	James Harrison, Alexander catlin	USA
Rubber (Latex foam)	1928	Dunlop Rubber Co.	Britain
Rubber (Vulcanised)	1841	Charles Goodyear	USA
Rubber (Waterproof)	1823	Charles Macintosh	Britain
Safety pin	1849	Walter Hunt	USA
Ship (Steam)	1775	I C Perier	France
Ship (Turbine)	1894	Hon Sir C Parsons	Britain
Steam engine	1698	Thomas Savery	Britain
Steam engine (Piston)	1712	Thomas Newcomen	Britain
Steam engine (Condenser)	1765	James Watt	Britain
Steel (Stainless)	1913	Harry Brearley	Britain
Stethoscope	1819	Laennec	France
Submarine	1776	David Bushnell	USA
Super computer	1976	J H Van Tassel	USA
Tank	1914	Sir Emest D Swington	Britain
Tape recorder	1899	Fessenden Poulsen	Denmark
Telegraph	1787	M Lammond	France
Telegraph code	1837	Samuel F B Morse	USA
Telephone (Cellular)	1947	Bell Labs	USA
Telephone (Imperfect)	1849	At tonio Meucci	Italy
Telephone (Perfected)	1876	Alexander Graham Bell	USA
Telescope	1608	Hans Lippershey	Netherlands
Television (Mechanical)	1926	John Logic Baird	Britain
Television (Electronic)	1927	P T Farnsworth	USA
Television (Colour)	1928	John Logie Baird	Britain
Transistor	1948	Bardeen, Shockley and Brattain	USA
Transistor radio	1948	Sony	Japan
Uranium Fission (Atomic reactor)	1942	Szilard Fermi	USA
Vacuum cleaner (Elec.)	1907	Spangler	USA
Vidoe tape	1956	Charles Ginsberg	USA
Washing machine (Elec.)	1907	Hurley Machine Co.	USA
Watch	1462	Bartholomew Mantredi	Italy
Wireless (Telegraphy)	1896	G Marconi	Italy

www.ingramcontent.com/pod-product-compliance
Ingram Content Group UK Ltd.
Pitfield, Milton Keynes, MK11 3LW, UK
UKHW020049200726
13845UKWH00055B/610

9 789352 036035